AIR, WATER AND SOIL POLLUTION SCIENCE AND TECHNOLOGY

AIR POLLUTION

EFFECTS AND DANGERS

AIR, WATER AND SOIL POLLUTION SCIENCE AND TECHNOLOGY

Additional books and e-books in this series can be found on Nova's website under the Series tab.

AIR, WATER AND SOIL POLLUTION SCIENCE AND TECHNOLOGY

AIR POLLUTION

EFFECTS AND DANGERS

JORGE ESTEBAN COLMAN LERNER
EDITOR

nova
science publishers
New York

Library of Congress Cataloging-in-Publication Data

ISBN: 978-1-53619-544-6

Published by Nova Science Publishers, Inc. † New York

CONTENTS

PREFACE

Air pollution from anthropogenic sources has been and continues to be considered a serious environmental problem affecting public health. Its seriousness lies in the fact that it raises the level of pollutants in places where their impact on human health is most likely. Therefore, air pollution is a major threat to public health worldwide. According to the World Health Organization (WHO) more than two million premature deaths per year are attributable to the effects of outdoor and indoor air pollution. More than half of this burden of disease is borne by populations in developing countries.

Epidemiological studies in several countries have shown that chronic exposure to traffic-related pollutants and chemicals and petrochemicals, such as particulate matter (PM), metals and polycyclic aromatic hydrocarbons (PAHs) associated with PM, volatile organic compounds (VOCs) and inorganic gases (SO_2, NOx, O_3) produce adverse effects on development and affect lung function; along with an increase in respiratory diseases, often expressed in terms of diagnosed asthma or chronic obstructive pulmonary disease, and cardiovascular effects and increased mortality.

For all this, it is important to characterize and quantify air pollution and to recognize and assess the associated health effects, and different tools can be used individually or together. These tools are presented in this book, initially in introductory form (Chapters 1 to 3) and then in applied form in different case studies (Chapters 4 to 10) for the assessment of air pollution and associated risk.

Chapter 1 - Air pollution from anthropogenic sources has been and continues to be considered a serious environmental problem affecting public health. Its seriousness lies in the fact that pollutant levels rise in places that increase the likelihood of impact on human health. Therefore, air pollution is a major threat to public health worldwide. According to the World Health Organization (WHO), more than two million premature deaths per year are attributable to the effects of both indoor and outdoor air pollution. In addition, more than half of this disease burden is supported by the populations of developing countries. Epidemiological studies in several countries have shown that chronic exposure to traffic-related pollutants, chemicals and petrochemicals, such as particulate matter (PM), metals and polycyclic aromatic hydrocarbons (PAHs) associated with PM, volatile organic compounds (VOCs) and inorganic gases (SO_2, NOx, O_3) produce adverse effects on development and affect lung function; along with an increase in respiratory disease, often expressed in terms of a diagnosis of asthma or chronic obstructive pulmonary disease, cardiovascular effects and increased mortality. In this sense, it is important to characterize and quantify air pollution, recognizing and evaluating the associated health effects. an adequate way to present these results (from public management) is the elaboration of an air quality index (AQI) that represents, in a simple way, the relationship between pollution levels and its possible health effects. To achieve this objective it is necessary to design and implement a monitoring network of atmospheric pollutants. In its design it is necessary to have clear objectives to be accomplished, which must be clear and concise, as well as the resources available to be carried out. The network alone is not sufficient to develop an integral program of air pollution control, but must be complemented by other tools including an inventory of gaseous emissions and the use of air pollutant dispersion modeling tools.

Chapter 2 - Air pollution is a known hazard to human and environmental health. Susceptibility, differential exposure, differential preparedness, and coping ability are not uniform within populations. For this reason, the differential risk assessment is to have a good description of the vulnerabilities associated with the population studied. Also, it has levels of air pollution throughout the study region. Air quality models are presented here as efficient tools for the determination of hazardous atmospheres and threat zones. Also, these are useful for the identification of different sources and their relative contributions.

Chapter 3 - The World Health Organization (WHO) reports that 7 million people died because of exposure to air pollution in 2012, doubling previous estimates and confirming that air pollution constitutes the most important environmental health risk worldwide. The most affected populations being those living in developing countries. This report synthesizes the abundant epidemiological evidence that correlates chronic exposure to pollutants of urban and industrial origin, such as particulate matter (including adsorbed metals and polycyclic aromatic hydrocarbons), toxic and irritant gases (nitrogen and sulfur oxides, and ozone), in addition to other volatile substances such as benzene and perchloroethylene. All of them exert adverse effects on the development and lung function, on the cardiovascular system. Also, produce an increase in mortality, and one excess of lifetime cancer risk.

Chapter 4 - To ensure the prevention of air pollution and to avoid short-term health effects or long-term effects on the environment, it is necessary that AEIs are approached as a preventive rather than a reactive tool to work on existing issues. To assess air quality linked to the mobile source, it is first necessary to define the objective of the inventory, the scope and the particularities of the site. In this work, the results of the emissions estimation are compared with the Top-Down and Bottom-Up methodological approaches, with three methodologies in increasing order of complexity - IPCC 2006, COPERT 4 and IVE - that analyze a heavy transport fleet whose driving pattern belongs to the collection of urban solid waste (USW) in an intermediate city by 2016. The pollutants included in the Air Emission Inventory are CO_2, CH_4, N_2O, PM_{10}, CO, HC, SO_x, and NO_x. The main results obtained are quite similar for the IPCC and COPERT 4, while they show great differences when compared with the results of the IVE if the particularities of the place for fuel consumption are not considered. The most notable finding is the difference in CO_2 values estimated with IVE because the IPCC 2006 considers that it is not necessary to make estimates with high levels of information when it comes to this pollutant. The values of PM and SO_2 emitted during start-up are significant but do not contribute to the total emissions of pollutants. The work also concludes that the implementation of an AEI should be systematic and cyclical, and subject to improvements in planning and methodologies used; when the inventory is carried out periodically, it changes from being a diagnostic tool to one of prevention, being able to intervene with actions aimed at continuous improvement of the objectives of reducing air pollution.

Chapter 5 - The transport of hazardous materials presents a strong inherent risk, in particular when industrial buildings are located in urban areas. In this chapter, the authors presented a methodology for risk assessment in order to prevent disasters. In the first stage, the authors proposed the use of geographic information systems to detect conflict areas. In the identified hotspots on the hazardous materials transport route, the authors performed a simulation with ALOHA and analyzed the vulnerability by an index of social vulnerability. Finally, this information was integrated using a simple formula for risk estimation. As a specific case, the authors applied it to methanol transport in La Plata, Argentina. By implementing this strategy, the authors detected areas associated with higher risks, which may coincide either with both areas of high or low threat, understanding that vulnerability deserves a mitigation approach to prevent future disasters.

Chapter 6 - Activities carried out in gas stations –mainly filling processes– expose the surrounding population to high levels of contamination due to the emissions originated in their normal operations. This work analyzes the impact of these emissions as diffused sources in the city of La Plata (Argentina) and assesses the risk of contracting cancer (LCR) from exposure to the benzene estimated concentrations. To achieve this goal, the AERMOD modeling system was used as a dispersion assessment tool. The results show that, although the estimated concentrations do not represent a risk for the surrounding population, they reach 7.2-8.5% of the limit concentration recommended by the United States Environmental Protection Agency (corresponding to a LCR of 1 $x10^{-6}$), regarding chronic exposure to benzene in the adult population. However, the levels may be alarming when considering other significant sources in the region that have not been included in the modeling, such as the petrochemical industrial area and the emission of gases coming from motor vehicles.

Chapter 7 - In many countries, mostly - but not exclusively - developing countries, only small and/or medium databases obtained by discrete monitoring are available to analyze some characteristics of atmospheric pollution. Ambient Volatile Organic Compounds are pollutants that affect public health, mainly in urban regions close to a petrochemical complex. In this context, the source apportionment is addressed across to different statistical tools oriented to receptors, applied to small and medium databases obtained by discrete monitoring, so that valuable information is acquired in spite of the technical constraint above mentioned. Through this application, it is possible to characterize the relevant emission sources, estimating the profiles of the sources and their contributions by principal component analysis with multiple linear regression (PCA – MRL). Finally, the most probable areas where the pollutants under study originated are estimated from the back trajectories of the contributions of the present sources by means of a Nonparametric Trajectory Analysis (NTA). As a case study, the analysis of ambient Volatile Organic Compounds (VOCs) levels in Gran La Plata (Argentina) is addressed. This region lacks continuous monitoring and is also characterized by the presence of a petrochemical area and high traffic. This study shows that the highest concentrations are near the petrochemical area but that the data collected at the receptor sites can be explained with two main sources, emission associated to the industry (70.6%) and the traffic (28.0%). Then, the integration of PCA-MRL and NTA allowed access to graphic and numerical information of simple interpretation, useful for stakeholders. Furthermore, the estimated lifetime cancer risk (LCR) in Gran La Plata shows that conglomerate inhabitants are not probably exposed to this risk.

Chapter 8 - It is recommended that each population has its own reference values for studies of respiratory function, due to the ethnic, age, social, geographical and climatic characteristics of the different regions. Previous studies from the National University of La Plata show high levels of particulate matter and volatile organic compounds in La Plata and in areas near the Ensenada Petrochemical area, which correlate chronic exposure to pollutants linked to vehicular traffic and the chemical and petrochemical industry with adverse effects on development and lung function. Based on these recommendations, and the risk factors presented for respiratory function in the authors' region, it is proposed to establish values of forced spirometry parameters in adolescents from 13 to 17 years of age in the city of La Plata, following the recommendations of the American Thoracic Society y la European Respiratory Society (ATS/ERS) based on anthropometric variables. A cross-sectional, observational, and descriptive study was carried out, in which 171 spirometries were analyzed in healthy young people between 13 and 17 years of age selected in the city of La Plata during 2018. The correlation between variables was analyzed to identify predictive variables of spirometry. To establish the prediction equations for the spirometric parameters, regression models were adjusted according to sex. The highest correlation was found with height and weight. Predictive equations for forced vital capacity (FVC) and forced expiratory volume in the first second (FEV1) were developed, separately for gender. In conclusion, it has been possible to establish regional reference equations, which show differences with respect to those proposed by the international bibliography, confirming the importance of having the authors' own estimated.

Chapter 9 - Air pollution in urban environments is a serious problem affecting human health and the environment. A wide range of tools are needed to identify air pollution problems and to test the effectiveness of any air quality action plan. This document describes a new decision support system for linking existing air quality tools through a modular methodology which allows the air quality and health effects health impact assessment. The system is designed to incorporate multi-source, multitype and multi-scale databases, queries and visualization, and thus take advantage of ground-based stations, remote sensing technology, and atmospheric chemical transport models. Also, allows linking the air quality with health impact assessment through tools proposed by the WHO as environmental burden disease and lifetime cancer risk.

Chapter 10 - Low air quality is a public health problem with severe global repercussions. This work presents a data science approach for obtaining relevant information from an urban monitoring network assessment. The long-term concentration of carbon monoxide (CO), nitrogen dioxide (NO_2), and particulate matter (PM_{10}) were examined at three stations (Córdoba, Centenario, and La Boca) in Buenos Aires for the period 2009-2017. The methodology allowed the evaluation of the monitoring network, the analysis of extreme events, and local meteorology's influence on the recorded levels. The description of the hourly, daily, and seasonal behavior of pollutants and the determination of concentration trends using the Mann-Kendall trend test were also included. The analyzed pollutants have reached high maximum values, which poses a risk to the exposed population's health. All stations showed a significant reduction in NO_2 concentrations, but only two stations had a reduction in PM_{10}. The authors observed significant increases in CO levels. These results provide evidence of air quality degradation and provide critical information for regional government policy development.

In: Air Pollution: Effects and Dangers ISBN: 978-1-53619-544-6
Editor: J. Esteban Colman Lerner

Chapter 1

ATMOSPHERIC POLLUTION

Daniela Giuliani[1], Daniela Mellado[1,*]
and Jorge E. Colman Lerner[2]
[1]Centro de Investigaciones del Medioambiente (CONICET-UNLP), La Plata, Argentina
[2]Centro de Investigación y Desarrollo en Ciencias Aplicadas (CONICET-UNLP-CICPBA), La Plata, Argentina

ABSTRACT

Air pollution from anthropogenic sources has been and continues to be considered a serious environmental problem affecting public health. Its seriousness lies in the fact that pollutant levels rise in places that increase the likelihood of impact on human health. Therefore, air pollution is a major threat to public health worldwide. According to the World Health Organization (WHO), more than two million premature deaths per year are attributable to the effects of both indoor and outdoor air pollution. In addition, more than half of this disease burden is supported by the populations of developing countries. Epidemiological studies in several countries have shown that chronic exposure to traffic-related pollutants, chemicals and petrochemicals, such as particulate matter (PM), metals

* Corresponding Author's Email: dmellado@quimica.unlp.edu.ar.

and polycyclic aromatic hydrocarbons (PAHs) associated with PM, volatile organic compounds (VOCs) and inorganic gases (SO_2, NOx, O_3) produce adverse effects on development and affect lung function; along with an increase in respiratory disease, often expressed in terms of a diagnosis of asthma or chronic obstructive pulmonary disease, cardiovascular effects and increased mortality. In this sense, it is important to characterize and quantify air pollution, recognizing and evaluating the associated health effects. an adequate way to present these results (from public management) is the elaboration of an air quality index (AQI) that represents, in a simple way, the relationship between pollution levels and its possible health effects. To achieve this objective it is necessary to design and implement a monitoring network of atmospheric pollutants. In its design it is necessary to have clear objectives to be accomplished, which must be clear and concise, as well as the resources available to be carried out. The network alone is not sufficient to develop an integral program of air pollution control, but must be complemented by other tools including an inventory of gaseous emissions and the use of air pollutant dispersion modeling tools.

1. Atmospheric Pollution

Atmospheric pollution means the presence of any physical, chemical or biological agent in the atmosphere, or combinations thereof in places, forms and concentrations which are or may be harmful to the health, safety or welfare of the population, prejudicial to animal and plant life, or prevent the use and enjoyment of the properties and places of recreation [1]. In this sense, a pollutant can be a chemical substance or an energy form (such as noise, heat, or radioactivity). Daily experience shows that air pollution changes constantly with human activities and weather conditions, and consequently, this state cannot be considered as static. The sweeping of a polluting fog after the passage of an air front and the changes in the associated atmospheric emissions modifications in the regional economic activity are examples that consist of non-staticity of the atmospheric pollution.

1.1. Historical Background

Air pollution and efforts to control it are not a recent phenomenon. The first activities related to its control date back to the 13th century, when King Edward I of England banned the burning of certain highly polluting coals in London, leading to the first pollution control regulations [2]. During the reigns of Richard II (1377-1399) and Henry V (1413-1422), the use of coal in England was regulated and restricted. One of the first publications on the subject known was the book published by royal order of Charles II: "Fumifugium: The inconvenience of the air and smoke of London dissipated together with some remedies humbly proposed" written by John Evelyn in 1661 [3].

The use of coal to generate energy was an important factor in the Industrial Revolution at the end of the 18th century, which constituted the basis of our current technological society, although unfortunately and at the same time, the starting point of pollution and degradation of the environment. In the middle of the 19th century the exploitation of oil began. Its refining and the automobile industry grew exponentially during the 20th century, along with derived industries, and the manufacture of steel and rubber. Since World War II, the sophistication of technological development increased and accelerated air pollution almost unnoticed. The cases of pollution after the Second World War worsened considerably, when the world increased energy consumption, as well as the extraction, production and use of various substances both natural and artificial, exceeding the capacity of ecosystems to regenerate, assimilate and/or degrade them.

On the other hand, natural processes including forest fires, decomposition of organic matter, sandstorms and volcanic eruptions have always polluted the air. Through precipitation, oxidation and absorption into the oceans and soil, the atmosphere has been able to clean itself. However, even though the proportion of gases and particulate matter recognized as pollutants and emitted by anthropogenic sources is lower than natural sources, the rate of discharge to the atmosphere of pollutants associated with humans in densely populated regions often exceeds the natural clean-up capacity. Because of the above, most air pollution is man-made.

Interest in air pollution in Latin America and the Caribbean began in the 1950s, when universities and health ministries made the first measurements of air pollution. In 1965, the Pan American Health Organization (PAHO) began establishing research programs on water and air pollution, as well as collaborating with Member Governments in the development of appropriate control policies [2].

Some of the most famous air pollution episodes of the 20th century (Table 1) happened in the Meuse Valley (Belgium), in Donora (Pennsylvania) and in London (England). These events coincided with a particular weather condition: thermal inversion. In which, the dispersion of the pollutants was made difficult by causing both the surface air masses and the pollutants emitted into the atmosphere to concentrate around their emission sources. In the Meuse Valley episode, which occurred in December 1930, the atmosphere of an industrial area in Belgium was covered with a thick fog for 3 days, so hundreds of people became ill and 60 died. Soon after, in 1931, a thick fog covered the Manchester and Salford area in England for 9 days leading to the death of 592 people. In 1948, in Donora, a small town in Pennsylvania, where chemical plants and steel mills were located, it was covered by a fog for 4 days, making almost half of its 14,000 inhabitants sick and causing 20 deaths. Ten years later, Donora residents who had been seriously ill during the episode showed higher rates of illness and also died earlier than the average resident [3].

Table 1. Most famous episodes of pollution in the 20th century

Episode	Year	Excess mortality	Suggested causes
Meuse Valley	1930	60+	SO_2, fluorides, H_2SO_4
Donora	1948	20	SO_2, PM
Londres	1952	4000+	SO_2, PM, H_2SO_4

In the 1952 London episode, the potential for air pollution became apparent: a fog covered the city from December 5th to 8th. Ten days later, it was learned that the total number of deaths in the main region of London was 4,000 higher than the average. Statistics indicated that almost all inhabitants who had died unexpectedly had a clinical history of bronchitis, emphysema or cardiac disorders; making them more vulnerable [3].

1.2. Pollution in the Current Times

Ambient (outdoor) air pollution in both cities and rural areas was estimated to cause 4.2 million premature deaths worldwide per year in 2016; this mortality is due to exposure to small particulate matter of 2.5 microns or less in diameter ($PM_{2.5}$), which causes cardiovascular and respiratory disease, and cancers.

People living in low- and middle-income countries disproportionately experience the burden of outdoor air pollution with 91% (of the 4.2 million premature deaths) occurring in low- and middle-income countries, and the greatest burden in the WHO South-East Asia and Western Pacific regions. The latest burden estimates reflect the very significant role air pollution plays in cardiovascular illness and death. More and more, evidence demonstrating the linkages between ambient air pollution and the cardiovascular disease risk is becoming available, including studies from highly polluted areas.

WHO estimates that in 2016, some 58% of outdoor air pollution-related premature deaths were due to ischaemic heart disease and strokes, while 18% of deaths were due to chronic obstructive pulmonary disease and acute lower respiratory infections respectively, and 6% of deaths were due to lung cancer. Also, The International Agency for Research on Cancer (IARC), in the press report 221/13, has categorized the polluted ambient air (extramural) as a proven carcinogen, category IA.

In our region, these reports echo the abundant international epidemiological evidence linking chronic exposure to pollutants associated with the petrochemical industry and vehicular traffic - such as particulate matter (PM), metals, polycyclic aromatic hydrocarbons (PAHs) and volatile organic compounds (VOCs) - with adverse effects on development and lung function, on the cardiovascular system, and increased mortality

[1-7]. Within this framework, the agreement signed by Argentina together with Latin American countries is particularly relevant. It is called "Plan de Acción Regional de Cooperación Intergubernamental en Materia de Contaminación Atmosférica para América Latina y el Caribe" (*Regional Action Plan for Intergovernmental Cooperation on Air Pollution in Latin America and the Caribbean*) [4], it gives rise to the "Plan de Acción Nacional sobre Contaminación Atmosférica" (*National Action Plan on Air Pollution*), which includes among its objectives the establishment of a national air quality database.

The maps developed by WHO detailing deaths associated with environmental pollution (extramural) and intramurals worldwide in 2012 are shown in Figure 1. The study involved more than 100 experts who analyzed 14 regions of the planet with the aim of knowing which factors most affect health. The map details the contamination of each country on a chromatic scale from light to dark green, associating the intensity of the color with greater contamination. The study analyzed both the presence of large and small particles in the environment, and their impact on global air pollution.

In addition, a worldwide study conducted in 2013 reveals troubling data about air pollution and human mortality from diseases related to this environmental problem. Vehicle traffic, industry or wood burning are some of the activities that cause more deaths in countries like China and India. The report prepared by a group of researchers in Canada [5] shows in an illustrated way (Figure 2) the places where most deaths have occurred in 2013. These deaths are classified into two specific groups: extramural pollution (cars, industry, agriculture) and intramural pollution (coal stoves or wood burning).

China and India are the countries that concentrate more than half of these deaths with 1.6 and 1.4 million deaths, respectively. In broad terms, Asia concentrates the most alarming figures in this area, with 64 percent of deaths worldwide. Countries such as Indonesia and Pakistan also stand out within this air pollution problem. The remaining 36 percent is distributed in the rest of the world, topped by the developed countries, followed by Eastern Europe, sub-Saharan Africa, the Middle East and North Africa, and Latin America and the Caribbean. Countries such as the United States, Russia and Nigeria are in the middle position, with deaths of around 100,000 people. At the end of the line in pollution, and with an enviable environmental concern, are countries like Canada, Australia, Iceland or the rest of the Nordic countries.

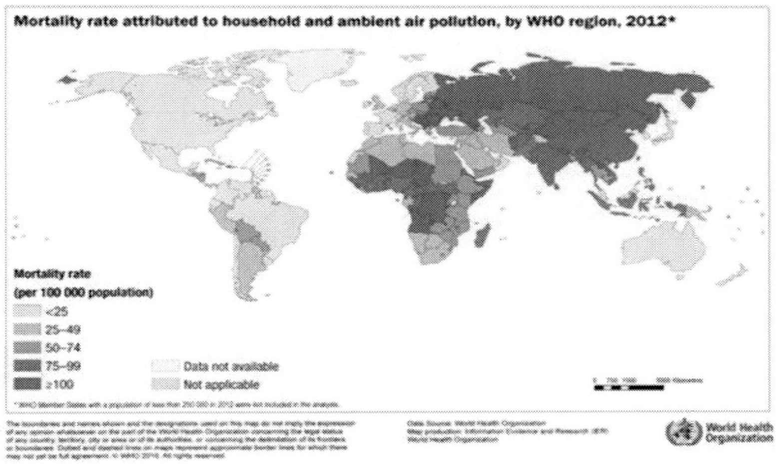

(a)

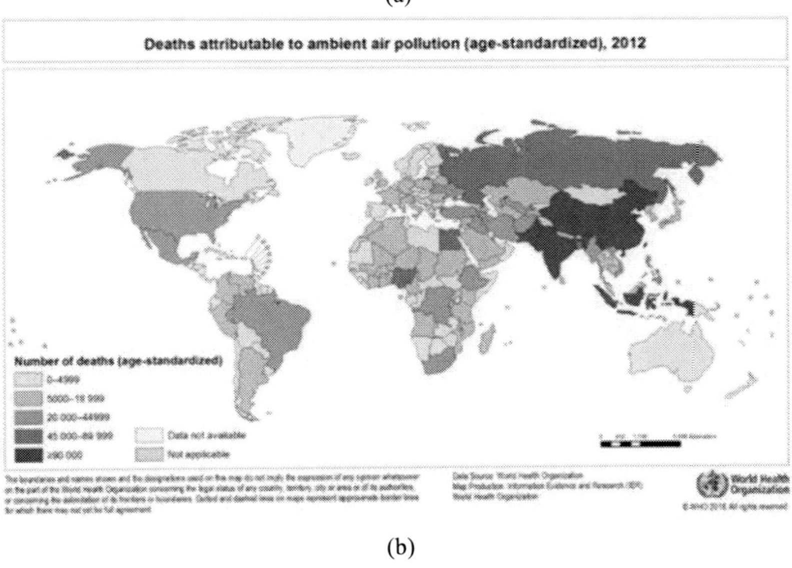

(b)

Figure 1. Maps of mortality rate (a) and deaths (b) attributable to air pollution [4].

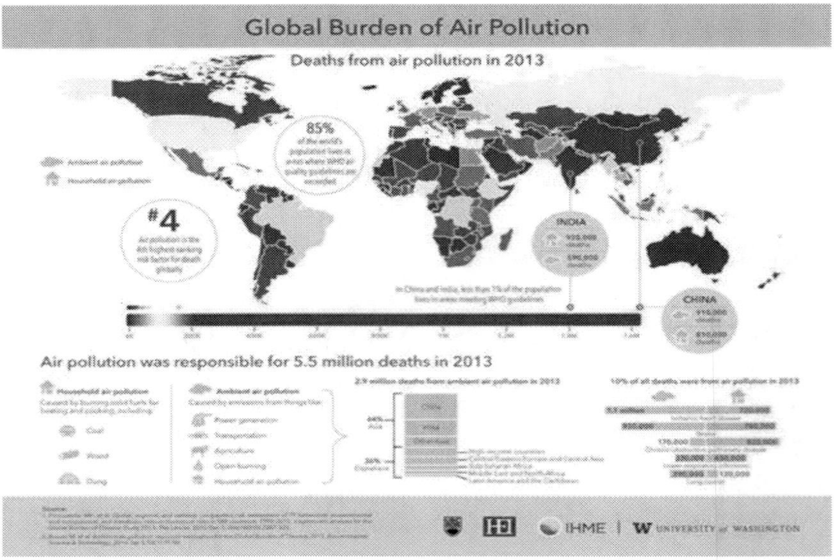

Figure 2. Global burden of air pollution [5].

In China a total of 910,000 deaths were caused by outdoor pollution and 100,000 fewer by indoor pollution. On the other hand, in India, deaths in enclosed spaces exceed by 330 thousand the deaths from outdoor pollution. The report indicates that these contrasts are due to different practices in each country. In China, the burning of coal in the open air is the fundamental cause of the poor state of the air, causing 366 thousand deaths with this practice alone. If no action is taken to control and reduce the practice, the number of deaths could exceed one million people by 2030. In contrast, practices such as burning wood for cooking or heating are common in India, causing thousands of families to be exposed and living together to particles that are harmful to their health.

The WHO says that over the past 20 years, air pollution has declined in higher-income countries and highlights the actions taken by Western Europe, North America and Japan to combat it. However, it indicates that globally the levels of pollution have increased thanks to the contribution of countries such as China and South Asia.

1.3. Air Quality

The air quality of a region is strongly influenced by meteorological parameters such as ambient temperature, wind magnitude and direction, cloud cover, and rain/snow/hail, etc., including the nature of the soil, biota and the different activities that take place in the territory.

To define the air quality of a region, an air quality index (AQI) is established on a scale from good to bad [6]. When the AQI value is increased, an increasing percentage of the population may be affected by the adverse effects of the pollutants studied. The main problem is that there is no unified AQI value, so countries develop their own air quality indexes based on existing national standards.

1.3.1. Urbanization and Pollution

The world's population lives mostly in urban areas [7], concentrated in mega-cities with populations over 10 million. Mega cities are not only centers of economic growth but also important sources of air pollutants and precursors of climate change [8].

Between 1950 and 2005 the world population increased from 2.5 to 6.5 billion people, therefore, this figure is expected to increase to 9.1 billion by 2050. Interestingly, the increase in the world's population was accompanied by a percentage increase in the number of people living in urban areas (Figure 3). Although the percentages of the urban population are higher in developed countries than in developing countries, South America is an exception, with the urban population currently accounting for 83% of the total population (Figure 4). These dramatic increases in the total population, and especially the urbanized population, have been accompanied by technological growth and economic development, changes in land use, irrational use of resources, and a high demand for energy and transport. These changes to improve the quality of life of citizens have had a dramatic impact on the quality of the environment. As significant examples, it is worth mentioning that anthropogenic and biogenic emissions have altered the composition of the atmosphere at different scales.

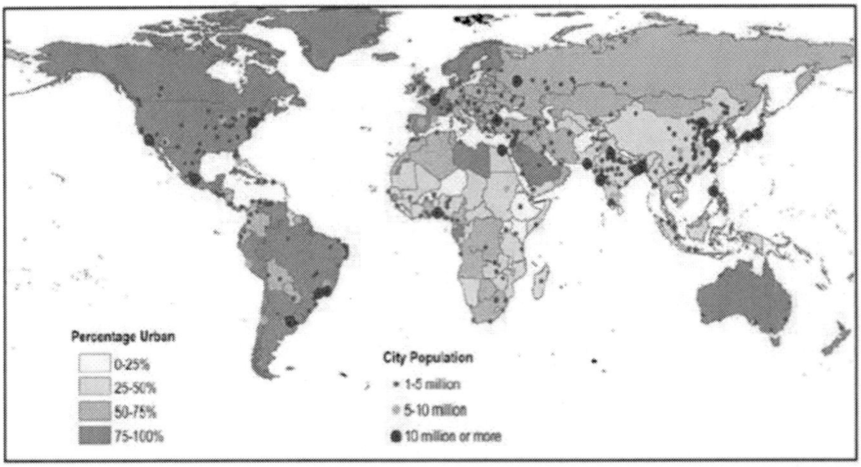

Source: United Nations Department of Economic and Social Affairs, https://www.un.org/en/development/desa/news/2010.html.

Figure 3. Agglomeration in urban centers for the year 2009 (proportion of urban population to world population: 50.1%).

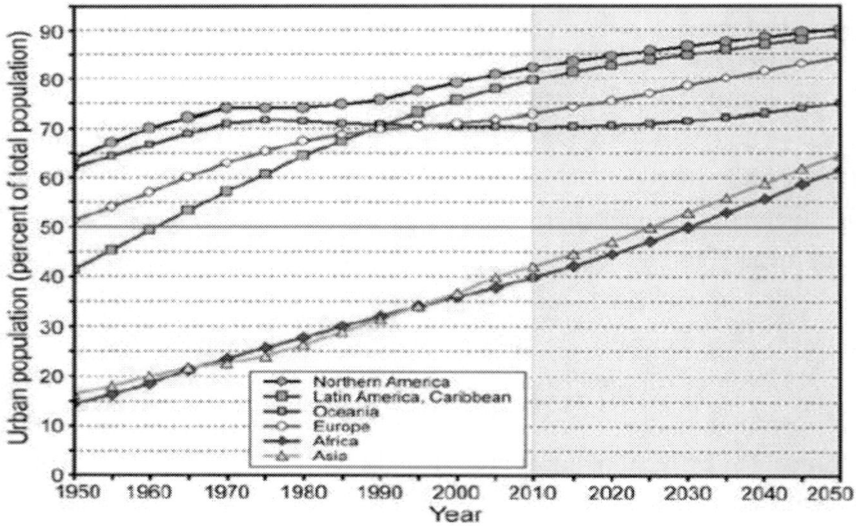

Source: United Nations Department of Economic and Social Affairs, https://www.un.org/en/development/desa/news/2010.html).

Figure 4. Urban population in large geographical areas (percentage of total population).

In this regard, the attempt to understand the impact of urbanization on atmospheric chemistry is of great interest in these times.

The geographic scale and intensity of the impact of megacities on air pollution and climate are determined not only by the magnitude of emissions but also by their geography and regional meteorology. These factors affect the degree to which pollutants are trapped, such as in cities located in valleys surrounded by mountains, or those characterized by persistent atmospheric inversions investments. However, it should be noted that most of the world's megacities are located in coastal areas, where a mixture of urban and marine air masses is prevalent. Clearly, each urban area must be treated as a unique setting for the study of the impacts of urbanization on atmospheric chemistry.

1.3.2. Air Quality Indexes

An "AQI" air quality index is a number used by government agencies to communicate to the public how polluted the air is at a given location. As the AQI increases, an increasing percentage of the population is likely to experience adverse health effects.

Different countries have their own air quality indexes that correspond to their respective national air quality standards. Some of these are the Air Quality Health Index (Canada), the U.S. EPA Air Quality Index, and the United Kingdom Daily Air Quality Index [9-11].

The calculation of the AQI requires an air pollutant concentration over a specified averaging period obtained from a monitor or model. Together, the concentration and time represent the dose of the pollutant in the air. The health effects corresponding to a given dose are established by epidemiological investigations.

Pollutants vary in potency, and consequently, the function used to convert that concentration into an AQI value varies according to the pollutant. Air quality index values are generally grouped into ranges. Each of these ranges is assigned a descriptor, a color code, and a standardized public health notice.

The AQI value may increase due to increased air emissions (e.g., traffic during peak hours or when a forest fire develops) or due to lack of dilution of air pollutants as a result of weather parameters. Stagnant air, often caused by an anticyclone, temperature inversion, or low wind speeds, allows air pollution to remain in a local area leading to high concentrations of the pollutants, thus favoring chemical reactions among air pollutants.

For a day when the AQI is expected to rise due to particulate matter pollution or increased concentration of some other pollutant, a public health agency or organization could:

- Advise sensitive groups such as the elderly, children and those with respiratory or cardiovascular problems to avoid outdoor exercise.
- Encourage voluntary measures to reduce air emissions such as promoting the use of public transportation.
- Recommend the use of masks to prevent fine particles from entering the lungs.

Because most of the pollutants we emit do not have an associated AQI, many countries monitor tropospheric ozone, particulate matter, sulfur dioxide, carbon monoxide, and nitrogen dioxide, calculating air quality indices for these pollutants only. The definition of AQI in a particular nation reflects the discourse surrounding the development of national standards governing air quality in that region. For example, the most widely used air quality index in the United Kingdom is the Daily Air Quality Index recommended by the Committee on the Medical Effects of Air Pollutants (COMEAP). This index contains ten points grouped into 4 bands (see Table 2): low, moderate, high and very high. Then, each of the bands includes clear advice to communicate to both the groups at risk and the general population. The index is based on the concentrations of five pollutants: ozone, nitrogen dioxide, sulfur dioxide, $PM_{2.5}$ and PM_{10} (Table 3). The breakpoints between the index values are defined by pollutant, while the overall index is calculated as the maximum index value.

Table 2. AQI values the United Kingdom

Band	Value	Advice to groups at risk	Advice to the general public
Low	1–3	Enjoy your regular outdoor activities	Enjoy your regular outdoor activities
Moderate	4–6	Adults and children with lung problems, and adults with heart problems, who experience symptoms, should consider reducing strenuous physical activity, particularly outdoor.	Enjoy your regular outdoor activities
High	7–9	Adults and children with lung problems, and adults with heart problems, should reduce strenuous physical exertion, particularly outdoors, and especially if they experience symptoms. People with asthma may find that they need to use their reliever inhaler more often. Older people should also reduce physical exertion.	Anyone experiencing discomfort such as eye pain, coughing or sore throat should consider reducing their activity, especially outside.
Very High	10	Adults and children with lung problems, adults with heart problems, and seniors should avoid strenuous physical activity. People with asthma may find that they need to use their reliever inhaler more often.	People should reduce physical exertion, especially outdoors and particularly if they experience symptoms such as a cough or sore throat.

Table 3. AQI levels for each pollutant

AQI	O_3, 8-hour average (μg m^{-3})	NO_2, hour average (μg m^{-3})	SO_2, 15-min average (μg m^{-3})	$PM_{2.5}$, 24-hour average (μg m^{-3})	PM_{10}, 24-hour average (μg m^{-3})
1	0-33	0-67	0-88	0-11	0-16
2	34-66	68-134	89-177	12-23	17-33
3	67-100	135-200	178-266	24-35	34-50
4	101-120	201-267	267-354	36-41	51-58
5	121-140	268-334	355-443	42-47	59-66
6	141-160	335-400	444-532	48-53	67-75
7	161-187	401-467	533-710	54-58	76-83
8	188-213	468-534	711-887	59-64	84-91
9	214-240	535-600	888-1064	65-70	92-100
10	≥ 241	≥ 601	≥ 1065	≥ 71	≥ 101

Table 4. AQI levels for each U.S. EPA pollutant

AQI	Level with respect to health	Color
0-50	Good	Green
51-100	Moderate	Yellow
101-150	Unhealthy for sensitive groups	Orange
151-200	Unhealthy	Red
201-250	Very unhealthy	Purple
251-300	Dangerous	Granate

The United States Environmental Protection Agency (U.S. EPA) has developed an Air Quality Index. This AQI is divided into six categories that indicate increasing levels of health concern. An AQI value above 300 represents hazardous air quality, while a value below 50 indicates good air quality (Table 4.). In this context, the AQI is based on the five pollutant "criteria" regulated by the Clean Air Act: ground-level ozone, particulate matter, carbon monoxide, sulfur dioxide, and nitrogen dioxide. The agency has established National Ambient Air Quality Standards (NAAQS) for each of these pollutants in order to protect public health. An AQI value of 100 generally corresponds to the NAAQS level for the pollutant defined in the Clean Air Act (1990). This law requires the U.S. EPA to revise its National Ambient Air Quality Standards every five years to reflect evolving information on health effects. Accordingly, the Air Quality Index is periodically adjusted to reflect these changes.

The air quality index is a linear, part-by-part function of pollutant concentrations, defined by equation 1. It can be seen that at the boundary between the AQI categories there is a discontinuous jump of one AQI unit:

$$I = \frac{I_{high} - I_{low}}{C_{high} - C_{low}}(C_{high} - C_{low}) + I_{low} \tag{1}$$

where:

I = the air quality index (AQI),

C_{low} = the concentration breakpoint which is ≤ C,

C_{high} = the concentration breakpoint which is ≥ C,
I_{low} = the index breakpoint corresponding to C_{low},
$I_{high}$the index breakpoint corresponding to C_{high}

If multiple pollutants are measured at a monitoring site then the AQI value is calculated as the largest among the pollutants, also known as "dominant", for the location. The U.S. EPA cut-off values used are shown in Table 5.

2. Atmospheric Chemistry

For many centuries "dilution was the solution", that is to say, putting higher and higher chimneys to cushion the impacts on human health in the surrounding areas. However, historical events have motivated the scientific community and the competent authorities to address studies on atmospheric chemistry and air pollutants, and to implement regulations to improve air quality.

In this context, it is known that most of the substances present in the troposphere are transformed there by physical and/or chemical mechanisms, so that only the least reactive species reach the stratosphere and intervene in the chemistry of this region. On the other hand, the water content of the stratosphere is very low, with the absence of precipitation, since most of the water vapor condenses before crossing the tropopause. For this reason, the removal of gaseous and particulate pollutants by precipitation, a mechanism for self-purification of the atmosphere, is not generally present in the stratosphere. Consequently, the materials that achieve enter the stratosphere by slow diffusion through the tropopause, vertical currents in the tropics, volcanic eruptions and storms, among others, remain in this layer for long periods of time, forming what is known as stratified layers [12].

Table 5. Cutoff values used by the U.S. EPA to calculate the AQI

O_3 (ppb)	O_3 (ppb)	$PM_{2.5}$ (μg m^{-3})	PM_{10} (μg m^{-3})	CO (ppm)	SO_2 (ppb)	NO_2 (ppb)	AQI	
Clow - Chigh (mean)	Clow - Chigh (mean)	Clow - Chigh (mean)	Clow - Chigh (mean)	Clow - Chigh (mean)	Clow - Chigh (mean)	Clow - Chigh (mean)	Ilow - Ihigh	Category
0-54 (8-hr)	-	0.0-12.0 (24-h)	0-54 (24-h)	0.0-4.4 (8-h)	0-35 (1-h)	0-53 (1-h)	0-50	Good
55-70 (8-hr)	-	12.1-35.4 (24-h)	55-154 (24-h)	4.5-9.4 (8-h)	36-75 (1-h)	54-100 (1-h)	51-100	Moderate
71-85 (8-hr)	125-164 (1-h)	35.5-55.4 (24-h)	155-254 (24-h)	9.5-12.4 (8-h)	76-185 (1-h)	101-360 (1-h)	101-150	Unhealthy for sensitive groups
86-105 (8-hr)	165-204 (1-h)	55.5-150.4 (24-h)	255-354 (24-h)	12.5-15.4 (8-h)	186-304 (1-h)	361-649 (1-h)	151-200	Unhealthy
106-200 (8-hr)	205-404 (1-h)	150.5-250.4 (24-h)	355-424 (24-h)	15.5-30.4 (8-h)	305-604 (24-h)	650-1249 (1-h)	201-300	Very unhealthy
-	405-504 (1-h)	250.5-350.4 (24-h)	425-504 (24-h)	30.5-40.4 (8-h)	605-804 (24-h)	1250-1649 (1-h)	301-400	Dangerous
-	505-604 (1-h)	350.5-500.4 (24-h)	505-604 (24-h)	40.5-50.4 (8-h)	805-1004 (24-h)	1650-2049 (1-h)	401-500	

2.1. Stratospheric Chemistry

Unlike the rest of the trace components, the ozone concentration normally increases in the first few kilometers of the surrounding stratosphere by an order of magnitude. In addition to the recognized properties of ozone in the stratosphere as a protective shield for living organisms, it determines the vertical temperature profile of the stratosphere and is responsible for stratospheric heating. It is also considered a very reactive gas that is involved in many chemical reactions. For these reasons, the chemistry of the stratosphere revolves around the chemistry of ozone and the effect that the different trace species present have on the concentration of the same. The complexity of the chemistry of this region, although extensive, is lower than that of the troposphere because the number of chemical species present is much lower.

2.2. Tropospheric Chemistry

The chemical species present in the troposphere, either because they are part of the natural composition of the layer or have been added to this matrix; once they are in this matrix they can experience chemical reactions in the emission zone, be transported or deposited on the earth's surface (dry or wet deposition). It is estimated that of the large number of species emitted from the earth's surface to the troposphere, those with a life span of less than one year are eliminated in the troposphere itself [12]. From the point of view of the chemical reaction, there are a series of factors that influence the reactivity that occurs in this part of the atmosphere. Then, although most of the high energy solar radiation is absorbed in the stratosphere, the solar radiation reaching the troposphere is capable of initiating certain photochemical reactions.

Composition is another decisive factor in tropospheric chemistry. In the lower part of the troposphere, hundreds of compounds appear in locally high concentrations emitted into the atmosphere by human action and naturally. Then, the composition varies and increases its complexity as an increasing diversity of compounds resulting from human activity appears.

Since the troposphere is characterized by an oxidizing medium, the tendency of the species found in the layer is to react by oxidizing. The main oxidizing chemical species present in the troposphere that will determine its chemistry are the hydroxyl radical (OH), the nitrate radical (NO_3^-) and ozone (O_3). The OH radical, of photochemical origin, will dominate the chemistry of the troposphere during the day while the NO_3^- radical, which undergoes photolysis, will determine the chemistry during the night. Although the majority of the compounds present a greater reactivity with the OH radical than with the NO_3^- radical, the greater concentration of the latter (108 molecules per cm^3 as opposed to 105 molecules per cm^3 for the OH radical) makes the reactive processes of the NO_3^- radical decisive in atmospheric chemistry.

Many are the processes that occur continuously in the troposphere. Among the most representative are:

1. Basic photochemical cycle of NO_2, NO y O_3.
2. Photolysis of O_3
3. Reactions of NO_x.

Which are governed by the following reactions:

1. Basic photochemical cycle (Eq 2-4) of NO_2, NO y O_3.

$$NO_2 + h\nu \rightarrow NO + O \tag{2}$$

$$O_2 + O + M \rightarrow O_3 + M \quad (M = N_2, O_2, CO_2) \tag{3}$$

$$NO + O_3 \rightarrow NO_2 + O_2 \tag{4}$$

2. Photolysis of O_3 (Eq 5)

$$O_3 + h\nu \rightarrow O^{\bullet} + O_2 \tag{5}$$

3. Reactions of NO_x (Eq 6 and 7). The main reactions in which NO_x is consumed during the day are those where nitric acid and nitrous acid are produced:

$$OH^{\bullet} + NO_2 \rightarrow HONO_2 \tag{6}$$

$$OH^{\bullet} + NO \rightarrow HONO \tag{7}$$

3. Atmospheric Pollutants

3.1. Particulate Matter (PM)

In the atmosphere there are various suspended substances such as dust particles, pollen, soot (coal), metals (lead, cadmium), asbestos, salts, small drops of sulfuric acid, dioxins, pesticides, among others. The term "aerosol" is usually used to refer to the part of these materials that is smaller, either in solid or liquid form. The solids that are part of the aerosol are usually referred to as particles.

Particulate matter (PM) is defined as all solid and liquid particles suspended in the atmosphere in sizes between 2×10^{-4} and 5×10^{2} µm. Because they are of different size and shape, PM has been classified according to the aerodynamic diameter, mainly grouped in fine and coarse fractions. Fine particles are those with an aerodynamic diameter less than or equal to 2.5 µm, referred to as $PM_{2.5}$. Another group of particles is made up of those whose aerodynamic diameter is less than or equal to 10 microns, which are called PM_{10}. Finally, in the classification of total suspended particles (TSP) refers to all particles surrounded by air in a given volume of undisturbed air [13].

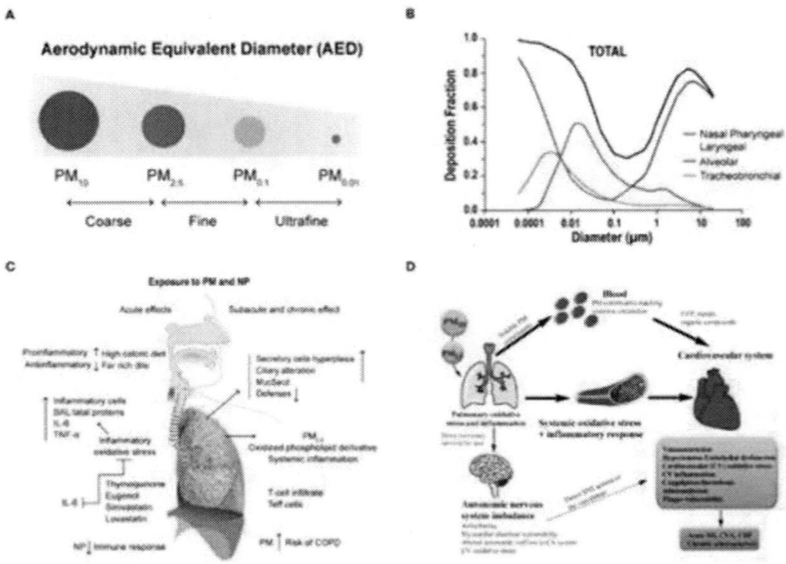

Figure 5. PM classification by aerodynamic diameter and its impact on the respiratory system [14].

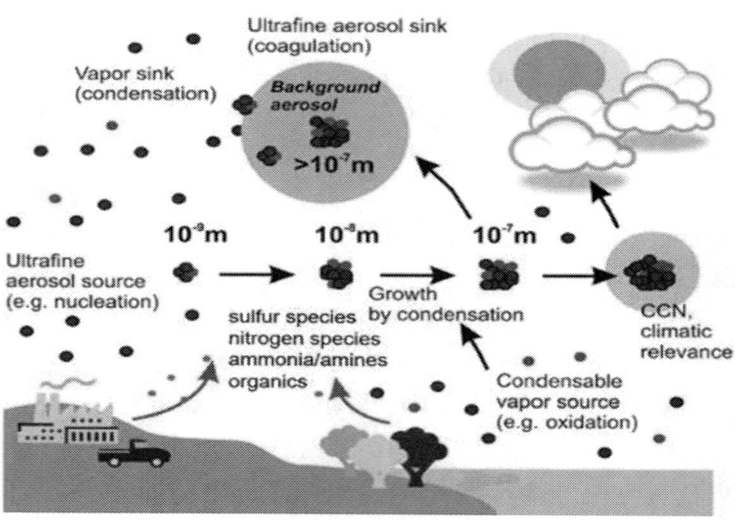

Figure 6. Sources of PM and aerosol atmospheric [15].

Figure 5 shows the classification of particles and their impact on the respiratory system. These particles are widely distributed in the atmosphere, and form a stable suspension in the air. The chemical composition of the particles depends on their origin and, generally, they constitute a mixture of diverse substances, among which we can mention: silicates, carbonates, sulfates, chlorides, nitrates, oxides, metals, coal, tar, resins, pollen, fungi, bacteria, etc. Many come from natural processes and are transported due to the action of the wind; examples include ashes from fires and volcanic eruptions, aerosols formed by the action of the wind on the oceans and the turbulence of the sea, dust from dry soils without plant cover and pollen, among others (Figure 6). In addition, particles can be originated by a wide range of technological processes such as wood and fossil fuel combustion and solid material processing that includes the stages of size reduction, drying and transport [3].

3.2. Sulfur Compounds

Sulfur dioxide (SO_2) and sulfur trioxide (SO_3) are the main sulfur oxides present in the atmosphere. Their main source of anthropogenic emission is the combustion of fossil fuels rich in sulfur, such as coal and fuel oil, representing about one third of the total SO_2 emitted to the atmosphere. The main sulfur component present in coal is pyrite, Fe_2S_3, which oxidizes to Fe_2O_3 and SO_2 during combustion. Other relevant industrial activities are oil refineries and sulphur ore foundries [16].

SO_2 is a non-flammable, non-explosive gas and colorless gas. In the air, SO_2 is partially oxidized into SO_3 (Eq 8) and, in the presence of high humidity, it is transformed into sulfuric acid (H_2SO_4) and its salts by means of atmospheric photochemical processes according to the following main reaction :

$$SO_2 + \frac{1}{2}O_2 \rightarrow SO_3 + H_2O \rightarrow H_2SO_4 \tag{8}$$

Another environmentally relevant sulfur compound is hydrogen sulfide (H_2S) which is produced from the reaction of sulfur with hydrogen in industrial or natural processes. In the presence of oxygen, this compound is oxidized to SO_2. In addition, it is a by-product generated in anaerobic biodegradation in the presence of sulphates. H_2S has a strong odor similar to a rotten egg, which is detected by humans at very low levels of concentration, (around 6 μg m^{-3}). At higher concentrations it is highly toxic and explosive so its emissions must be kept under strict control.

3.3. Nitrogen Oxides

The three nitrogen oxides commonly found in the atmosphere are nitrous oxide (N_2O), nitric oxide (NO), and nitrogen dioxide (NO_2). These are generically called NO_X. N_2O is a relatively unreactive gas and is a trace component naturally found in the atmosphere. NO is a colorless and odorless gas; while NO_2 is yellowish brown and has an unpleasant odor [16].

Nitrogen oxides are produced directly and indirectly by high temperature combustion processes. In such processes, N_2 present in the air is oxidized to form mainly NO, which is transformed into NO_2 by photochemical reactions (Eq 9). In urban regions, maximum NO_X concentrations occur in areas with the highest traffic due to car engine exhaust gases being rich in NO and poor in NO_2. Generally, only a small fraction of total NO_X is present as NO_2. During the initial mixing of the air with the hot exhaust gases from the cars, some of the NO emitted is oxidized through the following reaction:

$$2NO + O_2 \rightarrow 2NO_2 \qquad (9)$$

Much of the atmospheric nitrogen dioxide (Eq 10) is converted to nitric acid and nitrate salts. These salts form particulate matter and/or sediment, or are washed away by the rain.

$$4NO_2 + O_2 + H_2O \rightarrow 4HNO_3 \tag{10}$$

3.4. Hydrocarbons

Within the hydrocarbons, we can mention the volatile organic compounds (VOCs) that comprise an important group of air pollutants. Among these we can name benzene, toluene, xylenes, hexane, heptane, chloroform, trichloroethane and perchloroethane and cyclohexane. Exposure to VOCs is associated with respiratory effects and allergies [16], and these compounds are one of the main sources of tropospheric ozone, a known respiratory tract irritant. In addition, several VOCs can be classified as carcinogenic to humans such as 1,3-butadiene, benzene and perchloroethylene [17]. These compounds are important components in the urban environment arising from natural and anthropogenic processes, and are commonly present both indoors and outdoors. Fifty percent of total VOCs and more than eighty percent of outdoor benzene emissions are due to vehicle exhaust and evaporative losses. In addition to vehicle traffic, there are emissions from industrial sources, e.g., oil refineries and petrochemical plants that play a major role in air pollution [18].

Another important family of hydrocarbons are the so-called polycyclic aromatic hydrocarbons (PAHs) which are organic compounds formed by two or more condensed aromatic rings (Figure 7), and are derived from the combustion of organic material (burning of coal, oil, gasoline, garbage, tobacco, wood or other organic substances, such as roast meat). They can also be found naturally when they are released during forest fires and by volcanoes. Some of them are called "priority pollutants" by the United States Environmental Protection Agency because of their wide distribution in the environment and because of their mutagenic, carcinogenic and teratogenic properties.

The physical and chemical properties of low molecular weight PAHs (two or three aromatic rings) make them ideal candidates for dispersion through the atmosphere [19]. In contrast, higher molecular weights PAHs (four to seven aromatic rings) have a greater affinity for the more persistent particulate material in the environment [20-22]. In this sense, PAHs are persistent organic pollutants (POPs).

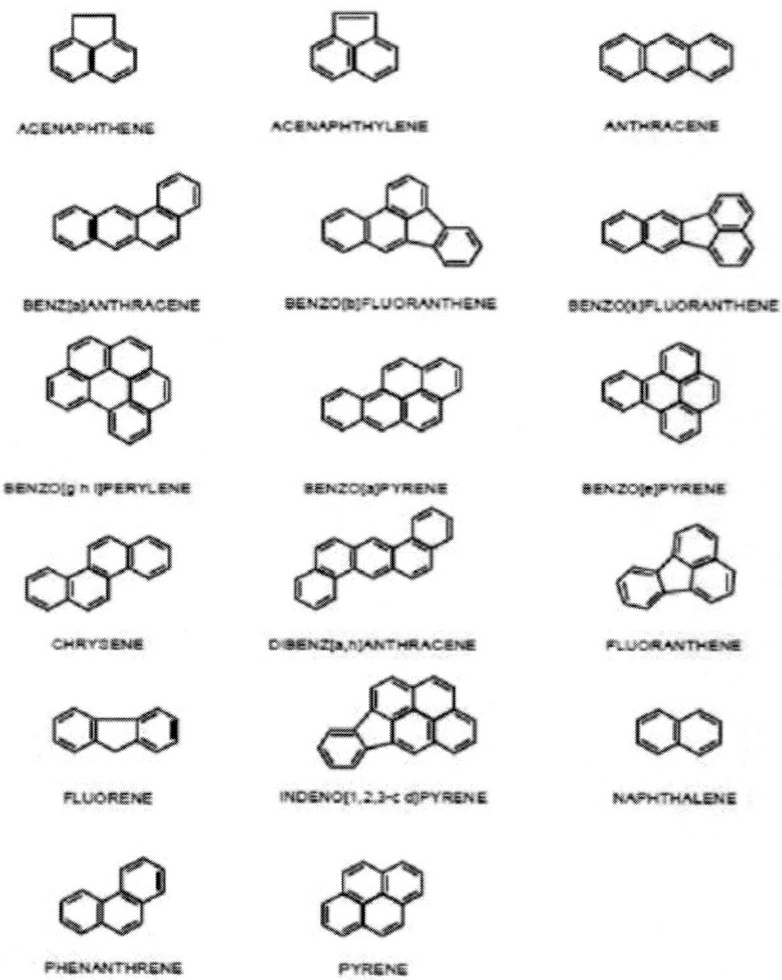

Figure 7. Structures of the main PAHs.

3.5. Photochemical Oxidants

Ozone (O_3) is one of the most abundant photochemical oxidants in the atmosphere. Generated by the formation reaction (Eq 11) between the oxygen molecule (O_2) and atomic oxygen (O), it requires the presence of a catalyst to take place:

$$O_2 + O \xrightarrow{cat} O_3 \tag{11}$$

In turn, ozone (O_3) is consumed (Eq 12) by reacting with nitrogen monoxide (NO), forming an oxygen molecule (O_2).

$$NO + O_3 \rightarrow NO_2 + O_2 \tag{12}$$

In the troposphere and in the presence of substances such as carbon monoxide (CO) and various hydrocarbons (RH), together with nitrogen oxides (NO_x) and solar radiation, ozone behaves as a photo-oxidant promoting the oxidation of these compounds. In this sense, these compounds (RH, CO, NO_x) are called precursors of tropospheric ozone formation.

For its part, atomic oxygen (O) is produced (Eq 13) due to the action of sunlight on nitrogen dioxide (NO_2). Atomic oxygen is highly reactive and its interaction with water generates hydroxyl radicals (OH), which play a fundamental role in complex atmospheric chemical processes. The formation of the hydroxyl is represented below:

$$O + H_2O \rightarrow 2OH^{\bullet} \tag{13}$$

Finally, hydroxyl radicals react with hydrocarbons (RH) to produce hydrocarbon free radicals (R •) in the presence of sunlight at wavelengths less than 0.38 μm (Eq 14-17). In turn, these substances react with nitrogen monoxide (NO) to form more nitrogen dioxide (NO_2). This sequence is represented in the following expressions:

$$NO_2 + h\nu \rightarrow NO + O\ OH^{\bullet} + RH \rightarrow R^{\bullet} + H_2O \tag{14}$$

$$R^{\bullet} + O_2 + NO \rightarrow RO^{\bullet} + NO_2 \tag{15}$$

$$RO^{\bullet} + O_2 \rightarrow HO_2^{\bullet} + R'CHO \tag{16}$$

$$HO_2^{\bullet} + NO \rightarrow HO^{\bullet} + NO_2 \tag{17}$$

The net effect of these reactions is that one molecule of hydrocarbon (RH) converts two molecules of nitrogen monoxide (NO) to nitrogen dioxide (NO_2) and produces one molecule of aldehyde. As can be seen for each molecule of nitrogen dioxide consumed, two are produced, which increases its concentration and decreases that of nitrogen monoxide (NO).

3.6. Carbon Oxides

Carbon oxides are made up of carbon dioxide (CO_2) and carbon monoxide (CO). CO_2 is generally not considered an atmospheric pollutant because it is a natural component of the atmosphere where it plays an important role in the greenhouse effect. However, at present there is great concern about the increasing emissions of this gas, as a product of combustion, which leads to an adverse effect on the earth's heat balance.

On the other hand, CO is a pollutant that comes mainly from the incomplete combustion of any type of fuel. It is also produced in large quantities by various natural sources such as volcanic gases, forest fires and the oxidation of environmental methane, among others. Vehicles with internal combustion engines are one of the main sources of CO emissions in urban areas. Fireplaces, boilers, water heaters, stoves and other fuel-burning appliances are also major sources of CO both outdoors and indoors.

3.7. Heavy Metals

Heavy metals are found in the atmosphere in the form of particles. Some metals and their compounds, such as mercury and tetraethyl lead, have a high vapor pressure, so they can appear in a gaseous state. This group of pollutants also includes the following metals: cadmium, chromium, copper, zinc and arsenic. In general, they are considered to be persistent and bioaccumulative in living organisms and, in some cases, also biomagnifiable.

In turn, arsenic and cadmium have become a concern in some cities or regions that are located near metal processing industries, such as mining operations and smelters. In contrast to specific gaseous compounds, the assessment of metal and metalloid compounds in ambient air is complicated by the fact that different chemical species may possess different toxicity. In ambient air, metals, metalloids, and their compounds are primarily found as part of particulate matter, although volatile chemical species may be the most toxic. They may be present in the non-soluble phase or as soluble ionic or organometallic compounds. Under certain environmental conditions, metallic compounds can be found in gaseous forms such as low weight organometallic molecules. As an example, arsenic compounds (As) most commonly found in ambient air include arsenic trioxide, arsenic oxide, arsenic acid (ortho-arsenic acid), monomethylarsonic acid, dimethylarsinic acid, and other trivalent arsenic compounds (H_3As), these species being the most toxic.

In the case of arsenic (As) compounds, it is known that increasing the organic structure of the compounds strongly reduces their toxicity, providing a natural way of detoxification. However, in the case of mercury (Hg), the toxicity of its chemical species increases from inorganic mercury to methyl and phenyl (organomercury) compounds. The latter can easily cross cell membrane barriers. In addition, the state of oxidation is also a key factor to be considered in the evaluation of the toxicity of mineral elements, just as, for example the Chrome (Cr), where Cr (III) compounds are less toxic than those associated with Cr (VI), and have a totally different effect on humans.

4. Classifications of Pollutants

The atmospheric pollutants are so numerous that it is difficult to group them together for study. Some authors do so according to their state of aggregation, such as solids (PM), liquids and gases, even considering that air quality can be degraded by other pollutants such as noise, heat, ionizing radiation and electromagnetic fields. Today, air pollution typically refers to a set of criteria pollutants, which were historically designated as such because of their impacts on human health and later, their effects on ecosystems. In this regard, most air pollutants can directly affect people's health, as they enter the body through the respiratory system or through skin contact. In other cases, the pollutant is transported from the air to the soil or water bodies, where it subsequently enters the food chain. The transport of pollutants from the atmosphere to the earth's surface occurs by sedimentation of air particles and/or by absorption and carryover of rain.

4.1. Primary and Secondary

Primary atmospheric pollutants are those substances that are emitted directly to the air from an emission source, and result in their harmful form; while secondary pollutants are not emitted directly to the atmosphere (Table 6). The latter are formed within the atmosphere through chemical and/or photochemical processes that primary pollutants undergo, and in several cases are the direct cause of the most pressing pollution problems, such as tropospheric ozone. Examples of primary air pollutants include CO, SO_2, and suspended particulate matter that can be generated by natural sources, such as resuspension of soil particles, or by human activity such as transportation, heating, or industrial processes.

On the other hand, secondary air pollutants are formed in the atmosphere as a result of reaction of hydrolysis, oxidation and/or photochemical oxidation, and examples include acid mists and photochemical oxidants. Other relevant secondary pollutants are: sulphuric acid (H_2SO_4) which is formed by the oxidation of SO_2, nitrogen dioxide NO_2 which is formed by the oxidation of the primary pollutant NO, and ozone O3 which is formed from oxygen O_2. As a result of the above, it can be concluded that air pollution can be avoided if we take as our main action the control of emissions from primary pollutant sources, paying special attention to anthropogenic sources, which will be drastically enhanced by the increase in human population and activities related to economic development.

Table 6. Primary and secondary pollutants depending on the type of compound

Type	Primary pollutants	Secondary pollutants
Compound of S	SO_2, SH_2	SO_3, H_2SO_4, MSO_4
Compound of N	NO, NH_3	NO_2, MNO_3
Organic compound of C	C1-C7 compound	Aldehydes, ketones, acids
oxides of C	CO, CO_2	------
Halogenated Compounds	HF, HCl	-------

4.2. Criteria and Non-Criteria

The expression “atmospheric criteria pollutants” refers to a set of pollutants that allow defining air quality. Their name is due to the fact that they have been the subject of evaluations on air quality standards in the United States, with the purpose of establishing permissible levels that protect the health, the environment and the well-being of the population [13]. This classification has been adopted by many countries today, and includes the following pollutants:

- particulate matter.
- tropospheric ozone.
- carbon monoxide.
- sulfur dioxide.
- nitrogen dioxide.
- lead.

The main difference between criteria and non-criteria pollutants is that the former have been studied extensively, generating vast information about their emission sources, environmental levels and health impacts, while non-criteria pollutants have begun to be studied more recently so the available information is not sufficient to develop exposure limits. Furthermore, criteria air pollutants have been used as indicators for several decades, estimating that, if they are within acceptable values, the rest of the pollutants should not appear in excess. However, this premise is not fully accepted at present, since there are chemical processes that release dangerous pollutants without emitting any of the criteria simultaneously [13].

Non-criteria air pollutants in small concentrations in the atmosphere can have significant acute and chronic health impacts. Non-criteria air pollutants include volatile organic compounds (VOCs) and polycyclic aromatic hydrocarbons (PAHs). The U.S. EPA lists 187 hazardous air pollutants, including substances that cause cancer and neurological, respiratory and reproductive effects. These pollutants are often called "toxic air pollutants." The risk from these particular pollutants depends on their intrinsic toxicity, concentration, and duration of exposure. Examples of hazardous air pollutants include benzene, which is found in gasoline; perchloroethylene, which is emitted from some dry cleaning facilities; and methylene chloride, which is used as a solvent/paint separator in industry. Other relevant examples on this list are dioxins, asbestos, toluene and metals such as cadmium, mercury, chromium and lead compounds.

4.3. Emerging

The emerging air pollutants can be defined in general terms as chemicals or microorganisms that are not commonly controlled in the environment, but have been detected in the air. However, in this section, we must include two different types of emerging pollutants: (i) pollutants that appeared in the atmosphere recently and (ii) pollutants that have been in the environment for a long time but concern for them has become recent. The growing interest in recent years is due in some cases to the availability of new, highly sensitive analytical methods suitable for explaining the relationship between molecular concentration in the air and adverse health effects.

The main groups of emerging pollutants are phthalates, pesticides, flame retardants and manufactured nanoparticles. Phthalate esters are widely used as plasticizers in the production of polymeric materials such as polyvinyl chloride (PVC). PVC and plasticizer are not chemically bonded and, therefore, the plasticizer can vaporize or leach PVC products. The extensive use of PVC as a building material is the main reason why indoor air can be polluted by phthalate esters. Some of the phthalate esters are known as teratogens. The case of flame retardants, such as phosphate esters, is similar to that of phthalate esters and these compounds can be leached into indoor air. Phosphate esters have been considered because of acute central nervous system toxicity.

The case of nanoparticles is quite complex: referring to all those particles that have at least one dimension that is less than 100 nm. In turn, nanoparticles can be subdivided into carbon-based nanoparticles, such as carbon nanotubes, nanohorns, nanowires, or fullerenes, and metal-based nanoparticles, such as noble metal nanoparticles, metal oxides, or quantum dots. In addition to the problems of the difficulty of sampling, the challenge in assessing the environment or the risks to human health of nanoparticles must be distinguished between toxicity due to the composition and size of the particles considered.

Carbon-based nanoparticles behave partly as heavy organic compounds and partly as small particles. On the other hand, metal-based nanoparticles must also be evaluated in terms of the metallic components that are dissolved in the different media considered, and in comparison, by the intrinsic problems related to the toxicity of the nanoparticle itself.

Today, the prolonged use of motor catalysts, along with the incorporation of nanomaterials into clothing, packaging, and industrial products has dramatically increased the presence of small particles in indoor and outdoor environments, and so far there is not much data available on their true distribution mobility and possible toxic effects on the respiratory system.

5. Emission Sources of Atmospheric Pollution

Air pollution includes elements of natural origin and emissions resulting from human activities. Although regulatory programs focus exclusively on air pollution from anthropogenic sources, it is important to understand that nature contributes to air pollution and, in some events, causes significant air quality problems.

5.1. Natural Emission Sources

Emissions of natural pollutants vary from place to place, according to seasonal, geological and meteorological conditions, and the type of vegetation present. Human activities can also contribute to creating conditions that increase the proportion of pollutants from natural sources, for example the deforestation/burning of soils without an immediate restoration. Natural emission sources [23] can be classified into biogenic (related to living organisms) and geogenic (related to geological processes).

5.2. Anthropogenic Emission Sources

The demands to improve the quality of life of the human being and the demographic growth in urban areas, together with the lack of public policies, constitute elements that lead to the sustained contamination of resources. In this regard, the pollutants that the human being releases into the atmosphere could be classified into four main groups:

Table 7. Classification of pollutants emitted into the atmosphere according to domestic activity [22]

Activity	Air pollutants emitted
Heating rooms	CO, CO_2, NOx, SOx, Soot, smoke
Cook	Solid, liquid and gaseous fats; particles and odors
Cleaning	Vapores de disolvente, polvo, pelusa, propelentes
Gardening	Pesticides, fertilizers
Painting	Solvent vapors
Laundry	Soap and detergent particles, lint

1. *Domestic sources:* They are the main cause of pollutant emissions in residential areas. Some examples are shown in Table 7.

Before the industrial revolution, domestic fires were the most important source of pollution in London; the problem was amplified by the fact that emissions came from low chimneys and homes had notoriously inefficient combustion and heating systems. Modern domestic combustion systems are more efficient but low emission discharges still predominate.

2. *Commercial sources:* A classic example of this is dry cleaners, where almost all of the solvent used evaporates and enters the atmosphere. Dry cleaning emissions are between 10 and 16% of lost solvent. Also, the widely used solvent in almost all small commercial cleaning machines is perchlorethylene, a chlorinated hydrocarbon [24]. Other major commercial establishments or

activities that release pollutants into the atmosphere include restaurants, hotels, auto shops, schools, printers, and paint shops.

3. *Industrial sources:* Industrial sources of air pollutants are the most notable because emissions are generally concentrated.
   - A common source of NO_x is combustion processes; other sources of these pollutants are nitrogen fertilizer and explosives factories.
   - SO_x are also produced in combustion processes, especially when using coal. Another source of SO_x is sulphide mineral refining processes, non-ferrous smelters, and oil refineries. H_2S is emitted in large quantities from paper mills, natural gas processing and cleaning plants, oil refineries, synthetic fiber industries and anaerobic wastewater treatment plants. The H_2S is oxidized within hours in the atmosphere to form SO_2 and H_2O.
   - CO in high concentrations is emitted in the production of cast iron and in other metallurgical processes, where the reduction of the presence of oxygen to a minimum is desirable. Combustion processes also generate CO, the concentration of the pollutant will depend on their efficiency.
   - Meanwhile, VOCs originate from a multitude of industrial processes that are generally related to the oil and natural gas industries, or to industries that use their derivatives, such as the paint industry. Likewise, the amount of VOCs emissions in the plants that manufacture plastics is considerable, most of which use petroleum derivatives or natural gas as base products, and on the other hand, in the manufacturing or processing facilities of rubber or synthetic materials.
   - Metal oxides (from the spray painting and coating industries), dust (from refinery catalysts), asbestos fibers (from the insulation industries), fibers (from the textile industry) and metal dust (from metal processing industries), among others, are classified as dangerous particles because they are very toxic or carcinogenic and their size can be in the respirable range.

- Finally, the largest industrial sources of PM emissions are the ashes from the combustion of coal, petroleum derivatives and waste, as well as PM generated in quarries and mining.

4. *Sources related to transport:* CO and VOCs are unwanted products of inefficient combustion, which would be eliminated if combustion were complete. Most of the VOC emissions come from the tailpipe. Most of the VOC emissions come from the exhaust pipe, and can be controlled by using catalytic reactors and injecting air into the engine exhaust ports, thus burning the hydrocarbons that could be emitted in this high temperature area. About 20% of uncontrolled VOC emissions from automobile engines come from the cylinder exhaust vent and evaporated lubricating oil, or crankcase, and from the carburetor vent to the atmosphere. These emissions can be controlled using a crankcase exhaust pipe connected to the engine air intake duct, and an absorption unit for evaporative losses. Starting in 1985, fuel injection systems began to take the place of carburetors; these systems provide more precise metering of fuel to the cylinders and a significant reduction in pollutant emissions. During combustion, the formation of NOx in vehicles takes place, increasing its production very quickly with the increase of combustion temperature, and to reduce this effect have changed the design of engines seeking to reduce the temperature in the cylinders, mainly by reducing the compression ratio of the engine, modifying the timing of ignition, introducing the recirculation of exhaust gases and performing combustion in two stages. In addition, another significant improvement is the use of oxygenated or high-octane fuels to reduce VOC emissions in the tailpipe, promoted today [24]. On the other hand, an atmospheric pollutant that caused great concern generated by the transportation sector was lead, fortunately this has been modified since unleaded gasoline is sold.

5.3. Stationary Sources

Under this classification are sources that operate at a fixed point, that is, the emission source does not move autonomously in time. Examples of this type of sources are industrial and domestic chimneys. National Law 20.284/73- Argentine Republic defines as stationary sources all those designed to operate in a fixed place, which do not lose their condition even if they are mounted on a transport vehicle in order to facilitate their displacement. Stationary sources can be classified according to the size and spatial distribution of the emissions in:

- *Point source:* a single fixed point from which pollutants can be released into the atmosphere continuously or instantaneously (e.g., during an accidental release).
- *Area source:* emission of air pollutants from a specific area, e.g., a city or a forest fire. In an emission inventory, area sources can refer to a whole country or a region.
- *Line sources:* generally refers to emissions from transportation along a line. Examples of this classification are emissions from road vehicles, railways running on tracks, aviation routes, ships navigating in canals, etc.

In particular, gas stations and dry cleaning establishments are often treated as area sources.

5.4. Mobile Sources

In this classification we have all those sources that can move autonomously, emitting pollutants in their trajectory. Examples of this type of source are cars, trains, trucks, buses, planes and ships, among others. National Law 20.284/73- Argentine Republic, defines them as those sources capable of moving between different points, by means of a propulsive element (engine) that generates and emits pollutants.

Motor vehicles are primarily responsible for air quality in most urban areas. Most of today's transportation systems are supplied by the combustion of fossil fuel, which causes different compounds to be emitted into the atmosphere. The internal combustion engines are one of the main sources of emissions of air pollutants such as CO, hydrocarbons, aldehydes, NOx, SO_2, suspended particles, Pb. The proportion in which these pollutants are emitted depends on a number of factors including the type of engine, the fuel used and the state of maintenance of the vehicle.

6. Effects of Air Pollution

6.1. Chemistry of Stratosphere

The Earth's atmosphere is divided into a series of layers characterized by how the temperature varies within it. The troposphere is the most interesting in ecology since it is part of the biosphere, where birds fly, clouds are generated, rain, wind, lightning and the most important vertical movements of the air masses, which have an impact on temperature change, are produced. The thickness of the troposphere is variable, about 9 km in the poles and about 18 km in the equator. In this layer, the air temperature decreases abruptly with altitude.

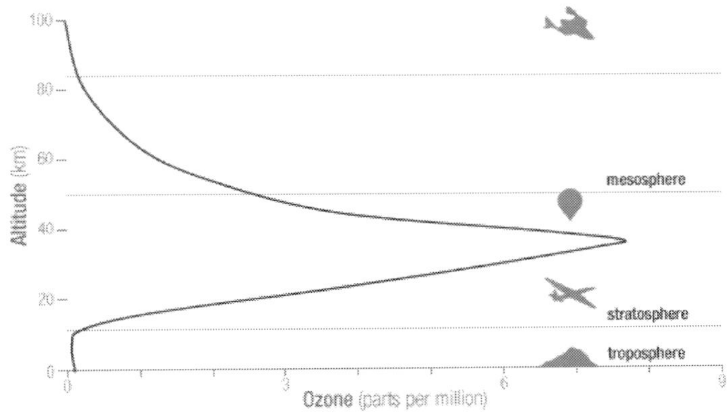

Source: [https://ozonewatch.gsfc.nasa.gov/facts/SH.html].

Figure 8. Distribution of ozone in the atmosphere.

The stratosphere is the layer that borders the troposphere, up to about 50 km from the earth's surface, decreasing the temperature as the height increases until it reaches about zero degrees centigrade. The vertical movements in this layer are much less important than horizontal movements, which can reach up to 200 km per hour. This is the reason that commercial airplanes usually fly in this area. Therefore, a pollutant can be diffused in the stratosphere very quickly, just like the CFCs (chlorofluorocarbons) that destroy ozone.

Ozone is mostly found between 15 and 35 km high (Figure 8). Its concentration in the atmosphere is measured in Dobson Units (DU) that is 1 DU is equivalent to the thickness of 0.01 mm of pure ozone under normal conditions of temperature and pressure (zero degrees centigrade and 1 atm). The average amount of ozone in the atmosphere is 350 DU, corresponding to a thickness of only 3.5 mm under normal conditions. Given the low concentration of ozone in the atmosphere, even in the area of higher concentrations, it would be more appropriate to speak of an ozone screen instead of an ozone layer [25].

6.1.1. Mechanism of Destruction of the Ozone Layer

Ozone is a gas that absorbs a large part of the highest energy solar radiation, which is very harmful to life on Earth. Solar radiation breaks the ozone molecule making it lose energy and thus becoming a harmless radiation for life on Earth. Ozone (Eq 18) is generated in the stratosphere from a first reaction that consists of the dissociation of the oxygen molecule by the action of UV radiation. The reaction that takes place is as follows:

$$O_2 + h\nu \rightarrow 2O \quad \Delta H^o = 495\, kJmol^{-1} \tag{18}$$

Once atomic oxygen is available, it reacts (Eq 19) with other surrounding oxygen molecules, allowing the formation of ozone through an exothermic process:

$$O_2 + O \rightarrow O_3 \quad \Delta H^o = -105\, kJmol^{-1} \tag{19}$$

The heat released in this process causes the temperature of the stratosphere to rise, causing the layer of air in this zone to always be cooler than the one immediately above it. Since this layer is less dense, it should tend to rise rapidly, but the force of gravity works in the opposite direction, making the vertical movements of air in this layer very slow. This creates air strata, hence the name stratosphere.

Ozone, in turn, can decompose following the reverse reaction, with dissociation energy so low that even infrared radiation could break it. The problem is that this absorption by ozone is carried out inefficiently, with very few molecules dissociating at these wavelengths. However, the absorption spectrum of ozone shows that it absorbs very well below 320 nm (Eq 20), which means that it does indeed absorb UV radiation very well, which is how we see it behaves in the atmosphere, as a primary UV filter and absolutely necessary for life on Earth.

$$O_3 + h\nu(\leq 320nm) \rightarrow O + O_2 \tag{20}$$

In addition, ozone is also destroyed (Eq 21) by the oxygen atoms themselves to regenerate molecular oxygen following the process:

$$O_3 + O \rightarrow 2O_2 \tag{21}$$

With all these processes in place, we can say that ozone is constantly being formed, destroyed and recomposed. Thus, the half-life of an ozone molecule at an altitude of 30 km is half an hour. This cyclic process in which ozone is involved is called the Chapman cycle (Figure 9).

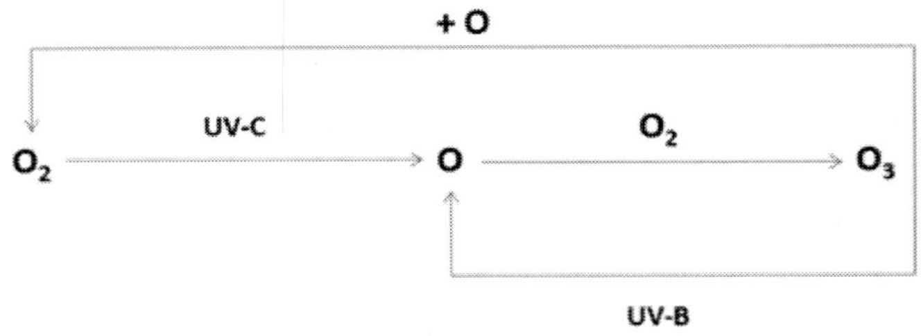

Figure 9. Chapman cycle. Non-catalytic ozone destruction and production reactions [23].

In practice, the observed ozone concentration is lower than the concentration predicted by Chapman's mechanism, so there are ozone destruction processes not taken into account by this mechanism [25].

There are molecules, both anthropogenic and of natural origin, that catalyze the destruction of the ozone layer:

- Hydrogen compounds: HO_X
- Nitrogen compounds: NO_X
- Halogens: Cl_X, Br_X, etc

Because they are catalytic processes, small amounts of catalyst produce the destruction of large amounts of ozone. The most important families of compounds in stratospheric chemistry are shown in Table 8.

Table 8. Most important family of compounds in the stratosphere

Symbol	Group	Components
O_x	Odd oxygen	$O + O_3$
NO_x	Nitrogen oxides	$NO + NO_2$
NO_y	Oxidized nitrogen	$NO + NO_2 + HNO_3 + N_2O_5 + ClONO_2 + NO_3 + HOONO_2 + BrONO_2$
HO_x	Hydrogen radicals	$OH + HO_2$
Cl_y	Inorganic chlorine	$Cl + Cl_2 + ClO + OClO + Cl_2O_2 + HOCl + ClONO_2 + HCl + BrCl$
ClO_x	Reactive chlorine	$Cl + ClO$
CCl_y	Organic chlorine	$CF_2Cl_2 + CFCl_3 + CCl_4 + CH_3CCl_3 + CFCl_2CF_2Cl$ (CFC113) + CF_2HCl (CFC22)
Br_y	Inorganic bromine	$Br + BrO + HOBr + BrONO_2$

The general mechanisms of ozone destruction in the stratosphere are given by (Eq 22 and 23):

- Upper stratosphere (30-50 km, higher O concentration):

$$\begin{array}{l} X + O_3 \rightarrow XO + O_2 \\ \underline{XO + O \rightarrow X + O_2} \\ O_3 + O \rightarrow 2O_2 \end{array} \qquad (22)$$

- Lower stratosphere (18-25 km, lower O concentration):

$$\begin{array}{c} X + O_3 \rightarrow XO + O_2 \\ \underline{XO + O_3 \rightarrow X + 2O_2} \\ 2O_3 \rightarrow 2O_2 \end{array} \qquad (23)$$

6.2. Chemistry of the Troposphere

A large number of species are emitted at the Earth's surface; those with half-lives of less than one year are destroyed in the troposphere. The troposphere behaves as a chemical reservoir, relatively distinct from the stratosphere. The transport of species from this layer to the stratosphere is much slower than the actual mixing within the troposphere.

Although solar radiation of more energetic wavelengths is removed in the stratosphere, the wavelength radiation penetrating the troposphere has sufficient energy to promote significant photochemical reactions. Another important factor in the chemistry of the troposphere is the relatively high concentration of water vapor.

As discussed in 3.5, ozone in the troposphere is generated from two main classes of precursors: volatile organic compounds (VOCs) and nitrogen oxides (NOx). The ozone formation process is initiated by the reaction of the OH radical with organic molecules. The subsequent reaction sequence is catalyzed by NOx, in an interwoven network of free radical reactions. Most of the direct emission of NOx to the atmosphere is in the form of NO, while NO2 is formed in the atmosphere by the conversion of NO. As in the stratosphere, the two nitrogen oxides are generally lumped together as NOx because the interconversion between NO and NO_2 is rapid (with a tropospheric time scale of ~ 5 min) compared to the time scale for organic oxidation (from one to several hours). NOx mixing ratios range from 5 to 20 ppb in urban areas, about 1 ppb in rural areas, and 10 to 100 parts per thousand in the remote troposphere.

In the remote troposphere, ozone formation is sustained by the oxidation of carbon monoxide (CO) and methane (CH_4), both reacting with the $OH^{\cdot}$ radical. CO and CH_4 are long-lived species, with an atmospheric life against the OH reaction of approximately 2 months and 9 years, respectively. Ozone formation in the urban and regional atmosphere is driven by shorter-lived VOCs emitted by anthropogenic and biogenic sources. These include alkenes, aromatics, and oxygenated organic species. Because two classes of precursor compounds are involved in the formation of tropospheric ozone, volatile organic compounds and nitrogen oxides, a key question is how changing the levels of VOCs and NOx affect the amount of ozone formed. Ozone formation is critically dependent on the level of NOx.

6.2.1. Photochemical Smog

Photochemical smog is the combined effect of the stagnation of the air mass with a strong emission of pollutants (emissions from diesel and / or gasoline engines, domestic heating and industrial activity). When under certain weather conditions a low upward (vertical) or horizontal air velocity is observed, prolonged for several hours and combined with a relative humidity greater than 60%, the so-called classic smog is formed, a toxic mixture of soot (black carbonaceous fatty particles), smoke (composed mainly CO, CO_2, SO_2, NOx), particulate matter, and fog (liquid aerosol that reduces visibility). The health effects of classic smog include catarrh and irritation of the mucous membranes of the upper airways and irritation of the skin from exposure to acid mist (SO2 and NOx dissolved in water vapor).

Photochemical smog (Figure 10) occurs when atmospheric pollutants form a gaseous mixture capable of producing photochemical reactions in the presence of sunlight (mainly UV rays), generating dangerous secondary pollutants from criteria pollutants. These include ozone (O_3), nitrogen dioxide (NO_2) and compounds known as PAN (peroxyacyl nitrates). When PANs are part of the fog, they are associated with the particles usually present, giving it a characteristic yellowish haze appearance. In the fog can also occur the reaction SO_2 + fog $(NH_4)_2SO_4$ forming particles with diameters greater than 10 microns that affect the nostrils, intermediate particles (5 microns) such as SO_2 affect the mucous membranes of the upper tract, mouth, nostrils, rhinopharynx and trachea, fine particles (less than 3 microns) such as O_3 and NO_2 are captured by macrophages in the pulmonary alveoli causing a medium conducive to superinfections [25].

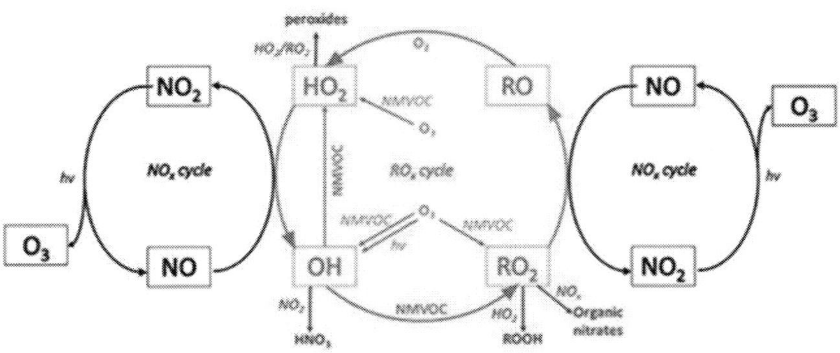

Figure 10. Mechanism of photochemical smog formation [26].

6.2.2. Acid rain

Rain is one of the parts of the water cycle: water evaporates from the earth's surface forming clouds of vapour, which on contact with a current of cold air condenses to form droplets that increase in size until they fall (precipitate). This is a natural occurrence that can be assimilated to the operation of distilling water in the laboratory. But while distilled water has pH 7, rainwater normally has a pH of 5.6 due to the presence of dissolved CO_2.

It is called acid rain to that whose pH is less than 5.6 and this can occur by different mechanisms of wet deposition of polluting compounds present in the atmosphere, generating the formation of acids in the presence of water. The following are the mechanisms of wet deposition:

- Aggregation: incorporation of substances into cloud droplets and crystals.
- Derrubio: sweeping of the material under the clouds during rain-snow precipitation.

The compounds that determine acid rain can be SO2 and NOx, either of natural or anthropogenic origin, which form their corresponding acids in aqueous solution: H2SO4 and HNO3.

6.2.3. SO_2 and Other Sulfur Compounds

In the atmosphere it is usual its presence in quantity of traces, but even in such low quantities they have great impact in the production of the acid rain at local level. The most common sulfur compounds present in the atmosphere are SO_2, COS (carbonyl sulfide), CS_2 (carbon disulfide), RSR (organic gaseous sulfides), and H_2S. They are incorporated into the atmosphere by volcanism (SO_2, COS, SO_4^{2-}, H_2S), wind erosion of sulfate-rich rocks, RSR formation in the photic zone of the ocean, or by biomass degradation (CH_3HS, COS, CS_2).

In addition to these natural sources, human activity generates 25% of the SO2 present in the atmosphere, as a result of the combustion of fossil fuels with high sulphur content. In the atmosphere these compounds participate in a series of complex kinetic reactions that can be classified into two large groups:

- Homogeneous oxidations: where other gaseous compounds in the atmosphere participate. These transformations occur in the presence of ozone, oxygen or radicals such as $HO^\cdot$, $HO_2^\cdot$.
- Heterogeneous oxidations: SO_2 dissolves into $H_2O(L)$ present as droplets in the atmosphere or is absorbed in aerosols (e.g., soot),

forming SO_3^{2-} and HSO_3^- which in the presence of cations such as Fe^{3+}, Mg^{2+} or Ca^{2+}, that function as catalysts, react with ozone or oxygen to form SO_4^{2-}.

Sulfuric acid in the atmosphere is found as a colloidal suspension (liquid) and its formation can occur by two different mechanisms:

- SO_2, hν in presence of O_2 or O_3, SO_3 and H_2O
- SO_2 in presence of radicals $HO^·$ y $HO_2^·$

6.2.4. NO_x and Others

Under the generic name of NO_x, environmentalists refer to these nitrogen-polluting oxides because they contribute to the formation of photochemical smog, acid rain and the greenhouse effect. In this context, the oxides of NO and NO_2 are mainly treated, along with NH_3 and other nitrogenous pollutants, which can be present in the atmosphere for the following reasons:

- *Volcanic emissions*: they can emit NO_x and NH_3 into the stratosphere and even enter it. If they accumulate in the troposphere they contribute to the greenhouse effect. Also, as NO_x reacts with different free radicals by different mechanisms, they can be transformed into atmospheric nitric acid.
- *Biological processes*: since nitrogen is an essential constituent of living beings, the nitrogen cycle is a fundamental part of the chemical balance of the planet.
- *Anthropogenic sources*: during hydrocarbon combustion processes at high temperatures in the presence of air, various nitrogen oxides can be generated, mainly NO.

In the case of cars and other means of transportation, these emissions are reduced by using catalytic converters for the exhaust gases, mainly Al_2O_3 with platinum, palladium or rhodium, which catalyze the oxidation of CO and hydrocarbons to CO_2 and the reduction of NO_x to nitrogen.

6.3. Greenhouse Effect and Climatic Change

Basically all the energy that reaches the Earth comes from the Sun. The absorption and loss of radiant energy by the Earth and the atmosphere are almost entirely responsible for the climate (on both global and local scales). The average temperature of the Earth remains fairly constant, indicating that the Earth and the atmosphere in general lose as much energy by radiation back into space as is received from the Sun. The quantification of input and output radiation energy constitutes the energy balance of the Earth (Figure 11).

The atmosphere, although it may appear to be transparent to radiation, plays a very important role in the Earth's energy balance. In fact, the atmosphere controls the amount of solar radiation that actually reaches the Earth's surface and, at the same time, controls the amount of outgoing terrestrial radiation that escapes into space.

The radiation spectrum is the radiant energy classified according to its wavelengths (λ) (Figure 12). The sun radiates across the entire electromagnetic spectrum, although most of the energy is concentrated near the visible portion of the spectrum, that is the wavelength band from 400 to 700 nm (0.4 to 0.7 microns).

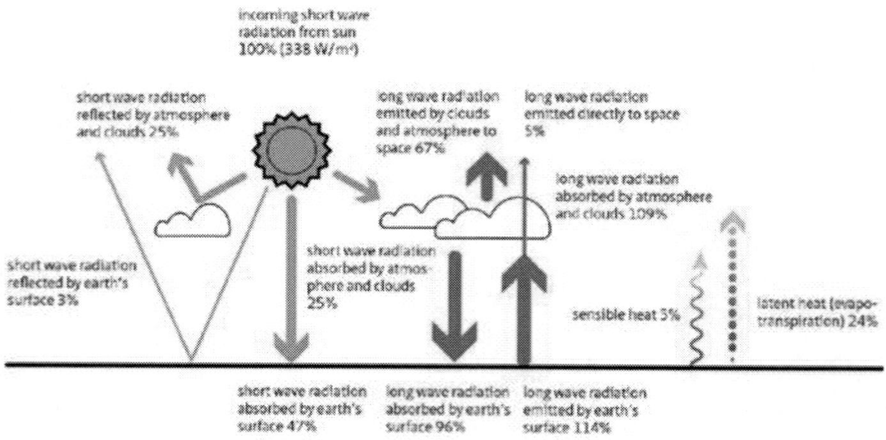

Figure 11. Energy balance of the earth [27].

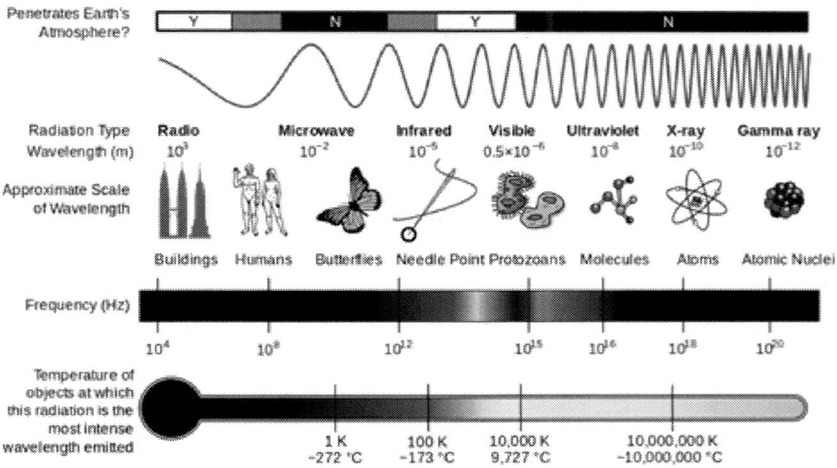

Figure 12. Electromagnetic spectrum.

Our interest is limited to the so-called optical region, which extends over the near ultraviolet, visible and near infrared, meaning, the wavelength range from 200 nm to 100 microns. This range covers most of the solar radiation and that emitted by both the surface and the atmosphere of the Earth.

The following three measurements are used to specify the location in the electromagnetic spectrum: wavelength (λ), frequency (v), and wave number ($v = \lambda^{-1}$). The frequency and the wavelength are related by $v = c/\lambda$, where c is the speed of light. In the ultraviolet and visible part of the spectrum, it is common to characterize radiation by its wavelength, expressed either in nanometers (nm) or micrometers (microns). In the infrared part of the spectrum, the wave number (cm^{-1}) is often used.

6.3.1. Greenhouse effect

Energy absorption by a given gas occurs when the frequency of electromagnetic radiation is similar to the molecular vibrational frequency of the gas. When energy is absorbed by a gas, it is transformed into internal molecular motion that produces an increase in temperature.

The atmosphere is a fluid formed by different types of gases and each of them behaves differently. Consequently, the absorbed energy is selectively performed for different wavelengths, being in some cases transparent for certain ranges of the spectrum. The atmosphere has a low absorption capacity in the visible spectrum, but it has a significant capacity to absorb ultraviolet radiation (short-wave radiation from the sun), and the main responsible for this phenomenon is ozone. Likewise, the atmosphere has a good capacity to absorb infrared or long-wave radiation that comes from the Earth, whose responsible are water vapor, carbon dioxide and other trace gases such as methane and nitrous oxide.

Gases that are good absorbers of solar radiation are important in the heating of the atmosphere, for example, the absorption of solar radiation from ozone provides the energy that heats the stratosphere and mesosphere.

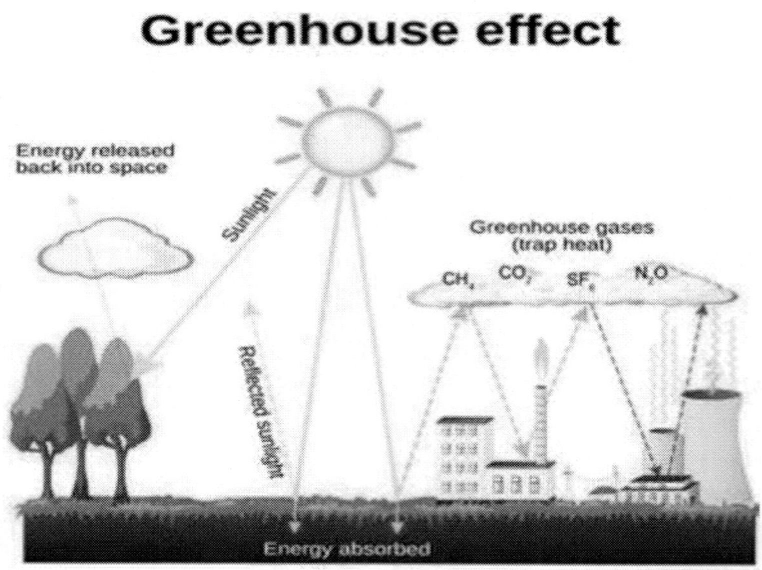

Source:
https://www.123rf.com/photo_66662245_stock-vector-greenhouse-effect-diagram-showing-how-the-greenhouse-effect-works-global-warming.html.

Figure 13. Diagram of the greenhouse effect.

The absorption of infrared radiation from the Earth is important in the energy balance of the atmosphere. This absorption by trace gases heats the atmosphere, promoting them to emit longer-wave radiation. Some of this radiation is released into space and some is radiated back to the Earth's surface (Figure 13). Two-thirds of the atmospheric radiant energy is directly returned to the surface, providing an additional source of energy in addition to direct solar radiation.

The net effect of this phenomenon allows the Earth to store more energy near its surface than it could store if the Earth had no atmosphere, and consequently, the temperature is higher, on the order of 33°C. This process is known as the natural greenhouse effect. Without the greenhouse effect, the average surface temperature would be approximately 18°C below zero and life on the planet would not be possible.

Consequently, gases in the atmosphere that absorb infrared radiation from the Earth, or outgoing radiation, are known as Greenhouse Gases (GHG), including carbon dioxide, water vapor, nitrous oxide, methane and ozone. These gases have molecules whose vibrational frequency is located in the infrared range of the spectrum.

Some GHGs emitted by human activities are carbon dioxide, nitrous oxide, methane and some halocarbons (such as CFCs, HCFCs, HFCs and PFCs), which together with tropospheric ozone (which is formed from carbon monoxide, nitrogen oxides and other volatile organic compounds), are good absorbers of infrared radiation. Halocarbons are especially noteworthy because many of them absorb energy at wavelengths that are not absorbed by carbon dioxide or water vapor (a region known as an atmospheric window).

Changes in the atmospheric concentration of GHGs and aerosols, solar radiation and soil surface properties affect the absorption, scattering and emission of radiation within the atmosphere and at the earth's surface. Positive or negative results in the energy balance due to these factors are expressed as radiative forcing, which is used to compare the influence of warming or cooling on the climate system [28-30].

Radiative forcing is a measure of the influence of altering the balance between incident solar radiation and outgoing infrared radiation on the atmosphere-Earth system, denoted by a change in net irradiance at the tropopause, and it is expressed in watts per square meter (W m^{-2}). Generally, radiative forcing values are for changes relative to conditions defined in 1970. These perturbations are due to internal changes or external forcings of the climate system, such as changes in the concentration of a GHG or in the radiation emitted by the sun.

A positive radiative forcing tends to warm the troposphere, while a negative forcing tends to cool it. The depletion of the ozone layer due to its destruction by the increase in halocarbon emissions since 1970 has represented a negative radiative forcing of the climate system, since ozone is a GHG. On the other hand, the increase in GHGs has produced a positive forcing. It is estimated that the global average net effect of human activities since 1750 has been a warming with a radiative forcing of 1.6 W m^{-2}.

The radiative forcing of GHGs and ozone is shown in Figure 14, and its magnitude is given by the product of their concentration and radiative efficiency (energy absorbed per unit concentration W m^{-2}*ppb). CO_2 has contributed the most to the positive radiative forcing with 1.66 W m^{-2}, followed by CH_4 with 0.48 W m^{-2}, tropospheric ozone with 0.35 W m^{-2}, N_2O with 0.16 W m^{-2} and some halocarbons. The radiative forcing of CO2 has increased by 20% during the last 10 years (1995-2005), being the largest observed or inferred change for a decade in the last 200 years.

Other agents that contribute to radiative forcing are changes in tropospheric aerosols, changes in surface albedo and in the radiation emitted by the sun. Regarding the latter, it is estimated that changes in solar radiation since 1750 have generated a radiative forcing of 0.12 W m^{-2}.

Aerosols of anthropogenic origin (mainly sulfates, organic carbon, black carbon, nitrates and dust) cause a direct radiative forcing of -0.5 W m^{-2} and an indirect forcing of cloud albedo of -0.7 W m^{-2}. Therefore, aerosols have an influence on cloud lifetime and precipitation.

Changes in surface albedo due to changes in the soil cover and deposition of carbon black aerosols on snow exert forcings of -0.2 W m^{-2} and +0.1 W m^{-2} respectively.

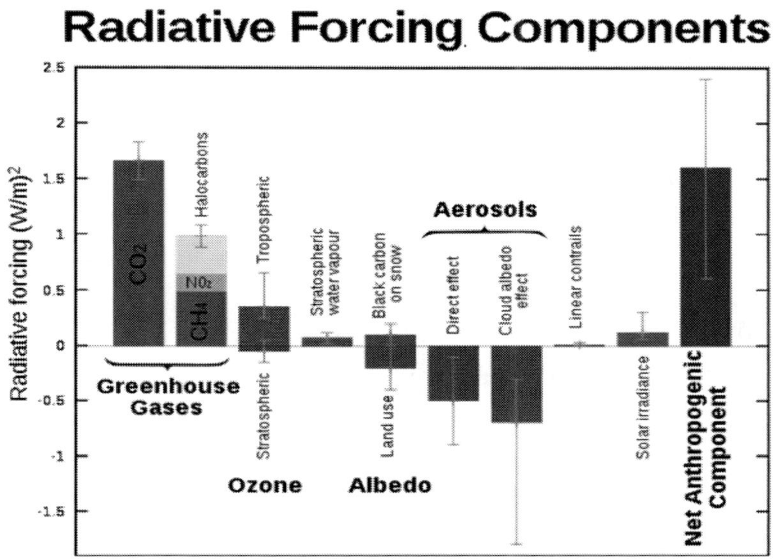

Source: https://energyeducation.ca/encyclopedia/Radiative_forcing.

Figure 14. Radiative forcing.

Table 9. Global warming potential (GWP) of GHGs

GHG	GWP for 20 years	GWP for 100 years
CO_2	1	1
CH_4	62	23
NO_X	275	296
HFC-23	9400	12000
HFC-125	5900	3400
HFC-134a	3300	1300
HFC-143a	5500	4300
CF_4	3900	5700
C_2F_6	8000	11900
SF_6	15100	22200

Among the halogenated compounds, CFCs have contributed 0.28 W m^{-2} to the positive radiative forcing since 1750, while HCFCs have contributed 0.033 W m^{-2} and HFCs 0.007 W m^{-2}. Increases in halocarbon gas concentrations between 1970 and 2000 have contributed about 30% of the increase in radiative forcing related to GHGs during this period.

Halocarbon molecules can be thousands of times more efficient as absorbers of energy emitted by the earth than a molecule of carbon dioxide, and small amounts of these gases can contribute significantly to the radiative forcing of the climate system.

The radiative effect of CO_2 and water vapor is to warm the surface climate and cool the stratosphere, while the radiative effect of halocarbons is to warm the troposphere and the stratosphere due to their absorption in the atmospheric window.

Table 9 shows the global warming potential (GWP) for various gases, which is an indicator of the radiative effect of a substance over a chosen time horizon, based on carbon dioxide. The highest GWP values occur for species that absorb the most radiation or have long half-lifetimes. The time horizon chosen is generally 100 years, in order to represent the future impact of the substance in the next 100 years.

6.3.2. Climatic Change

According to the United Nations Framework Convention on Climate Change (UNFCCC), climate change is understood as a change in climate attributed directly or indirectly to human activity that alters the composition of the global atmosphere and that is in addition to natural climate variability observed over comparable time periods. On the other hand, the Intergovernmental Panel on Climate Change (IPCC) defines climate change as any change in climate over time, due to natural variability or as a result of human activities.

From the meteorological point of view, climate change is the alteration of the prevailing conditions. External processes such as variations in solar radiation, variations in the Earth's orbital parameters (eccentricity, the inclination of the Earth's axis with respect to the ecliptic), movements of the Earth's crust and volcanic activity are factors of great importance in climate change. Internal processes of the climate system can also produce changes of sufficient magnitude and variability through interactions among its elements. The Earth's climate depends on the radiative balance of the atmosphere, which in turn depends on the amount of solar radiation entering the system and on the atmospheric concentration of some variable gases that exert a natural greenhouse effect (traces gases with radiative activity, clouds and aerosols). These radiative forcing agents vary both naturally and as a result of human activity, producing alterations in the climate of the planet (Figure 15).

Climate change is partly the result of increased GHG emissions. Climate variability occurs when a phenomenon generates abnormal climate behavior with a certain frequency, but it is a temporary and transitory phenomenon. However, there is a difference between climate variability (e.g., the El Niño phenomenon) and climate change. On the other hand, climate change denotes a process that is not temporary and can be verified over time by reviewing climate data (e.g., temperature).

An integrated approach to climate change takes into account the dynamics of the full cycle of interrelated causes and effects in all affected sectors. The following diagram (Figure 16) presents an integrated assessment framework for the consideration of anthropogenic climate changes as reported in the TAR (IPCC, Third Assessment Report).

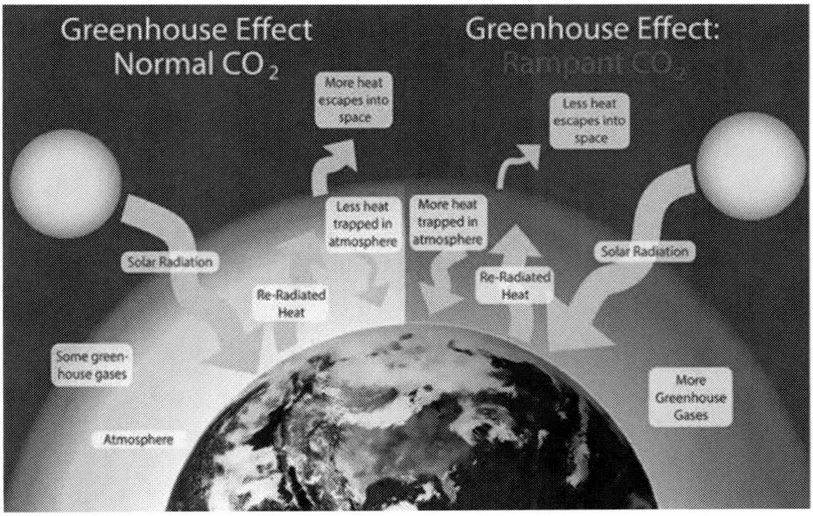

Source: https://energyeducation.ca/encyclopedia/Radiative_forcing.

Figure 15. Natural greenhouse effect and its formation leading to global warming (increase in global surface temperature).

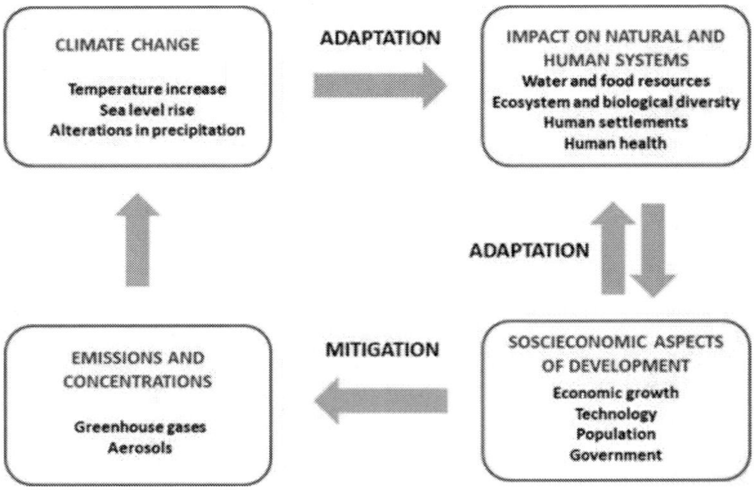

Figure 16. Integrated approach to climate change.

Decision making on climate change is essentially a sequential process that takes place under conditions of overall uncertainty.

Global warming can be understood in a simplified form as the gradual increase in the temperature of the planet as a consequence of the increase in the emission of certain GHGs that prevent the sun's rays from leaving the earth under normal conditions. A "thicker" layer of greenhouse gases retains more of the infrared rays and raises the temperature.

On the other hand, it is a term commonly used to define two phenomena: It is the observed phenomenon that shows on average an increase in the temperature of the Earth's atmosphere and oceans in recent decades. It is also a theory that predicts, from projections based on computer simulations, a future growth in temperatures. The majority scientific opinion on climate change says that "most of the warming observed over the last 100 years is attributable to human activity". Simulations seem to indicate that the main cause of the human-induced warming component is due to the increase in carbon dioxide. The temperature of the planet has been rising since the end of the 19th century, when the Little Ice Age ended. Global warming and the greenhouse effect are not synonymous. The greenhouse effect, enhanced by pollution, may, according to theories, be the cause of the observed global warming.

7. Tools for Air Quality Assessment

An air quality monitoring and assessment program must combine three fundamental tools [31]:

- Air pollutant monitoring
- Pollutant dispersion modeling
- Emissions inventory

Monitoring provides a scientific basis for the development of strategies and policies to improve air quality. It also allows the quantification of the effect of the measures adopted to implement them. However, any monitoring network, although well designed and operated, only gives a partial, but useful, picture of the behavior of air pollution variables in space and time for a city or region. The other tools are used to complete and clarify that picture, through interpolation calculations and mapping processes.

Figure 31 shows how these tools are integrated into an air quality monitoring and assessment program.

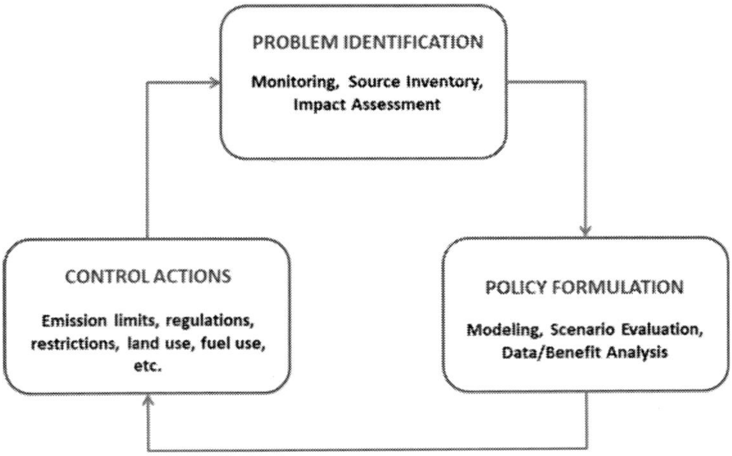

Figure 17. Tools to be integrated into an air quality monitoring and assessment program.

When initiating an air quality monitoring and assessment program, one must begin by defining the purpose of the monitoring tool, i.e., what pollution phenomena or effects are to be measured. The remaining tools - the modeling and the emissions inventory - should be integrated around these objectives.

Based on this first step and not before, the data quality objectives and aspects such as the number or density of sampling points and their location, among others, should be defined.

A clear definition of objectives allows us to maximize resources to achieve them (Figure 18). Objectives should be defined in a concise and clear manner, and should be congruent with economic realities and technical possibilities. Fuzzy, too restricted or overly ambitious monitoring objectives will result in cost ineffective programs and poor data utilization. In this circumstance it is not possible to make optimal use of available resources.

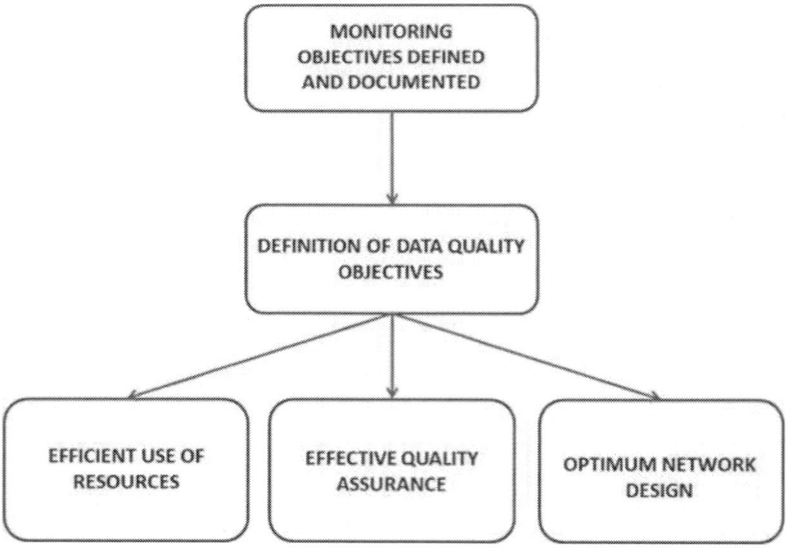

Figure 18. Optimal working diagram to maximize resources and meet the objectives of the monitoring network.

Among the general objectives recommended by the WHO [32] are the following:

- Determine exposure and assess the impact on the health of the population.
- To inform the public about air quality and increased threats.
- Determine compliance with national or international standards.
- Provide information for air quality management, traffic and land use planning.
- Identify trends in natural ecosystems.
- Identify and locate natural resources.
- Guide development policies and prioritize management actions.
- Develop and validate management tools (Geographic Information Systems and Models).
- Evaluate the impact of resources on areas and sites.
- Trend assessment to identify future problems or progress versus management and control objectives.

According to the U.S. EPA [31], the general objectives should be:

- Evaluate compliance with air quality standards and/or progress achieved in this regard.
- Activate control procedures to prevent or alleviate pollution episodes.
- Observe pollution trends in a region, including non-urban areas
- Provide databases for research and evaluation of urban, land use, and transportation effects.
- Provide a basis for planning; development and evaluation of control strategies; and development and validation of diffusion models.

Based on the above, U.S. EPA establishes four different types of monitoring:

- *Standards Compliance Monitoring:* performed with the primary objective of assessing compliance with air quality standards and/or progress achieved by applying control measures.
- *Monitoring of Emergency Episodes:* It is performed in order to activate control procedures to prevent or alleviate pollution episodes. Automatic monitoring networks are ideal for this case because they allow implementing control measures based on real time readings. It is recommended to take hourly averages in order to obtain useful and agile information. In order to maintain the equipment at the maximum operating point, personnel must be available at each station during the episodes, or the network must have a well-adjusted maintenance program to avoid any type of failure in the collection or transmission of data.
- *Monitoring of trends*: It is carried out to provide databases for research and evaluation of urban, land use and transportation effects. The monitoring of trends is characterized by a minimum number of stations deployed in the largest possible area, as long as the proposed objectives are met. The main objective of these programs is to determine the amount and nature of air pollution, along with the evolution of pollution levels with respect to geographic, socio-economic, climatological and other factors. The data collected are useful for planning epidemiological studies and for conducting preliminary surveys to provide background information on the need to conduct or expand studies on pollution in certain areas. When interpreting the data obtained from these surveys, it is necessary to keep in mind that although the monitoring sites have been carefully chosen, it is impossible to ensure that a site is not influenced by local factors such as topography, structures, sources of pollution too close to the site, and other variables. The effect of these factors cannot be accurately anticipated, but should still be considered in the design. Comparison between pollution levels in various areas is valid only if the sites are representative of the conditions for which the study was designed.

- *Research Monitoring:* Air quality monitoring networks related to health effects are designed with dual-purpose stations: to determine pollutant concentrations over 24-hour periods and to determine long-term (greater than 24-h) pollution standards. The research requires that the monitoring sites be located to represent the concentrations to which the population groups under study are exposed. Therefore, monitoring sites should be located in small, well-defined residential areas within a community. Averaging times of measurements should be short enough to allow for studies on acute health effects that form the scientific basis for short-term standards. The frequency of monitoring, usually daily, should be sufficient to characterize air quality as a function of time. The monitoring system should be flexible and able to respond to emergency conditions, with immediate availability of data.

According to the above general objectives, the US-EPA suggests that monitoring networks should be designed with stations that pursue one of the following six specific objectives:

- Detect the highest expected concentrations in the area covered by the monitoring network.
- To find representative concentrations in areas of high population density.
- To show the impact of large sources on ambient pollution levels.
- Detect background concentration levels.
- Study the regional transport of pollutants over populated areas.
- Detect the impact on population and goods in rural and remote areas.

There is no basic regulation for the design of an air quality monitoring network (AQMN), since decisions on the number and location of monitoring sites are ultimately based on objectives and resource constraints. An AQMN is almost never oriented to a single objective; therefore, the sites selected may obey two or more objectives at the same time. However, it should be ensured that it is possible to contrast results among several sites, without duplicating efforts. The main design goal is to ensure the greatest amount of information with the least amount of effort.

REFERENCES

[1] *National Law, 20.284/73- Republica Argentina*. .http://servicios.infoleg.gob.ar/infolegInternet/anexos/40000-44999/40167/norma.htm.

[2] Pan American Center for Sanitary Engineering and Environmental Sciences, (1982). *Red Panamericana de Muestreo de la Contaminación del Aire – REDPANAIRE* [*Pan American Air Pollution Sampling Network - REDPANAIRE*]. Final Report 1967-1980. Lima. CEPIS. https://pesquisa.bvsalud.org/bvsvs/resource/pt/lil-146263.

[3] Wark, Kenneth & Warner, Cecil. (1990). *Air pollution: origin and control*. México. Limusa & Noriega, ISBN: 978-968-18-1954-5.

[4] *Secretariat of Environment and Sustainable Development, Resolution 1327/2014*. http://www.loa.org.ar/legNormaDetalle.aspx?id=26187.

[5] Brauer M, Lencar C, Tamburic L, Koehoorn M, Demers P, Karr C (2008) A cohort study of traffic-related air pollution impacts on birth outcomes. *Environ Health Perspect*, 116(5):680-6. doi: 10.1289/ehp.10952. Erratum in: Environ Health Perspect. 2008 Dec; 116(12):A519. PMID: 18470315; PMCID: PMC2367679.

[6] *UNEP*, 2015. https://wedocs.unep.org/rest/bitstreams/11114/retrieve.

[7] UNFPA, (2016). *United Nation Population Fund*. available, http://www.unfpa.org/pds/urbanization.htm, accedido Septiembre 2016.

[8] Parrish, D.D.; Kuster, W.C.; Shao, M.; Yokouchi, Y.; Kondo, Y.; Goldan, P.D.; de Goue, J.A.; Koike, M. and Shirai, T. (2009). Comparison of air pollutant emissions among megacities. *Atmospheric Environment*. 43(40):6435-6441

[9] *Air Quality Health Index Categories and Health Messages*, http://www.ec.gc.ca/cas-aqhi/default.asp?lang=En&n=79A8041B-1.

[10] *Daily Air Quality Index*, https://uk-air.defra.gov.uk/air-pollution/daqi?view=more-info.

[11] "*Air Quality Index (AQI) - A Guide to Air Quality and Your Health*", https://www.airnow.gov/index.cfm?action=aqibasics.aqi.

[12] Martínez, E & Díaz Y. (2004). *Atmospheric pollution. Science and Technology Collection*, Ediciones de la Universidad de Castilla – La Mancha.

[13] Sbarato, D. R., SbaratoV., Ortega J. E., (2007). *Prediction and evaluation of environmental impacts on the atmosphere.* Córdoba Argentina. Center for research and training in environmental health. Universidad Nacional de Córdoba. Editorial Brujas, https://www.editorialbrujas.com.ar/.

[14] Yang, J. W., Shen, Y. C., Lin, K. C., Cheng, S. J., Chen, S. L., Chen, C. Y., Kumar, P. V., Lin, S. F., Lu, H. E., & Chen, G. Y. (2020). Organ-on-a-Chip: Opportunities for Assessing the Toxicity of Particulate Matter. *Frontiers in bioengineering and biotechnology*, 8, 519. https://doi.org/10.3389/fbioe.2020.00519.

[15] Riipinen, I., Pierce, J. R., Yli-Juuti, T., Nieminen, T., Häkkinen, S., Ehn, M., Junninen, H., Lehtipalo, K., Petäjä, T., Slowik, J., Chang, R., Shantz, N. C., Abbatt, J., Leaitch, W. R., Kerminen, V.-M., Worsnop, D. R., Pandis, S. N., Donahue, N. M., and Kulmala, M. (2011). Organic condensation: a vital link connecting aerosol formation to cloud condensation nuclei (CCN) concentrations, *Atmos. Chem. Phys.,* 11, 3865–3878, https://doi.org/10.5194/acp-11-3865.

[16] Zaror, C. A. (2000) *Introduction to environmental engineering for the process industry,* Universidad de Concepción, Santiago de Chile.

[17] Zhang H. B., Luo Y. M., Wong M. H., Zhao Q. G., Zhang GL (2006) Distributions and concentrations of PAHs in Hong Kong soils. *Environmental Pollution* 14(1): 107-114.

[18] Morillo E., Romero A. S., Madrid L., Villaverde J., Maqueda C. (2008) Characterization and Sources of PAHs and Potentially Toxic

Metals in Urban Environments of Sevilla (Southern Spain). *Water Air Soil Pollution* 187: 41-51.

[19] Daly G. L., Lei Y. D., Castillo L. E., Muir D. C. G., Wania F. (2007) Polycyclic aromatic hydrocarbons in Costa Rican air and soil: A tropical/temperate comparison. *Atmospheric Environment* 41: 7339-7350.

[20] Ma L., Chu S., Cheng H., Wang X., Liu X., Xu X. (2005) Polycyclic aromatic hydrocarbons contamination in subsoil from outskirts of Beijing, People's Republic of China. *Geoderma* 129: 200-210.

[21] Cai Q. Y., Mo C. H., Wua Q. T., Katsoyiannis A., Zeng Q. Y. (2008) *The status of soil contamination by semivolatile organic chemicals (SVOCs) in China: A review. Science of the Total Environment* 389: 209-224.

[22] Maliszewska B. K., Smreczak B., Klimkowicz A. (2009). Concentrations, sources, and spatial distribution of individual polycyclic aromatic hydrocarbons (PAHs) in agricultural soils in the Eastern part of the EU: Poland as a case study. *Science of Total Environment* 407: 3746-3753.

[23] Instituto Nacional de Ecología (2005). Guidance on the development and use of emission inventories. México. *Secretary of Environment and Natural Resources.*

[24] Henry J. G. and Heinke G. W. (1999). *Environmental Engineering,* 2º Edición. Editoral Pretince Hall, México, 777p.

[25] *Atmospheric Chemistry and Physics: From Air Pollution to Climate Change,* Second Edition, by John H. Seinfeld and Spyros N. Pandis. Copyright (2006) John Wiley & Sons, Inc.

[26] Fitzky Anne Charlott, Sandén Hans, Karl Thomas, Fares Silvano, Calfapietra Carlo, Grote Rüdiger, Saunier Amélie, Rewald Boris. The Interplay between Ozone and Urban Vegetation—BVOC Emissions, Ozone Deposition, and Tree Ecophysiology. *Frontiers in Forests and Global Change*, VOLUME=2, YEAR=2019, DOI=10.3389/ffgc.2019.00050.

[27] Ty – Jour Au - Esch, Marjolein P. Y. (2015). Designing the Urban Microclimate. A framework for a design-decision support tool for the

dissemination of knowledge on the urban microclimate to the urban design process VL – 5, DO - 10.7480/abe.2015.6, JO - A+BE: *Architecture and the Built Environment*, ER.

[28] Bates, D. R., and Nicolet, M. (1950). The photochemistry of atmospheric water vapor, *J. Geophys. Res*. 55, 301-327.

[29] Farman, J. C., Gardiner, B. G., and Shanklin, J. D. (1985) Large losses of total ozone in Antarctica reveal seasonal ClCx/NOx interaction, *Nature* 315, 207-210.

[30] Molina, L. T., and Molina, M. J. (1987) Production of Cl_2O_2 from the self-reaction of the ClO radical, *Journal of Physical Chemistry*, 91, 433-136.

[31] *Quality Assurance Handbook for Air Pollution Measurement Systems.* Volume II. Ambient Air Quality Monitoring Program. United States, Environmental Protection Agency, 2017. en https://www3.epa.gov/ttnamti1/files/ambient/pm25/qa/Final%20Handbook%20Document%201_17.pdf.

[32] World Health Organization, (2000). Developed by WHO Expert Group of: Sustainable Development and Healthy Environment (SDE), Department of Protection of the Human Environment (EHA), Environment and Occupational Health Program (PHE). "*Air Quality Guidelines.*" Ginebra.

In: Air Pollution: Effects and Dangers ISBN: 978-1-53619-544-6
Editor: J. Esteban Colman Lerner

Chapter 2

AIR QUALITY MODELS AS TOOLS FOR RISK ANALYSIS

Santiago Aguilar Ferraro, Micaela B. Del Sole, Daniela Mellado and Erica Y. Sanchez*
Centro de Investigaciones del Medioambiente (CONICET-UNLP), La Plata, Argentina

ABSTRACT

Air pollution is a known hazard to human and environmental health. Susceptibility, differential exposure, differential preparedness, and coping ability are not uniform within populations. For this reason, the differential risk assessment is to have a good description of the vulnerabilities associated with the population studied. Also, it has levels of air pollution throughout the study region. Air quality models are presented here as efficient tools for the determination of hazardous atmospheres and threat zones. Also, these are useful for the identification of different sources and their relative contributions.

* Corresponding Author's Email: yaninasanch@quimica.unlp.edu.ar.

1. General Description of the Risk

Risk can be understood as a product of exposure and vulnerability, as the human capacity to be harmed by or to cope with exposure. The risk value defines the consequence degree on an economic, social, and environmental level. This variable does not identify causal mechanisms to prevent or predict the future occurrence of undesirable events; however, it provides sufficient information to identify areas that merit attention. We can state with certainty that reaching a zero risk level is impossible. However, its understanding can provide a basis for the development of land-use strategies and controls [1].

Social and biological factors, among other factors, influence coping and exposure. In this sense, vulnerability encompasses characteristics that can modify the risk by influencing both the exposure and coping capacity [2].

Susceptibility, or sensitivity, is often used as a synonym for vulnerability to explain how people differ in their ability to experience the effects of exposure. In this context, sensitive groups or persons are identified by a higher likelihood of manifesting adverse effects at a given dose [3].

Air pollution is a major environmental risk for health and accounts for an estimated 4,2 million deaths per year [4]. Large cities are often accompanied by industrial activity in their surroundings and sometimes within them. Industrial development is directly related to productivity and quality of life and is considered the primary measure of progress. However, it is also associated with concerns such as pollution and disasters that may occur. With prejudice, it also exposes the surrounding population to undesirable levels of pollution [5].

The quality of air in the urbanized environments of industrial areas faces numerous and complex factors that modify it and make it dynamic. The presence of multiple sources located hundreds of meters, along with constructions that increase turbulence and affect the dispersion of pollutants, is one example [6-9].

There are several methodologies used to assess ambient air quality, especially to pollution levels. Usually, they are applied to determine compliance with standards and to reduce exposure levels of the population. The World Health Organization (WHO) highlights air monitoring, modeling, and emissions inventory as the three principal tools in air quality management. As mentioned in Chapter 1, these assessment tools are interdependent in scope and application. Therefore, they should be conceived as interrelated components in any comprehensive approach to assessing exposure or determining compliance with air quality criteria [10].

In the framework of risk management, all three tools are useful to estimate the impacts of a threat defined as air pollution. The risk or severity of health outcomes associated with exposure to ambient air pollution is not uniform within populations. The estimation of differential risks requires having a good description of the vulnerabilities of the population, as well as the levels of air pollution.

The identification of only the areas that are reached by a higher level of threat (higher concentrations of a pollutant in the air), or only the regions that present greater vulnerability (greater social inequality and lower response capacities), shows an incomplete vision to address efficient risk management. In this sense, the risk matrix is a technique that has been tested and validated to quantify the associated risk [1, 11]. It allows for the identification and assessment of risks, also considering the impact that could be generated by the combination of different levels of threat and vulnerability.

Complementarily, risk can be expressed in terms of exposure rates from the population health approach, as a vulnerability variable, and health risk recommended by the WHO. The latter will be discussed in more detail in chapter 3.

1.1. The Vulnerability Framework for Air Pollution

As Ge et al. [12] state in their work, the concept of vulnerability emerged within the geography and the natural hazards research field.

However, over time researchers have concluded that vulnerability is a state that exists in a system before it encounters a hazard. They emphasize that the threat impacts could be magnified by certain social factors such as poverty, low levels of education, social services, poor public infrastructure, and risk perception [13-15].

Generally, the vulnerability has two components: 1) external risks and stresses to which an individual or household is subject, and 2) internal abilities that offer the tools means for coping without causing damage or loss. Vulnerability is generally associated with a particular hazard; however, people in a community may not be equally vulnerable to that hazard [3, 16].

Kasperson et al. [17] and Gupta et al. [18] considered four main variables that affect the physical and social characteristics of the population, making it vulnerable. These authors highlight as main variables susceptibility, differential exposure, differential preparedness, and coping capacity. In this sense, in agreement with these authors, we emphasize that society's vulnerability to air pollution is complex because of its multi-dimensional connections that generate differential exposure and susceptibility, as well as adjusting and coping capacity [19].

In particular, social vulnerability taps on a broad range of susceptibilities at the individual and community level: lack of access to resources and life-lines, insufficient information and well-being, and certain beliefs and customs [20-25]. The underlying causes of social vulnerability are economic, demographic, and political processes that affect the allocation and distribution of resources among different social groups. Also, these reflect the power relations that are generated between them [1, 15, 26].

Some aspects of social vulnerability which contribute to the severity of risk are presented below [1]:

- The limited access to economic and cultural resources produces an unequal development of the social group capacities. At the same time, the conditions of poverty restrict people to access communication technologies. Consequently, individuals are not well prepared to respond to an emergency, both in terms of financial and communication means.

- The limited access to basic services (education, public health, potable water) increases the vulnerability of the population, both in previous instances and after a possible release. Although these characteristics do not cause this kind of unfortunate event, they condition the response of the population, the management system and the subsequent recovery. Additionally, building infrastructure quality and overcrowding are key factors in response decisions.
- Individuals who need care and attention, such as children and elders, increase the level of vulnerability of the population. In the same way, the high diversity of minority social identities is exposed to social and economic marginality. It should be noted that the degree of gender violence and unequal access to economic, cultural, social and political resources generate a greater vulnerability of women compared to men.

Finally, the vulnerability framework is useful to develop a local vulnerability index to examine the role of exposure, susceptibility, and the coping mechanism in determining vulnerability to air pollution.

1.2. The Threat Framework for Air Pollution

Air pollution is a major environmental health threat to humans, especially for children in whom respiratory function is still developing [27].

The potential for air pollution depends on both the meteorological potential and the complexity of the terrain. Particularly for the development of high concentrations of pollutants in a given area [16].

In this context, the threat can be quantified from pollution levels, estimated by dispersion models, which use emission and meteorological data to run the simulations. Models are a powerful tool for interpolating, predicting, and optimizing control strategies, but their quality depends on the availability of reliable emission data. It is also relevant to consider that these models are appropriate for local sources and specific topography conditions and that they are compatible with the available database on emissions and meteorology [5, 8-9, 28-31]. Finally, air monitoring data contribute to the validation of the simulations.

In the following section, we look in more detail at different software used for air quality modeling. These can be implemented as tools to assess threats and risks.

2. Air Quality Models

Air quality models (AQMs) use mathematical and numerical techniques to simulate the physical and chemical processes that affect air pollutants as they disperse and react in the atmosphere [32]. AQMs are useful for studying various types of air pollutants. Also, to provide the possibility to reveal the contributors of air pollutants [33]. These models help to identify source contributions to air quality problems and assist in the design of effective strategies to reduce harmful air pollutants [32].

Some models called dispersion models are useful to characterize primary pollutants that are emitted directly into the atmosphere, according to the classification of the U.S. EPA [32]. Photochemical models characterize secondary pollutants formed as a result of complex chemical reactions within the atmosphere. Last but not least, receptor models help to determine the relative contributions from different sources. A more detailed description of this type of model is explained below.

2.1. Dispersion Models

Dispersion models (DM) use mathematical formulations to represent the atmospheric processes that disperse pollutants emitted from a source. These models allow the prediction of concentrations at downwind receptor locations [32]. The meteorological factors directly infer in the dispersion of the pollutants. Consequently, these factors influence the determination of the dimension and characterization of the risk and threat zone. When contaminants are not readily dispersed, they tend to increase their concentration. The speed of the wind plays a fundamental role in its dilution, determining how quickly contaminants mix with the surrounding air. Also, the type of atmospheric stability may favor or limit the upward vertical motions. The type of atmospheric stability and the wind speed are the main factors of the dispersion, but we must remember that these variables depend on other factors [1].

DMs allow us to describe the dynamics of dense or neutral pollutant clouds. These can be classified in a range from simple to sophisticated, from modeling in simple algebraic expressions to those that require numerical methods to solve a system of differential equations system, which mathematically translates into direct proportionality to the computational cost [35]. The appropriate type of model depends on the properties of the problem, the scale of the study scenario, the required precision, the available input data, and the desired results [34].

DMs can provide answers to a multitude of contextualized questions in the design of new emission sources. Also, for chemical emergencies and the estimation of impact areas. In this sense, some software that use dispersion models offer graphic and numerical outputs that describe the estimated concentrations for each coordinate within the study domain. This type of information is useful to quantify the level of threat and finally to estimate the level of risk.

The following is an introduction and description of two software programs that implement dispersion models to estimate the concentrations to which the population may be exposed in the study region. These software are applied in some chapters in cases of study, to assess the threat and finally to estimate the risks.

2.1.1. ALOHA Software

ALOHA (Areal Locations of Hazardous Atmospheres) is free software, specially designed for use in chemical accident response, but also this is useful for emergency planning and training. It was developed by the United States Environmental Protection Agency (U.S. EPA) and the National Oceanic and Atmospheric Administration (NOAA). This software has a friendly interface with indications and messages of warning and caution. It was designed with the possible non-expertise of the users in mind.

The primary purpose of ALOHA is to provide an estimate of the spatial extent of hazards associated with short-term accidental releases of volatile and flammable chemicals. It allows quantifying the impact on human health due to various situations such as the inhalation of toxic chemical vapors, the thermal radiation from chemical fires, and the effect of the pressure wave from vapor-cloud explosions [36].

ALOHA is limited to assessing the hazards of chemicals that become airborne. It incorporates models for estimating the rate of release and vaporization of chemicals from containment. Besides, it links these "source strength" models to a dispersion model to estimate the spatial extent of toxic clouds, flammable vapors, and explosive vapor clouds.

The software has an extensive library of chemical properties that facilitates its application. Also, it has the option of loading new substances and chemical properties. The results of the model are in both text and graphic format. ALOHA also makes use of meteorological information to estimate the dispersion of pollutants.

ALOHA calculations represent a compromise between accuracy and speed of execution. It was designed to deliver good results quickly enough and to be useful in response to chemical emergencies. At present, ALOHA is applied by many government agencies, consultancies, and universities to quickly estimate the dispersion of pollutants and their impact on the population [37].

The Figure (1) below illustrates an outline of the stages and data required to simulate with ALOHA.

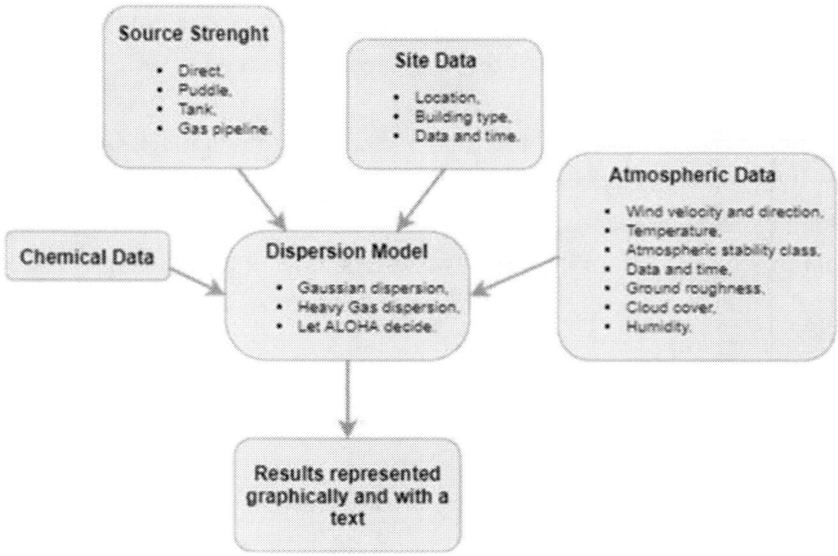

Figure 1. Outline of the stages and data required by ALOHA.

2.1.1.1. Sources

The rate at which a chemical becomes airborne is critical to estimate the size and duration of a toxic or flammable cloud. Due to this, ALOHA employs a variety of source strength models that allow predicting source strength for four general classes of chemical releases:

- Direct: It describes an instantaneous or continuous release of chemical vapors into the air from a single point. It is the only option that allows the simulation of an elevated release.
- Puddle: It simulates a constant area puddle, which may contain a boiling or unboiled liquid.
- Tank: It simulates the leakage or loss from a cylindrical or spherical tank at ground level with a single hole or leaking valve. The tank may contain a liquid, pressurized gas, or gas liquefied under pressure. Users can choose between two options for entering the chemical into the atmosphere: 1) directly an orifice or tank opening or 2) evaporation from a puddle formed by the spillage of the substance.

- Gas pipeline: It describes a pressurized pipeline containing gas, which may or may not be connected to a storage vessel.

The direct source is exemplified in the work undertaken by Pak and Kang [38] where a Cl_2 release is simulated from a one-ton cylinder. This is located close to residential areas. A case of puddle source can be found in Chakrabarti and Parikh [39]. This paper modeled different releases of United Nations Class-3 hazardous materials (benzene, toluene, p-xylene, methanol, and acetone) from road tankers along diverse routes. Instances for source tanks are explained in Bariha et al. [40] and Sanchez et al. [1]. The first work has modeled a leak of liquefied petroleum gas (LPG) from a truck tanker close to the residential area. The second has simulated a leak of ammonia from a storage tank in a meat-packing industry, located in the residential and urban context. A case of gas pipeline source is shown in Shao & Duan [41], they analyze the accident scene of a natural gas power plant. This paper simulated a pipeline leakage accident of natural gas.

The software limits the duration of any source to one hour. Nevertheless, it allows simulating an instant release with one minute of emission duration.

2.1.1.2. Dispersion

Air dispersion models are essential for predicting hazard zones associated with toxic or flammable gas clouds. These models are used to predict how the concentration of a pollutant, once released into the air, varies with time and position. ALOHA provides information about the concentration and duration of exposure but does not resolve the probability that an exposed individual will be injured [36].

The software uses airborne dispersion models that predict the concentrations of pollutants downwind of the source of a spill. The physical characteristics of the spilled material are also taken into account. Besides, it considers the weather conditions, site characteristics, and discharge type.

ALOHA includes two semi-empirical air dispersion models: the Gaussian model, appropriate for pollutant clouds that are not directly affected by gravity [42], and the Heavy Gas model, appropriate for pollutant clouds with densities greater than ambient air and significantly affected by gravity [43].

2.1.1.3. Receptors

ALOHA uses a graphical interface for both data entry and display of results. It allows visualizing to graphically the area where there is the possibility of exposure to toxic vapors, flammable atmospheres, overpressure from a vapor cloud explosion, or thermal radiation from fires, represented as threat zones. Threat zones indicate the area within which exposure at a ground level exceeds the level of concern specified by the user (for example, PAC, ERPG, AEGL, TEEL, among others) at some point after the start of a release; it is a record of the maximum exposure expected over time. In some scenarios, the user can also see the time dependency of the exposure at specific points.

Figure 2. Hazardous atmospheres estimated by ALOHA for a methanol release from a tanker truck [44]. Image edited in QGIS software.

The graphical outputs of the software, under the *.kml extension, can be imported from any geographic information system (GIS) and displayed as one more layer on the map of the scenario in question (Figure 2).

2.1.1.4. Strengths

ALOHA was designed to provide a rapid response in emergencies. Therefore, it has several features that facilitate its application in such situations. The most relevant are mentioned below.

- Few readily available input data are required regarding local atmospheric conditions, chemical identity, and spill scenario details. Besides, to minimize input data requirements, ALOHA includes an extensive database of chemical properties and geographic data.
- It is designed so that in situations where uncertainty is unavoidable, it will err in favor of overestimating rather than underestimating threat distances. It is conservative in its results.
- The software interface is intuitive, fast, and easy to use by the user, minimizing operator errors.
- It can estimate the rate of chemical emission and volatilization for many common accident scenarios, such as toxic vapor clouds, flammable vapor clouds, explosive vapor clouds, BLEVEs (Boiling Liquid Expansion Vapor Explosions), jet fires, and pool fires.
- It runs quickly on small computers that are easily transportable and affordable for most users.
- It is designed to predict the hazards associated with spills on a scale of 102 to 105 meters, with durations of up to one hour, which is typical of transportation accidents.
- The results are presented graphically with the possibility of being exported to a GIS.
- It is free to access.

2.1.1.5. Limitations

ALOHA cannot be more precise than the information that is provided to work. Even when the best possible input values are supplied, ALOHA can be unreliable in certain situations. Neither can model certain release types. Although accurately input information can be provided, ALOHA results may be unreliable when the following conditions exist:

- Very low wind speeds
- Very stable atmospheric conditions
- Wind shifts and terrain steering effects.
- Concentration patchiness, particularly near the source

ALOHA does not account for the effects of

- Chemical reactions
- Particulates
- Complex terrain
- Chemical mixtures
- Hazardous fragments

2.1.2. AERMOD Modeling System

AERMOD (American Meteorology Society-Environmental Protection Agency Regulatory MODel) is a modeling system, which was created by the American Meteorology Society together with the United States Environmental Protection Agency (U.S. EPA). It is a powerful software used to simulate the dispersion of pollutants in the air and is recommended by the U.S. EPA to estimate the impact of pollutant sources on environmental air quality levels at horizontal distances of up to 50 kilometers [45]. It is a model designed for the assessment of continuous emissions instead of short-term point emissions. It determines the distribution of pollutants in a plume, horizontally and vertically, following a normal steady-state behavior called Gaussian distribution [46].

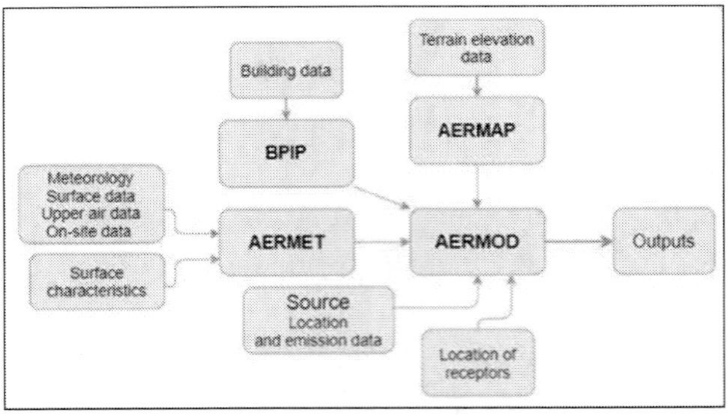

Figure 3. Block diagram of the AERMOD modeling system and pre-processors.

AERMOD has two pre-processing modules that are regulatory components for modeling. They must be used to configure the input data. First, the AERMET is a meteorological data preprocessor incorporating air dispersion based on planetary boundary layer turbulence structure and scale concepts, both which must be configured. Secondly, AERMAP is a terrain data preprocessor that incorporates the complex terrain using digital elevation models. It assigns the elevation height to the emission sources and each receptor of the grid [47, 48].

This modeling system incorporates a non-regulatory preprocessor module called BPIP to contemplate the effect of building descent. It is the effect that multiple buildings can have on the proximity of the sources during dispersion of the pollutants [48]. Figure 3 shows the AERMOD block and preprocessor diagram.

2.1.2.1. Emission Sources

AERMOD allows the simulation of the dispersion of contaminants from different types of sources, such as point, area, volumetric, and line sources. At the same time, there is the possibility of evaluating each type of source at different emission heights.

The main types of sources that AERMOD can model are described below, detailing the minimum input requirements for each of them [49]:

- *Point sources* are used to represent releases from sources such as chimneys and isolated vents. The inputs are the source location and height, the emission rate (g s^{-1}), the gas outlet temperature (°C), the stack internal diameter (m), and the gas outlet velocity (m s^{-1}).

- *Area sources* are used to model low-level or ground-level releases with no plume rise, such as lagoons or fires. The minimum requirements for this type of source are location coordinates, emission rate (g s^{-1}), source width (m), and source length (m). These last two are useful to calculate the area (m^2) of the emission source and the areal emission rate (g m^{-2}). Also, it is necessary to specify the angle of orientation of the polygon from the North.
- *Volume Sources* are useful to model releases from industrial sources, such as building roof monitors, multiple vents, conveyor belts, and coke stacks. The inputs are the source location and height, emission rate (g s^{-1}), length of each side (m), and height (m).
- *Line sources* are very similar to area sources. These are useful to model low-level or ground-level releases with no plume rise, such as storage piles, slag dumps, lagoons, and vehicle emissions on a highway. The inputs are the source location and height, source emission rate (g s^{-1}), endpoint coordinates of the line source, and the source width (m).

A large part of the literature simulates emissions from point sources, as illustrated by Silva [50]. This author simulates the dispersion from domestic heating equipment in Constitución city, Chile. Sosa-Pérez [51] exemplifies another case of application of the AERMOD for point sources. It models the three most important facilities of the industrial pole located in Mariel city, Cuba.

The author Gallego Martínez [52] models the area sources simulating the dispersion of various sources in the mining industry of the Cerro de San Pedro city, Mexico.

Diez et al. [53] illustrated an application case combining point and area sources to analyze the dispersion of pollutants from a cement industrial complex in the city of Malagueño, Argentina.

Heist et al. [54] model volumetric sources in their work. They analyze the dispersion of pollutants emitted by vehicle traffic on the road, considering volume and area sources.

Finally, the use of line sources is exemplified by Haobing et al. [55] which also simulates vehicular traffic on a high-volume highway in Gwinnett County, Georgia (United States).

2.1.2.2. Dispersion

AERMOD uses various parameters of the atmospheric boundary layer to model the dispersion of pollutants in the air. Many of these parameters are not observed but are estimated from other variables that are more easily measured. To make these estimates, data such as wind speed and direction and temperature observed near the surface, cloud cover, among other variables are required.

These meteorological input variables are obtained from a weather meteorological data observation station near the emission source and processed through the AERMET module. This preprocessor relates the meteorological data observed with other parameters such as Bowen's radius, surface roughness, and albedo. These depend on the land use in the study area (urban, rural, and forest, among others). AERMET creates two input files in a format acceptable to AERMOD: 1) The surface data file (*.SFC) provides a horizontal representation of the meteorology, and 2) the profile data file (*.PFL) reproduces a vertical characterization of the meteorological conditions [45].

2.1.2.3. Receptors

The model estimates the concentration in the centers of each grid from the generated receptor grid. The dispersion of the contaminants is graphically visualized with isoconcentration plumes.

The receptor grid can cover the entire modeling domain or have a smaller domain. The user must specify the location, the spacing between each point on the grid, and the grid type (polar or cartesian), to generate it.

2.1.2.4. Terrain

AERMOD can estimate the concentrations of air pollutants in all types of terrain. It is essential to give the elevation of the terrain in the study area when simulating the dispersion of the pollutants. Thus, the user can import and process the terrain height data through AERMAP terrain pre-process. The topography is configured in the modeling domain. The user must assign a representative terrain height to each receptor and source. Several standardized terrain data formats can be selected such as SRTM30, NTF, DTM, CDED, DEM, GTOPO30, and XYZ [55, 56]. However, it should be emphasized that only the SRTM and GTOPO30 formats are suitable for study areas other than the USA and can be downloaded automatically from the AERMAP interface. On the other hand, it is possible to generate a terrain file specific to the study area involved, Digital Elevation Model (DEM) or XYZ format, through other external pre-processors included in the AERMOD modeling system.

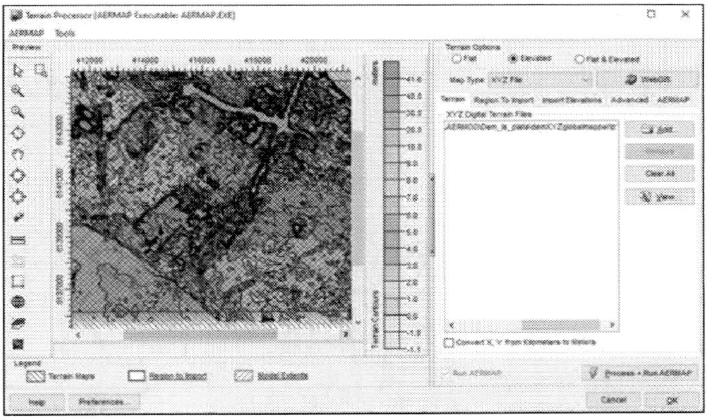

Figure 4. Terrain preprocessor. applied in the Gran La Plata.

Figure 4 shows the interface of the terrain pre-processor. Here, the emissions from a gas station in the Gran La Plata region were modeled. The terrain topography was imported in XYZ format.

2.1.2.5. Outputs

AERMOD uses its graphical interface for data entry and display of results. It is possible to calculate the maximum and the average concentration of the contaminant, both for each specified period of the year (1h, 4h, 6h, 8h, 12h, 24h, monthly, annual) as for each point receptor within the selected mesh. Isolines are drawn from the concentrations in each receptor of the mesh. These can be superimposed on cartography for the estimation of the affected surface. That is to say, it allows to export the concentration isolines in format .kml or .shp and then to post-process with some system of geographic information. It is then possible to calculate the affected areas and even the risk that contaminants may have in a specific population [58]. Besides being able to export the results in *kml or .shp format, they can be exported in text format. The last one has the concentrations in each receptor of the grid.

Finally, it should be emphasized that AERMOD is a complex model, so its proper use requires experience. The time needed to run the model will depend on the input data: i.e., number of receptors, number of years contained from the database of meteorological parameters, number of emission sources, and amount of pollutants. In addition, the running time is affected by the selected analysis period (i.e., hourly, daily and annual averages) and the required statistical treatments (i.e., calculation of percentiles, maximum value and average value). The duration of the simulation can take from several hours to a few days.

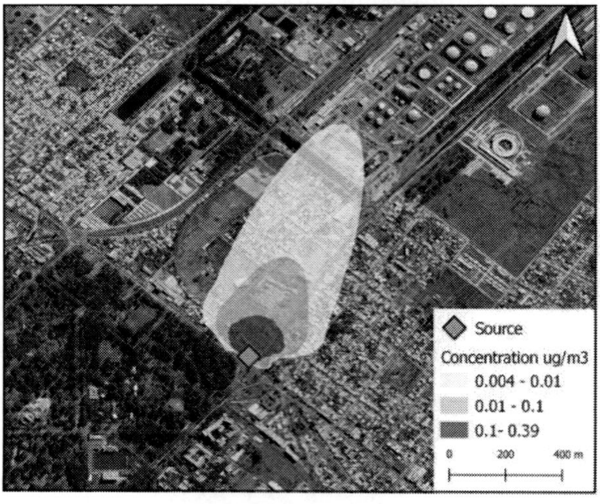

Figure 5. Benzene dispersion from an areal emission source

Figure 5 shows an example of execution with AERMOD, where the dispersion of benzene from a gas station was simulated as an area source.

2.1.2.6. Strengths

AERMOD can be used for multiple projects either by evaluating past and potential emissions since it allows us to evaluate a variety of emission scenarios by modifying the parameters of the emission sources. AERMOD also allows to evaluate the weight of the meteorological variations and to analyze the seasonal dynamics of a pollutant.

The main characteristics of the modeling system are [35, 59-72]:

- It allows the evaluation of regulatory or normative aspects, as well as its use for research purposes in the fields of air pollution and its effects on health.
- It is designed to work on a small scale (1m-50 km).
- It is the air quality dispersion model recommended by the U.S. EPA.
- Easy to implement model with an intuitive interface.
- No need for large computing resources (processors or computer memory)
- Its resolution (minimum distance between adjacent grids/receptors) has a spatial range of up to 1m.
- It allows working with all types of terrain (from complex, mountainous to flat terrain).
- Enable to simulate the dispersion of multiple substances simultaneously.
- It has a complete graphical analysis of the dragging by buildings (building downwash effect).
- Capable of processing terrain information in a wide variety of formats (DEM, SRTM, GTOPO30, NTF, DTM, CDED and XYZ formats).
- It allows us to elaborate graphs of direction, speed and frequencies of the winds.

- Once the execution of the model is finished, the results are automatically displayed in the main interface in a graphic form.

2.1.2.7. Limitations

Like ALOHA and other dispersion modeling software, AERMOD cannot be more accurate than the information provided to it to function. Within its limitations are [57]:

- It is a detailed model that requires a large amount of input data to be able to simulate.
- It only allows working on a local scale reaching up to 50 km from the emission source.
- It is a steady-state model, so it does not allow the evaluation of changes in the dispersion of pollutants as a function of time.
- The model is not valid for those emissions that undergo chemical reactions.
- The model has greater uncertainty in areas where weather conditions vary significantly, such as very complex terrain, boxed valley bottoms, or coastal areas.
- The model assumes that the plume follows a straight line.

2.2. Receptor Models

The Receptor Models (RM) are mathematical or statistical procedures that seek to respond to the source apportionment (SA) in a region, with the aim of identifying and quantifying the sources of air pollutants from observations made at a site receptor (immission data). These sites are chosen by the decision maker to be representative of the problem to be solved. With the purpose of planning, controlling, and mitigating air pollution, great efforts were dedicated to the development of different tools capable of interpreting air pollution, such as RM. The results obtained are used both in mitigation plans and in a public health study. Finally, the choice of the model is linked to the problem to be solved and the available data, this leads to the need to design complementary studies for a better understanding of their results [73-75].

Within the RM universe, models can be classified as either classic or hybrid, depending mainly on the use of local (non-historical) weather information for their simulation. Unlike classics that interpret the types of emissions from characteristic sources of the region, hybrid models combine meteorology with measurements of atmospheric pollutants to locate the regions with the greatest contribution of average emissions.

In terms of management, the RMs seek to respond to the Source Apportionment (SA), which represents a different concept for each model. Within the classics, SA is understood to be the identification and characterization of the sources of emission of atmospheric pollutants observed at the receptor site. On the other hand, in hybrids, the term is reserved for the identification and quantification of the geographic regions that contribute the highest emissions of a single pollutant to the receptor site [76].

Strength:

The application of RM as a complementary air quality management tool is explained under the following strengths [73, 76]:

- It does not depend on emissions inventories but makes use of prior knowledge about the relevant sources in the study region.
- It does not require complex chemical and meteorological processors.
- They are extensible to pollutants that do not present chemical or physical alterations during transport.
- Appropriate for local and regional scales.
- They offer graphical outputs that are easily interpretable by their users.
- They have a low computational cost.

2.2.1. Classic RMs

Currently, they are applied by different government agencies, consultancies and universities to quickly estimate the SA. They are procedures capable of identifying and characterizing the sources of emissions, quantitatively or qualitatively, of the air pollutants observed at the receptor site. Using common characteristics of the pollutants, the procedures study the joint effect of these pollutants through multiple linear regressions. Based on the resolution of mass balances, these models make use of concentrations of pollutants characterized in their associated chemical species to obtain profiles of the sources and their contributions. Understanding by "profile" the average composition of the chemical species emitted by the source, and by "contribution," the average concentrations emitted by the source that are observed in each of the measurements [73, 74].

The data to be modeled (concentrations) must be entered in matrix form, keeping the chemical species in its columns and the measurements in the rows. The choice of the participating species is made according to the objectives of the study, the characteristics of the region and the expected emission sources. In this instance, the species considered as source tracers or markers should be included because they help the user to interpret the profiles obtained. Other factors that determine the number of species available are the human, technical, and financial resources available.

Chemical profiles are of utmost importance in SA studies because they allow the identification of the sources that contribute the most to local / regional air pollution. In this way, the decision maker can interpret which are the conflicting sources that are controllable when there is a non-compliance with the established limits.

For a better understanding of the resulting profiles, the chemical characterization of the most relevant sources in the region must be known. The characterizations must come from experimental tests or previous studies. If there are no profiles obtained directly from the source, a sampling campaign can be designed with monitoring stations oriented to the source of interest, as long as, during the sampling period, no influence of emissions from other sources is observed. Anyway, there are repositories that have source profiles provided by the U.S. EPA (SPECIATE database) [77], the European Commission Joint Research Center (SPECIEUROPE database) [78], among others.

2.2.1.1. Classic Models Available

Within the variety of classic RM, the following three models stand out for their acceptance both in management and in academic fields:

- Principal Component Analysis (PCA)
- Positive Matrix Factorization (PMF)
- Chemical Mass Balance Sheet (CMB)

The last two, developed by the U.S. EPA, are available on the agency's website (https://www3.epa.gov/scram001/receptorindex.htm). These models require varying degrees of knowledge about the number of sources at a given site as well as the source profiles. Also, they present a friendly graphic interface for both data entry, selection of model parameters and the results obtained.

In particular, the CMB requires a database of chemical profiles in order to correctly distribute the concentrations measured by mass balance. In contrast, the PMF generates the source profiles internally from the environmental data. Also, PMF and CMB produce similar results in terms of the correlation of most sources, the magnitude of their contributions and the profiles obtained in the receptor sites [79, 80]. A typical graphic output of the PMF model can be seen in Figure 6. With respect to PCA, it is a technique widely used as RM due to its ability to reduce available information (relationships between concentrations of different contaminants), providing as a qualitative output two matrices that resemble source profiles and their contributions. This technique usually achieves solutions with a smaller number of sources compared to the models offered by the agency [74, 81-83]. Further detail of the PCA model is developed in chapter 6 of this book.

Since particulate matter is one of the most widely studied contaminants, several authors have implemented RM to respond to regional problems, managing to identify relevant sources and their impact on the site [85-88]. To a lesser extent, its use has been extended to SA of volatile organic compounds [89, 90].

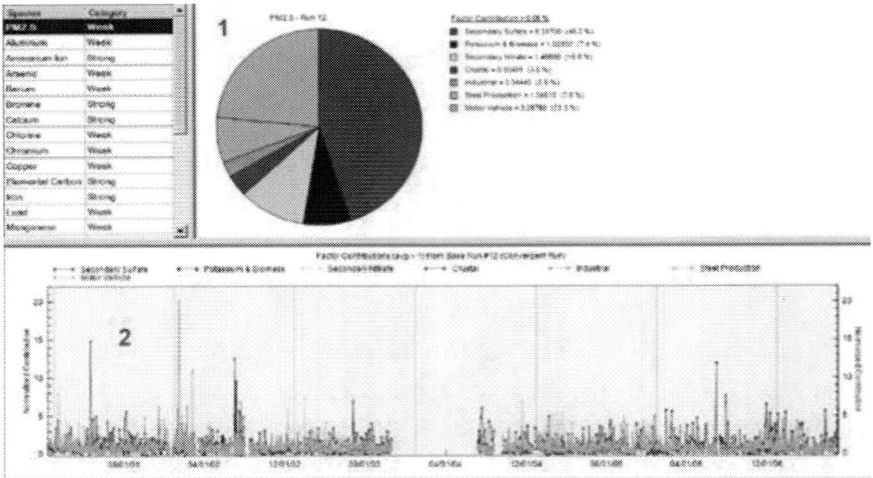

Figure 6. Example of the application of the PMF model [84].

2.2.1.2. Limitations of Classic RMs

As statistical procedures, classical RMs are designed to meet a set of natural physical constraints resulting from the mass principle of each atmospheric pollutant [73, 74]. The constraints to be considered are:

- The reproducibility of the original data by the model.
- The profiles and contributions of the resulting sources must be non-negative. This physical restriction implies that a source cannot be emitting a negative concentration, because RMs do not contemplate reactions between contaminants. Similarly, contributions must be positive because the model does not consider deposition or phase change.
- The sum of the chemical species mass contributions obtained by the model for each source cannot exceed its total observed mass.
- Its statistical nature requires large volumes of pollutant databases.

The third constraint defines the spatial domain of application, providing reliable results for local or regional scales. In addition, a better approximation of reality requires (i) a good knowledge of the processes that take place in the region, (ii) a good command of the chemical nature of the emissions from the sources, and (iii) skills in the use of computer tools.

The limitations of classical RMs are that i) source profiles must be stationary or quasi-stationary, ii) with constant emission rates, iii) modelled species must not react or change phase during transport from the sources to the receiving site, and finally, (iv) the data must be representative of the geographical region to be studied. In practice, failure to comply with the assumptions translates into the inability to approximate the study scenario, that is, the profiles and contributions of sources will not be able to explain the real physical system. Or, the solution may be to over/underestimate the number of sources, profiles and their contributions [73, 74].

2.2.2. Hybrid RMs

This RM classification seeks to identify and quantify the geographic regions that originate the emissions, which are perceived as contributions in the receptor site. In other words, it estimates the geographic areas with the highest probability of emission source presence. They are considered hybrid because they use concentrations of a single contaminant observed at a receptor site and the predominant role of wind in their execution. When applying the model, a great knowledge of the emission sources in the region is required to see how close the result of the model is to reality [91-94].

These models require as input data the concentrations and winds, both as a function of time, with a similar temporal resolution and taken at the same receptor site. In addition, the approximation to the concentrations associated with calm winds must be interpreted by the user, since they cannot be modeled, minimizing the solution error. Also, the periods in which the modeling is done must be previously defined to respond to a real problem. The most commonly used periods are daily, weekly, associated to working days, according to weather stations, among others [76].

Generally, the most complex hybrid models start from the estimation of back-trajectories of the pollutant measurements (concentrations). With origin in the monitoring center, the points of the trajectories are approximated by the data of the wind direction and speed. In this way, a data matrix is obtained that articulates the observed concentrations with their respective geographical coordinates as a function of time. Finally, some hybrid models such as Nonparametric Trajectory Analysis (NTA), estimate the average spatial concentrations by means of a nonparametric regression. Their graphical representation served on a map (e.g., GIS), allows visualizing the areas in which the highest average values are presented, which correspond to the regions of greatest contribution to the receptor site [92]. In summary, this type of models offers graphic and/or numerical outputs that give the estimated average concentrations for each coordinate within the studied domain. This type of information is useful to identify the regions that contribute the most emissions to the receptor site during the analysis period. A graphical output of the NTA model is shown in Figure 7. Further detail of this model and its application is developed in chapter 6 of this book.

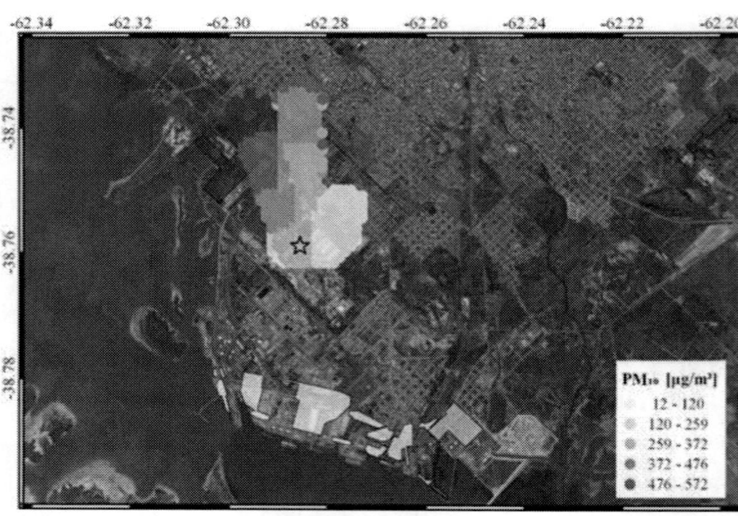

Figure 7. Example of the application of NTA Model [95].

2.2.2.1. Hybrid Models Available

Unlike classical RMs, the methods and models available for spatial analysis of contaminants are varied. The following are some examples of the most commonly used:

- Pollution rose (PR)
- Conditional bivariate probability function (CBPF)
- Nonparametric Trajectory Analysis (NTA)
- Hybrid single-particle lagrangian integrated trajectory (HYSPLIT)

In increasing complexity i) the PR is an extension of the wind rose, in which the frequency of average concentrations for a range of directions can be observed, ii) the CBPF relates the concentrations as a function of the probability of the wind direction and wind speed variables, iii) NTA estimates the average concentrations in space through a nonparametric regression of the observed data trajectories for a local scale, finally, iv) the HYSPLIT makes use of complex meteorological models to estimate the back-trajectories of the pollutants at higher scales.

The PR, CBPF and NTA models are available for use through libraries in commercial and free software. Therefore, their implementation is relatively simple, and they are selected by the different authors to identify mainly sources in coexistence with urban environments (traffic, resuspended dust, and other point-inventoried sources) [91, 93, 94, 96-98]. The HYSPLIT, designed by the National Oceanic and Atmospheric Administration (NOAA) of the United States, is the only one available for modeling at higher scales (regional to global) in its online format. This model has the particularity of being selected to study important sources that go beyond the territories of jurisprudence [99-101]. The interest lies in identifying the regional emission sources (through their georeferencing) and the directions of the prevailing winds.

2.2.2.2. Limitations of Hybrid RMs (Local Scale)

- Previous knowledge in statistical tools is required to implement the models for local or regional scales. Currently, there are many libraries that facilitate the design of the corresponding algorithms, with the exception of the PR that has specific commercial software.

- Most hybrid models assume that there is no loss of mass by deposition, nor chemical reaction of the contaminant during transport.
- The representativeness of the scenario has as a critical point the interpolation between the pollutant data and the meteorological parameters, which must have at least the same temporal resolution, the latter being representative of the monitoring site.
- Regarding the model, the estimated back-trajectories can be affected by the topography of the place, as well as by the presence of obstacles between the monitoring site and the study area.

ACKNOWLEDGMENTS

The authors would like to thank the National Agency for Scientific and Technical Promotion (ANPCyT), the National Scientific and Technical Research Council (CONICET) and the National University of La Plata, which have financed this work.

REFERENCES

[1] Sanchez, E. Y., Represa, S., Mellado, D., Balbi, K. B., Acquesta, A. D., Lerner, J. C., and Porta, A. A. 2018. "Risk analysis of technological hazards: simulation of scenarios and application of a local vulnerability index." *Journal of hazardous materials*, 352, 101-110.

[2] Kasperson, J. X., Kasperson, R. E., 2001. *International Workshop on Vulnerability and Global Environmental Change*. Stockholm Environment Institute, Stockholm.

[3] Makri, A., and Stilianakis, N. I. 2008. "Vulnerability to air pollution health effects." *International journal of hygiene and environmental health*, 211(3-4), 326-336.

[4] World Health Organization, WHO. 2020. "*Ambient air pollution*." (n.d). Accessed November 20 from https://www.who.int/teams/environment-climate-change-and-health/air-quality-and-health/ambient-air-pollution.

[5] Alhajraf, S., Al-Awadhi, L., Al-Fadala, S., Al-Khubaizi, A., Khan, A. R., and Baby, S. 2005. "Real-time response system for the prediction of the atmospheric transport of hazardous materials." *Journal of Loss Prevention in the Process Industries*, 18: 520–525.

[6] Perloff, H. S. 1973. "*La calidad del medio ambiente urbano* [*The quality of the urban environment*]." Colección de Urbanismo Oikos-Tau, Barcelona.

[7] Li, F., and Zhou, T. 2019. "Effects of urban form on air quality in China: An analysis based on the spatial autoregressive model." *Cities*, 89, 130-140.

[8] Qiu, G., Song, R., and He, S. 2019. "The aggravation of urban air quality deterioration due to urbanization, transportation and economic development–Panel models with marginal effect analyses across China." *Science of the Total Environment*, 651, 1114-1125.

[9] Sorte S., Arunachalam S., Naess B., Seppanen C., Rodrigues V., Valencia A., Borrego C., & Monteiro A. (2019). Assessment of source contribution to air quality in an urban area close to a harbor: Case-study in Porto, Portugal. *Science of the Total Environment*, 662, 347-360.

[10] World Health Organization, WHO. 2006. "WHO Air quality guidelines for particulate matter, ozone, nitrogen dioxide and sulfur dioxide: global update 2005: summary of risk assessment" (No. WHO/SDE/PHE/OEH/06.02). *World Health Organization*. Accessed November 20. https://apps.who.int/iris/bitstream/handle/10665/69477/WHO_SDE_PHE_OEH_06.02_eng.pdf.

[11] Calixto, E. 2016. "*Gas and oil reliability engineering: modeling and analysis*." 2nd ed. Gulf Professional Publishing.

[12] Ge, Y., Dou, W., Zhang, H. 2017. "A New Framework for Understanding Urban Social Vulnerability from a Network Perspective." *Sustainability* 9.10: 1723.

[13] Topuz, E., Talinli, I., and Aydin, E. 2011. "Integration of environmental and human health risk assessment for industries using hazardous materials: A quantitative multi criteria approach for environmental decision makers." *Environment International* 37.2: 393–403.

[14] Ramírez-Camacho, J. G., Carbone, F., Pastor, E., Bubbico, R., and Casal, J. 2017. "Assessing the consequences of pipeline accidents to support land-use planning." *Safety Science*, 97: 34–42.

[15] Wisner, B., Blaikie, P. M., Blaikie, P., Cannon, T., and Davis, I. 2004. "*At risk: natural hazards, people's vulnerability and disasters*." 2nd ed. Routledge: London.

[16] Wright, C., and Diab, R. 2011. "International Perspectives: Air Pollution and Vulnerability: Solving the Puzzle of Prioritization." *Journal of Environmental Health*, 73.6: 56-64.

[17] Kasperson, R. E. 2002. Personal communication to Michael Callahan. *Chair of the Cumulative Risk Technical Panel, as part of an EPA Science Advisory Board Consultation on the draft Framework for Cumulative Risk Assessment.*

[18] Gupta, S., Das, S., and Murty, M. N. 2019. "Quantifying Air Pollution Vulnerability and its Distributional Consequences." *Ecology, Economy and Society–the INSEE Journal*, 2.1: 93-125.

[19] U.S. EPA. 2003. "*Framework for Cumulative Risk Assessment*." EPA/600/P-02/001F. Washington, DC: U.S. Environmental Protection Agency, Office of Research and Development, National Center for Environmental Assessment.

[20] Martins, V. N., e Silva, D. S., and Cabral, P. 2012. "Social vulnerability assessment to seismic risk using multicriteria analysis: The case study of Vila Franca do Campo (São Miguel Island, Azores, Portugal)." *Natural Hazards*, 62.2: 385–404.

[21] de Oliveira Mendes, J. M. 2009. "Social vulnerability indexes as planning tools: beyond the preparedness paradigm." *Journal of Risk Research*, 12.1: 43-58.

[22] Barata, R. B., Ribeiro, M. C. S. D. A., and Cassanti, A. C. 2011. "Social vulnerability and health status: a household survey in the

central area of a Brazilian metropolis." *Cadernos de Saúde Pública*, 27: s164-s175.

[23] Zhou, Y., Li, N., Wu, W., Wu, J., and Shi, P. 2014. "Local spatial and temporal factors influencing population and societal vulnerability to natural disasters." *Risk analysis*, 34.4:614-639.

[24] Cutter, S. L., Boruff, B. J. and Shirley, W. L. 2003. "Social Vulnerability to Environmental Hazards." *Social Science Quarterly*, 84: 242-261.

[25] Yenneti, K., Tripathi, S., Wei, Y. D., Chen, W., & Joshi, G. 2016. "The truly disadvantaged? Assessing social vulnerability to climate change in urban India." *Habitat International*, 56:124-135.

[26] Represa, N. S., Sanchez, E. Y., y Porta, A. A. 2018. "Estudio de la vulnerabilidad social en Argentina mediante el uso de SIG: Construcción de un índice de aplicación local." *Anuario do Instituto de Geociencia*, 41.2:351-357. ["Study of social vulnerability in Argentina through the use of GIS: Construction of a local application index" Yearbook of the Institute of Geosciences-UFRG, 41.2:351-357].

[27] Liu, L., and Zhang, J. 2009. "Ambient air pollution and children's lung function in China." *Environment International*, 35:178–186.

[28] Guariso, G., Maione, M., and Volta, M. 2016. A decision framework for Integrated Assessment Modelling of air quality at regional and local scale. *Environmental Science & Policy*, 65: 3–12. https://doi.org/10.1016/j.envsci.2016.05.001.

[29] Casares, J. J., Rodríguez, R., Maceira, P., Souto, J. A., Ramos, S., Costoya, M. A., Sáez, A., Vellón, J. M., 2005: Inventario, Análisis y Proyección de las Emisiones Atmosféricas Industriales de Galicia [Inventory, Analysis and Projection of Industrial Atmospheric Emissions of Galicia]. Ed. *Servizo de Publicacións e Intercambio Científico*, *Universidade de Santiago de Compostela*, Santiago de Compostela, España.

[30] Casares, J. J., et al. 2005. "Inventario, análisis y proyección de las emisiones atmosféricas industriales en Galicia." *Universidade de Santiago de Compostela. Ed. Servizo de Publicaciones e*

Intercambio Científico, Santiago de Compostela, España. ["Inventory, analysis and projection of industrial atmospheric emissions in Galicia." *University of Santiago de Compostela. Publications and Scientific Exchange Service*, Santiago de Compostela, Spain].

[31] Dios, M., Souto, J. A., and Casares, J. J. 2010. "Emissions inventory analysis for an urban (industrial)-rural (agricultural) environment." C. A. Brebbia, S. Hernandez, E. Tiezzi, eds. *The Sustainable City VI: Urban Regeneration and Sustainability.* Southampton: WIT Press, p. 383-392.

[32] United States Environmental Protection Agency, US EPA. "*Managing Air Quality - Air Quality Modeling.*" (n.d). Accessed November 20 from https://www.epa.gov/air-quality-management-process/managing-air-quality-air-quality-modeling.

[33] Karroum, K., Lin, Y., Chiang, Y.-Y., Ben Maissa, Y., El Haziti, M., Sokolov, A., and Delbarre, H. 2020. "A Review of Air Quality Modeling." *MAPAN*, 35.2:287–300. https://doi.org/10.1007/s12647-020-00371-8.

[34] Inanloo, B., & Tansel, B. (2015). Explosion impacts during transport of hazardous cargo: GIS-based characterization of overpressure impacts and delineation of flammable zones for ammonia. *Journal of environmental management,* 156, 1-9.

[35] De Visscher A. 2014 "*Air Dispersion Modeling. Foundations and Applications.*" New Jersey: John Wiley & Sons.

[36] Jones, R., W. Lehr, D. Simecek-Beatty, R. Michael Reynolds. 2013. "ALOHA ® (Areal Locations of Hazardous Atmospheres) 5.4.4: Technical Documentation." *U. S. Dept. of Commerce, NOAA Technical Memorandum NOS OR&R 43.* Seattle, WA: Emergency Response Division, NOAA. 96 pp. Accessed November 20. https://response.restoration.noaa.gov/sites/default/files/ALOHA_Tech_Doc.pdf.

[37] Carrari, C. L., Aparicio, L. V., Tonelli, S. M., and Bandoni, J. A. 2004. "Utilización de modelos de dispersión atmosférica para la estimación de dosis de exposición." *Mecánica Computacional,*

1743-1760. ["Use of atmospheric dispersion models for exposure dose estimation." *Computational Mechanics*, 1743-1760].

[38] Pak, S., and Kang, C. 2021. "Increased risk to people around major hazardous installations and the necessity of land use planning in South Korea." *Process Safety and Environmental Protection*, 149: 325–333.

[39] Chakrabarti, U. K., and Parikh, J. K. 2013. "A societal risk study for transportation of class-3 hazmats – A case of Indian state highways." *Process Safety and Environmental Protection*, 91.4: 275–284.

[40] Bariha, N., Mishra, I. M., and Srivastava, V. C. 2016. "Fire and explosion hazard analysis during surface transport of liquefied petroleum gas (LPG): a case study of LPG truck tanker accident in Kannur, Kerala, India." *Journal of loss prevention in the process industries*, 40:449-460.

[41] Shao, H., and Duan, G. (2012). "Risk quantitative calculation and ALOHA simulation on the leakage accident of natural gas power plant." *Procedia Engineering*, 45: 352-359.

[42] Lozano, A., Muñoz, Á., Antún, J. P., Granados, F., and Guarneros, L. 2010. "Analysis of hazmat transportation accidents in congested urban areas, based on actual accidents in Mexico." *Procedia - Social and Behavioral Sciences*, 2.3:6053–6064.

[43] Law, W. P., Erain, N., Ramli, N. I., and Gimbun, J. 2018. "Assessment of chlorine leak dispersion around Gebeng industrial area and potential evacuation route." *Atmospheric Research.*

[44] Del Sole, M. B. (2019). *Simulación de incidentes en el transporte de materiales peligrosos en Gran La Plata* [*Simulation of incidents in the transport of hazardous materials in Gran La Plata*] (grade dissertation, Universidad Nacional de La Plata).

[45] United States Environmental Protection Agency, US EPA . 2019. "*User's Guide for the AERMOD Meteorological Preprocessor (AERMET)*" (No. EPA-454/B-19-028). Accessed November 20 https://www.epa.gov/sites/production/files/2020-09/documents/ aermet_userguide.pdf.

[46] Cimorelli, A. J, et al. 2004. "*AERMOD, description of model formulation. Staff Report*," EPA-454/R03-004, United States Environmental Protection Agency, US EPA. North Carolina, USA Accessed November 20. https://www3.epa.gov/scram001/7thconf/aermod/aermod_mfd.pdf.

[47] Espinoza Cruz, C. J., 2018. *"Determinación de la dispersión geográfica de los gases NOx, CO, SO2 y partículas (PM10), emitidos por una central térmica aplicando el modelo AERMOD."* Título Profesional, Universidad Nacional Tecnológica de Lima Sur. Repositorio Institucional de la Universidad Nacional Tecnológica de Lima Sur. ["*Determination of the geographical dispersion of NOx, CO, SO2 gases and particles (PM10), emitted by a power plant using the AERMOD model*" Professional Thesis, National Technological University of Southern Lima. Institutional Repository of the Universidad Nacional Tecnológica de Lima Sur]. Accessed November 20. http://repositorio.untels.edu.pe/handle/UNTELS/272.

[48] Porta, A. A. et al. 2018. Modelos de calidad de aire [Air quality models]. En, *Monitoreo y modelado de contaminantes atmosféricos. Efectos en la salud pública* (pp. 182 - 224). Editorial de la Universidad Nacional de La Plata (EDULP).

[49] United States Environmental Protection Agency, US EPA. 2019. "*User's Guide for the AMS/EPA Regulatory Model* (*AERMOD*)" (No- EPA-454/B-19-027) Accessed November 20 https://www3.epa.gov/ttn/scram/models/aermod/aermod_userguide.pdf.

[50] Silva, A. and Arcos, D. 2011. "Aplicación del programa AERMOD para modelar dispersión de PM10 emitido por equipos de calefacción a leña en la ciudad de Constitución [Application of the AERMOD program to model dispersion of PM10 emitted by wood heating equipment in the city of Constitución]." *Obras y Proyectos* 9: 4-10. Accessed November 20 http://revistas.ucsc.cl/index.php/obrasyproyectos/article/download/186/181/.

[51] Sosa-Pérez, C. 2018. "Gestión de la contaminación del aire en un municipio industrial. Caso de estudio: Mariel [Management of air

pollution in an industrial municipality. Case study: Mariel]." *Revista Cubana de Meteorología,* 24.1: 75-94.

[52] Gallego Martinez, E. 2010. *"Inventario de emisiones de partículas y modelado de su transporte a partir de fuentes de área de una industria minera."* Tesis de Maestría, Universidad Autónoma de San Luis Potosí. [*"Inventory of particle emissions and modeling of their transport from area sources of a mining industry"* Title of doctoral dissertation or master's thesis]. Accessed November 20 https://ninive.uaslp.mx/xmlui/bitstream/handle/i/3715/MCA1INV01001.pdf?sequence=1.

[53] Diez, S C., Abril, G. A., Piccioni, M. N. and Britch, J. 2013 "Dispersión de material particulado emitido desde un complejo industrial: comparación de resultados [Dispersion of particulate matter emitted from an industrial complex: comparison of results]." En Puliafito, Allende, Paniagatti (Eds.) *Contaminación Atmosférica e Hídrica en Argentina.* pp. (449-513). Argentina: Universidad Tecnológica Nacional.

[54] Heist, D., Isakov, V., Perry, S., Snyder, M., Venkatram, A., Hood, C., Stocker, J., Carruthers, D., Arunachalam, S., & Owen, R. C. 2013. "Estimating near-road pollutant dispersion: A model inter-comparison." *Transportation Research Part D: Transport and Environment, 25*:93–105. https://doi.org/10.1016/j.trd.2013.09.003.

[55] Liu, H., Xu, X., Rodgers, M. O., Xu, Y. "Ann," & Guensler, R. L. 2017. "MOVES-Matrix and distributed computing for microscale line source dispersion analysis." *Journal of the Air & Waste Management Association*, 67.7:763–775. https://doi.org/10.1080/10962247.2017.1287788.

[56] United States Environmental Protection Agency, US EPA. 2018. "*User's Guide for the AERMOD Terrain Preprocessor* (*AERMAP*)." (No. EPA-454/B-18-004). Accessed November 20 https://www3.epa.gov/ttn/scram/models/aermod/aermap/aermap_userguide_v18081.pdf.

[57] ATM-PRO. "*ISC-AERMOD View*" (n.d.). Accessed November 20 from http://www.atmpro.be/product.php?item=isc-aermod_view &onglet=general.

[58] Dirección General de Biodiversidad y Calidad Ambiental del Ministerio para la Transición Ecológica. (2019). "*Análisis de herramientas de evaluación de la difusión y comportamiento de agentes químicos en el marco de la normativa de responsabilidad medioambiental.*" Gobierno de España. ["*Analysis of tools to evaluate the diffusion and behaviour of chemical agents within the framework of environmental responsibility regulations*"]. Accessed November 20 https://www.miteco.gob.es/es/calidad-y-evaluacion-ambiental/temas/responsabilidad-mediambiental/herramientasde evaluaciondifusionycomportamientoagentesquimicos_251019_tcm30-177407.pdf.

[59] Grisales Vargas, S. C., Cubaque López, H. A., Agudelo Cadavid, R. M., Cárdenas Monsalve, J. J., Ramírez Vélez, A. M., and Piñeros Jiménez J. G. 2016. "Anexo 5, Modelo AERMOD dispersión contaminantes en el Valle de Aburrá." Universidad de Antioquia. ["*Annex 5, AERMOD Model of Pollutant Dispersion in the Aburrá Valley.*" Universidad de Antioquia] Accessed November 20 http://www.udea.edu.co/wps/wcm/connect/udea/9375a71f-7fbc-41ce-9806-a27a60921ba6/Anexo+5+Modelo+AERMOD++dispersion ++contaminantes+AMVA_compressed.pdf?MOD=AJPERES&CVID=n0baPr8.

[60] Holmes, N. S., and Morawska, L. 2006. "A review of dispersion modelling and its application to the dispersion of particles: An overview of different dispersion models available." *Atmospheric Environment*, 40.30:5902–5928. https://doi.org/10.1016/j.atmosenv.2006.06.003.

[61] Michanowicz, A. R. 2015. "*Hybrid Dispersion/ Land Use Regression Modeling for Improving Air Pollutant Concentration Estimates.*" Doctoral Dissertation, University of Pittsburgh. Institutional Repository at the University of Pittsburgh. Accessed November 20 http://d-scholarship.pitt.edu/23582/.

[62] Abril, G. A., Diez, S. C., Pignata, M. L., and Britch, J. 2016. "Particulate matter concentrations originating from industrial and urban sources: Validation of atmospheric dispersion modeling results." *Atmospheric Pollution Research*, 7.1:180–189. https://doi.org/10.1016/j.apr.2015.08.009.

[63] Gulia, S., Kumar, A., and Khare, M. 2015. "Performance evaluation of CALPUFF and AERMOD dispersion models for air quality assessment of an industrial complex." *Journal of Scientific & Industrial Research* 74: 302-307.

[64] Mokhtar, M. M., Hassim, M. H., and Taib, R. M. 2014. "Health risk assessment of emissions from a coal-fired power plant using AERMOD modelling." *Process Safety and Environmental Protection*, 92.5:476–485. https://doi.org/10.1016/j.psep.2014.05.008.

[65] Michanowicz, D. R., Shmool, J. L. C., Tunno, B. J., Tripathy, S., Gillooly, S., Kinnee, E., and Clougherty, J. E. 2016. "A hybrid land use regression/AERMOD model for predicting intra-urban variation in PM2.5." *Atmospheric Environment*, 131:307–315. doi:10.1016/j.atmosenv.2016.01.045.

[66] Venkatram, A. (2008) "Introduction to aermod" *Modeling Dispersion*, 2008; 1–24. http://tools.envirolink.govt.nz/assets/Uploads/AERMOD-Introduction-AERMOD-2008.pdf.

[67] Environmental M for the. Atmospheric dispersion modelling [Internet]. *Good practice guide for atmospheric dispersion modelling*. 2017. Available from: https://www.mfe.govt.nz/publications/air/good-practice-guide-atmospheric-dispersion-modelling.

[68] United States Environmental Protection Agency, US EPA. "*Air Quality Dispersion Modeling—Preferred and Recommended Models.*" Accessed November 20 https://www.epa.gov/scram/airquality-dispersion-modeling-preferred-and-recommended-models.

[69] Tartakovsky, D., Broday, D. M., and Stern, E. 2013. "Evaluation of AERMOD and CALPUFF for predicting ambient concentrations of

total suspended particulate matter (TSP) emissions from a quarry in complex terrain." *Environmental Pollution,* 179:138–145. doi: 10.1016/j.envpol.2013.04.023.

[70] Seangkiatiyuth, K., Surapipith, V., Tantrakarnapa, K., and Lothongkum, A. W. 2011. "Application of the AERMOD modeling system for environmental impact assessment of NO2 emissions from a cement complex." *Journal of Environmental Sciences*, 23.6:931–940. doi: 10.1016/s1001-0742(10)60499-8.

[71] Caputo, M., Giménez, M., and Schlamp, M. (2003). "Intercomparison of atmospheric dispersion models." *Atmospheric Environment*, 37.18: 2435–2449. https://doi.org/10.1016/S1352-2310(03)00201-2.

[72] Vardoulakis, S., Fisher, B. E. A., Pericleous, K., and Gonzalez-Flesca, N. (2003). "Modelling air quality in street canyons: A review." *Atmospheric Environment*, 37.2: 155–182. https://doi.org/10.1016/S1352-2310(02)00857-9.

[73] Belis, Claudio A., et al. (2014). "European guide on air pollution source apportionment with receptor models." *European Commission, Joint Research Centre. Institute for Environment and Sustainability.*

[74] Hopke, Philip K. (2016). "Review of receptor modeling methods for source apportionment." *Journal of the Air & Waste Management Association* 66.3: 237-259.

[75] United States Environmental Protection Agency, U.S. EPA. "*Air Pollutant Receptor Modeling.*" Accessed November 17. https://www.epa.gov/scram/air-pollutant-receptor-modeling.

[76] Mellado, Daniela. (2020). "Estudio de zonas críticas de emisión con modelos de receptores en regiones urbanas con entornos industrializados." PhD diss. Universidad Nacional de La Plata. ["*Study of critical emission zones using receptor models in urban regions with industrialized environments*." PhD diss., National University of La Plata].

[77] Simon, Heather, et al. (2010). "The development and uses of EPA's SPECIATE database." *Atmospheric Pollution Research* 1.4: 196-206.

[78] Pernigotti, Denise, Claudio A. Belis, and Luca Spano (2016). "SPECIEUROPE: The European database for PM source profiles." *Atmospheric Pollution Research* 7.2: 307-314.

[79] Begum, Bilkis A., Swapan K. Biswas, and Philip K. Hopke. (2007). "Source apportionment of air particulate matter by chemical mass balance (CMB) and comparison with positive matrix factorization (PMF) model." *Aerosol and Air Quality Research* 7.4: 446-468.

[80] Loska, Krzysztof, and Danuta Wiechuła. (2003). "Application of principal component analysis for the estimation of source of heavy metal contamination in surface sediments from the Rybnik Reservoir." *Chemosphere* 51.8: 723-733.

[81] Geng, Fuhai, et al. (2009). "Analysis of VOC emissions using PCA/APCS receptor model at city of Shanghai, China." *Journal of atmospheric chemistry* 62.3: 229-247.

[82] Shi, Guo-Liang, et al. (2014). "A comparison of multiple combined models for source apportionment, including the PCA/MLR-CMB, Unmix-CMB and PMF-CMB models." *Aerosol and Air Quality Research* 14.7: 2040-2050.

[83] Thurston, George D., and John D. Spengler. (1985). "A quantitative assessment of source contributions to inhalable particulate matter pollution in metropolitan Boston." *Atmospheric Environment* 19.1: 9-25.

[84] Norris, G. A., et al. (2014). "Positive Matrix Factorization (PMF) 5.0 Fundamentals and User Guide Prepared for the US Environmental Protection Agency." *USEPA,* Office Res Dev, Washington. DC.

[85] Błaszczak, Barbara. (2018). "The Use of Principal Component Analysis for Source Identification of PM2.5 from Selected Urban and Regional Background Sites in Poland." *E3S Web of Conferences*. Vol. 28. EDP Sciences.

[86] Dai, Qili, et al. (2019). "A size-resolved chemical mass balance (SR-CMB) approach for source apportionment of ambient particulate matter by single element analysis." *Atmospheric Environment* 197: 45-52.

[87] Landis, Matthew S., et al. (2017). "Source apportionment of ambient fine and coarse particulate matter at the Fort McKay community site, in the Athabasca Oil Sands Region, Alberta, Canada." *Science of the Total Environment* 584: 105-117.

[88] Zhang, Yanjun, et al. (2017). "Review of receptor-based source apportionment research of fine particulate matter and its challenges in China." *Science of the Total Environment* 586: 917-929.

[89] Cai, Changjie, et al. (2010). "Characteristics and source apportionment of VOCs measured in Shanghai, China." *Atmospheric Environment* 44.38: 5005-5014.

[90] Song, Sang-Keun, et al. (2019). "Source apportionment of VOCs and their impact on air quality and health in the megacity of Seoul." *Environmental Pollution* 247: 763-774.

[91] Feinberg, Stephen Neil, et al. (2019). "Examining spatiotemporal variability of urban particulate matter and application of high-time resolution data from a network of low-cost air pollution sensors." *Atmospheric environment* 213: 579-584.

[92] Henry, Ronald C. (2008). "Locating and quantifying the impact of local sources of air pollution." *Atmospheric Environment* 42.2: 358-363.

[93] Henry, Ronald C., Subha Mohan, and Shahram Yazdani. (2019). "Estimating potential air quality impact of airports on children attending the surrounding schools." *Atmospheric Environment* 212: 128-135.

[94] Wei, Peng, et al. (2019). "Solar-powered air quality monitor applied under subtropical conditions in Hong Kong: Performance evaluation and application for pollution source tracking." *Atmospheric Environment* 214: 116825.

[95] Mellado. Daniela et al. (2020). Location of areas of emission of pollutants when poor urban air quality is detected. *International Journal of Environment and Health* 1743-4955.

[96] Argyropoulos, G., et al. (2017). "Source apportionment of PM10 and PM2. 5 in major urban Greek agglomerations using a hybrid source-receptor modeling process." *Science of the Total Environment* 601: 906-917.

[97] Cahill, Thomas A., et al. (2016). "Transition metals in coarse, fine, very fine and ultra-fine particles from an interstate highway transect near Detroit." *Atmospheric Environment* 145: 158-175.

[98] Grange, Stuart K., Alastair C. Lewis, and David C. Carslaw. (2016). "Source apportionment advances using polar plots of bivariate correlation and regression statistics." *Atmospheric Environment* 145: 128-134.

[99] Squizzato, Stefania, et al. (2017). "Urban air quality in a mid-size city-PM2.5 composition, sources and identification of impact areas: From local to long range contributions." *Atmospheric Research* 186: 51-62.

[100] Urbančok, Dejan, Anthony JR Payne, and Richard D. Webster. (2017). "Regional transport, source apportionment and health impact of PM10 bound polycyclic aromatic hydrocarbons in Singapore's atmosphere." *Environmental Pollution* 229: 984-993.

[101] Yang, Kun, et al. (2020). "Human activities and the natural environment have induced changes in the PM2.5 concentrations in Yunnan Province, China, over the past 19 years." *Environmental Pollution*: 114878.

In: Air Pollution: Effects and Dangers ISBN: 978-1-53619-544-6
Editor: J. Esteban Colman Lerner

Chapter 3

HEALTH EFFECTS OF ATMOSPHERIC POLLUTANTS

***Myriam Aguilar*[1], *María de los Angeles Gutiérrez*[1], *Natacha S. Represa*[1,2] *and Andrés Porta*[1,*]**

[1]Centro de Investigaciones del Medioambiente (CONICET-UNLP), La Plata, Argentina

[2]Geo-Environmental Cartography and Remote Sensing Group, Polytechnic University of Valencia, Valencia, Spain

ABSTRACT

The World Health Organization (WHO) reports that 7 million people died because of exposure to air pollution in 2012, doubling previous estimates and confirming that air pollution constitutes the most important environmental health risk worldwide. The most affected populations being those living in developing countries. This report synthesizes the abundant epidemiological evidence that correlates chronic exposure to pollutants of urban and industrial origin, such as particulate matter (including adsorbed metals and polycyclic aromatic hydrocarbons), toxic and irritant gases (nitrogen and sulfur oxides, and ozone), in addition to other volatile substances such as benzene and perchloroethylene. All of

* Corresponding Author's Email: aaporta@yahoo.com.ar.

them exert adverse effects on the development and lung function, on the cardiovascular system. Also, produce an increase in mortality, and one excess of lifetime cancer risk.

1. Health Effects of Polluted Air

Air pollution is currently considered a decisive factor of health. Indeed, numerous epidemiological studies have found a direct association between air pollution and a wide variety of adverse effects on the health of the exposed population. These range from subtle subclinical effects to early death. Some social groups, such as older adults, children, pregnant women, and people with an underlying disease (such as asthma or chronic obstructive pulmonary disease), tend to be more susceptible, developing health effects more quickly and more seriously when exposed to polluted air. The populations with high exposure include those who live in the vicinity of routes of intense vehicular traffic, near to sources of pollutant emissions, occupational groups (occupational exposure), and even those groups of high vulnerability due to specific socioeconomic conditions [1-5].

The associations between exposure to polluted air and adverse health effects are carried out through epidemiological studies, examining temporal and spatial variations in the levels of the main pollutants, whether they are acute or chronic events. These studies seek the expression of different health alterations in human populations through different strategies such as cross-sectional or cohort studies.

Other useful strategies are toxicological and controlled human studies. Acute physiological effects include induction of pulmonary and systemic inflammation, endothelial dysfunction, thrombosis, arrhythmia, and blood pressure, which can also contribute to long-term consequences. Chronic exposure leads to other subclinical level indicators, such as insulin resistance, various disorders of the respiratory and cardiovascular system, and others indirect effects such as diabetes, atherosclerosis, and obesity. These morbidities ultimately lead to a shorter life expectancy. Chronic exposure often leads to high-impact diseases such as asthma or chronic obstructive pulmonary disease, lung cancer, and myocardial infarction [2, 5-7].

A particularly serious situation is the indoor use of solid fuels (homes). Almost 3 billion people among the poorest in the world continue to depend on this type of fuel (coal, wood, animal excrement and agricultural residues) that they burn in low-efficiency and highly polluting stoves and kitchens to cook and heat their households, which causes annually about four million premature deaths of children and adults from respiratory diseases, heart disease and cancer. Along with the widespread use of kerosene stoves, stoves and lamps, these energy-consuming practices in homes cause many deaths and serious injuries from burns and poisonings. Currently, indoor air pollution from the burning of fuels is the main environmental health risk worldwide [8].

Specifically, the data for 2012 published by the World Health Organization (WHO) shows that exposure to contaminated air situations led to the death of 7 million people. These data confirm that air pollution constitutes the most important environmental risk for health in the world. More seriously, more than 50% of this disease burden falls on the populations of developing countries [9]. This risk situation is also analyzed from the perspective of the International Agency for Research on Cancer (IARC), who, in their press release 221/13, categorizes outdoor polluted air as proven carcinogenic, category IA [10].

1.1. Analysis of Possible Correlations

Efforts to understand and mitigate the effects of air pollution on human health and well-being have a rich and interesting history. In the 1970s and 1980s, largely attributed to the well-documented early increase in morbidity and mortality from extreme air pollution episodes, the association between cardiopulmonary disease and very high concentrations of particulate matter (PM) was accepted.

However, disagreement persisted on what levels of exposure and what type of PM affected human health. In the first half of the 1990s, several epidemiological studies reported health effects associated with unexpectedly low concentrations of environmental PM. These documents included relevant findings between daily changes in PM and daily mortality in various cities in the USA; prospective cohort investigations from Harvard University and the American Cancer Society reporting long-term exposure to PM associated with respiratory disease in children and cardiopulmonary mortality in adults; and a series of studies in the Utah Valley (USA) that reported that particulate pollution was associated with a wide range of health endpoints, including respiratory hospitalizations, lung function and respiratory symptoms, school absences, and mortality. Comparable results were also reported in other studies from Germany, Canada, Finland, and the Czech Republic. The convergence of these data revealed large evidence, and hence a reconsideration of the effects of PM polluted air at low concentrations, prompting much more research that continues today. The numerous critical studies published on air pollution and its effects on public health, led in turn to the adoption of a series of measures leading to the reduction of air pollution [1, 2, 5-7].

For these chronic exposure studies, three main groups are recognized [5]:

a) Studies of different occupational groups with exposures to specific pollutants, including traffic police officers of Bangkok (Thailand, [11]), USA truckers [12], and Brazilian taxi drivers [13]. In these cases, workers develop lung cancer, doubling the maximum level accepted by the WHO.
b) Studies of fixed populations (cohorts) with long-term changes in exposure to air pollution, and differences in alterations in health. For example, the reports from Harvard University and the American Cancer Society, clearly demonstrated that the reduction of air pollution in fine particles is associated with an improvement in global mortality.
c) Studies of populations exposed chronically to different levels of atmospheric pollutants. Such studies provide information on the exposure of each population and related health endpoints. For

example, studies of the prevalence of asthma and respiratory symptoms in children, in La Plata city (Argentina) and neighboring areas of a petrochemical pole [14], and in Texas, USA [15].

On the other hand, good correlations were found between mortality of cardiovascular and respiratory origin and vehicular traffic in the Netherlands [16]; and health effects of exposure to PM_{10} and $PM_{2.5}$ in China [17]. Two additional studies compare the results obtained between populations of different cities, in Latin America [18] and in India [19].

All the studies of long-term effects on health of air pollution have limitations. For example, the extrapolation of the concentration of pollutant to a whole population, which in general do not consider the interactions between the different pollutants regarding the effects (for example synergy) or with other environmental factors such as temperature and barometric pressure and precipitation; nor confounding factors in the evaluation of effects such as different conditions in terms of socio-economic condition, educational and cultural aspects, and underlying diseases. To consider the components of this last group, epidemiological questionnaires are applied, generally standardized, which allow their incorporation in the analysis [5, 12-17].

1.2. Beneficial Effects of Reducing Air Pollution

Improvements in air quality, through application of specific public policies, decrease the impact on public health. In the European Union, through the application of legislative measures aimed at improving air quality, obtain significant reductions in emissions for vehicular traffic: carbon monoxide (-80%), volatile organic compounds (VOCs) (-68%), nitrogen oxides (-40%) and $PM_{2.5}$ (-60%). Reductions in particulate levels had a beneficial effect on respiratory symptoms in adults with reductions in regular cough, chronic cough, or phlegm, wheezing and dyspnea [2, 5, 20, 21].

In the USA, national air pollution control policies, and in particular the Clean Air Act, have provided opportunities to evaluate the effectiveness of these efforts in relation to improving health, indicating that reductions in air pollution resulted in substantial reductions in mortality risk. For example, in New York City (USA), due to improvements in NO_x reduction, a 2-9% decrease in ozone levels was obtained, and consequently fewer hospitalizations for respiratory diseases, up to 11% lower in some counties, including the New York metropolitan area [2, 5, 22]. In Tokyo (Japan), the application of specific regulations to reduce diesel emissions resulted in a gradual decrease in atmospheric pollutant concentrations: NO_2 by 11.5% while $PM_{2.5}$ by 10.9%. These lower levels of pollutants were associated with decreased cerebrovascular mortality rates. Something similar could be observed in Delhi (India), where radical measures were adopted to improve air quality, including the transformation of all commercial vehicles into compressed natural gas, finding a significant improvement in respiratory health [2, 5, 23, 24].

The reunification of the former German Democratic Republic (East Germany) and the former Federal Republic of Germany (West Germany) in 1990 was accompanied by marked changes in the political environment and socio-economic structures. Air quality improved markedly after German reunification, with a clear reduction in pollution, mainly associated with the switch from coal to natural gas from home heating fuel in the former East Germany. These reductions in the levels of NO_2, carbon monoxide (CO), $PM_{2.5}$ and ozone (O_3), correlate with a clear decrease in mortality. Likewise, the differences in lung function parameters in children from both sectors (east and west) disappeared simultaneously with the difference in PM levels [2, 25, 26].

1.3. Pulmonary Physiological Response to Pollutant Exposure

The respiratory system is the main portal of entry for air pollution and behaves basically as a biochemically active tissue that involves mediators that induce local and systemic inflammatory response after exposure, affecting lung function in a matter of hours or after years of exposure, according to severity. Diseases caused include asthma, chronic obstructive pulmonary disease (COPD), pneumonia, and possibly tuberculosis [3, 5, 27].

Lung function is an important measure of chronic respiratory and systemic inflammation, as well as premature morbidity and cardiorespiratory mortality. There is strong evidence of adverse effects of air pollution on the growth of lung function, accelerating its decline in adults, increasing the risk of chronic respiratory diseases and causing death from cardiovascular disease. However, the number of reports showing the association between short-term exposure to air pollution and exacerbation of COPD and asthma is increasing, but there are a limited number of studies indicating the impact of long-term air pollution in the development of these diseases and related symptoms. Current challenges include using individual exposure assessment to clarify the role of traffic, and to exclude potential confusion at the community level. Additional research is also needed to identify the specific sources of contamination that are most dangerous, and the groups that are most susceptible [3, 5-7, 14-16, 27-29].

1.4. Physiological Mechanisms Associated with Exposure to Air Pollutants

Ambient air contains various pollutants that react with human tissues, such as oxidizing gases (O_3 or NOx), VOCs and PM. An additional action occurs in cells in contact with inhaled particles, carriers of toxic substances such as metals and polycyclic aromatic hydrocarbons (PAHs). This leads to tissue damage and the influx of inflammatory cells to injury sites, where when activated they generate and release large amounts of free radicals. In the absence of antioxidants, these free radicals attack the tissue and local components and cause cellular damage with the development of inflammatory foci. Numerous experimental studies have demonstrated this mechanism. This response can spread to other tissues, mediated by cytokines and chemokines, because of the indirect release of inflammatory mediators in the blood, or as a result of the translocation of particles or their chemical components in the systemic circulation with direct action at the site objective. Inflammation in the lungs, caused by deposited particles, may even be a key process that could mediate adverse effects on the cardiovascular system. These findings have been supported by population-based studies [3, 5, 27, 30, 31].

About the different effects observed for a similar exposure to air pollution, the key lies in two concepts: vulnerability and susceptibility. Although it is necessary to continue studies in this direction, there are some certainties. Regarding vulnerability, specific exposure situations (work environments, proximity to relevant sources such as industry or high traffic), socio-economic and quality of life conditions intervene. Respect to susceptibility, it first refers to age groups such as children, who breathe a greater volume of air in relation to their body weight, and the elderly, especially those with pre-existing heart or lung disease. A key characteristic, related to the inflammatory process, is the oxidant-antioxidant activity: individuals with higher pulmonary antioxidant defenses are less sensitive to atmospheric pollution [5, 27, 30].

2. Epidemiological Evidence and Case Studies

Air pollutants often investigated in epidemiological and toxicological studies include PM_{10} and $PM_{2.5}$, black carbon (non-mineral fraction of PM), ozone (O_3), nitrogen oxides (NO_x), sulfur dioxide (SO_2), and VOCs. However, the results obtained are particularly relevant in relation to the impact on health associated with exposure to PM in outdoor air [2, 5, 32, 33].

2.1. Exposure to Airborne Particulate Matter (PM)

The particulate matter originates both from primary emissions (combustion in stationary or mobile sources, marine aerosols, soil particles resuspended by the wind), and from the formation of secondary particles in the atmosphere. PM can be characterized in terms of the concentration by mass or number of the particles, and their chemical composition (black carbon fraction, PAHs, metals, anions, and cations). Epidemiological and toxicological evidence shows that the mass of PM comprises fractions with different types and degrees of health effects. Furthermore, the different particle sizes and their composition are related to the specific emission sources, which allows defining the specific contributions to the total PM determined. In this sense, black carbon is a particularly sensitive indicator of emissions from combustion processes such as industrial and vehicular traffic emissions [2-5, 32].

The adverse health effects of PM are particularly well documented. There is no evidence of a safe level of exposure or a threshold below which adverse health effects do not occur. Thus, for example, PM contamination creates a substantial burden of disease, reducing life expectancy by almost 9 months on average in Europe, with more than 80% of its urban population exposed to levels above the Quality Guidelines for Air. Asthma is a burden in developing and developed countries, with a significant impact on public health and the economy. According to the 2011 United States National Health Data Survey, asthma is identified as one of the most prevalent chronic diseases in children. There are a total of 10.5 million (14%) children who have been diagnosed with this illness. Asthma is a serious respiratory disease that affects nearly 25 million North Americans and costs the nation $ 56 billion per year. The prevalence of asthma in the USA increased from 7.3% in 2001 to 8.4% in 2010. As children are more susceptible to ambient air pollution than adults, they are disproportionately affected by asthma, as shown by higher asthma hospitalization rates for people younger than 18 years [28, 32-35]. Other studies carried out in Latin America and China indicates equivalent results [36, 37]. In all these analyzes it is evidenced that $PM_{2.5}$ always showed a positive effect in triggering the asthma rate, that is, the prevalence of asthma is mainly dominated by $PM_{2.5}$ concentrations in the study areas [33-37].

This specific pollutant can induce inflammation and hyperresponsiveness of the airways, two characteristic features of asthma, in addition to oxidative stress. Exposure to these pollutants is associated, unsurprisingly, with exacerbations and even the appearance of asthma. However, when PM levels are very high, such as those found in large cities in India and China, air pollutants can trigger irritant and inflammatory effects on neuroreceptors in the airways and epithelium. These last exposure levels are rarely found in North American or European countries [17-19].

Many epidemiological and toxicological studies show that $PM_{2.5}$ is a strong risk indicator associated with exposure to PM from various sources and in different settings, noting that exposure to this pollutant is associated with systemic inflammation, oxidative stress, alteration of the electrical processes of the heart and vascular dysfunction. Epidemiological studies show variations in cardiovascular biomarkers of inflammation, such as C-reactive protein and fibrinogen, and they are consistently associated with cardiovascular disease and subsequent death. Long-term exposure has also been associated with preclinical markers of atherosclerosis and with the progression of this pathology, finding associations between various markers of atherosclerosis, including intima-media thickness and coronary artery calcification, and long-term concentration of $PM_{2.5}$ [12, 30, 36-38].

Other studies show that the levels of exposure to PM2.5 determine the type of mortality, those of cardiovascular origin being predominant if the levels are low, while exposure to high levels gives rise to the prevalence of mortality from lung cancer. Long-term exposure to fine particulate matter has been described other than cardiovascular and respiratory diseases. Evidence suggests effects on diabetes, poor neurological development in children, and neurological disorders in adults [16, 29, 39-42].

On the other hand, birth cohort studies have reported significant associations between $PM_{2.5}$ exposure and respiratory infections and asthma in young children. Several studies have found an association between $PM_{2.5}$ and childhood bronchiolitis, with significant risk for hospitalization. A systematic review [32] reported significant associations between exposure to $PM_{2.5}$ outcomes and effects in neonates, including preterm delivery, low birth weight, and small size relative to gestational age [43-45].

Regarding the evidence of the short-term effects of PM2.5 and PM10 on mortality, morbidity, and physiological endpoints, all have been significantly increased since 2005. Several studies confirm a mean percentage of increase in mortality for an increase in 10 μg m-3 of PM2.5 in short-term exposure is 1.0%, with considerable regional variation worldwide, with corresponding increases in mortality from respiratory diseases (+ 1.5%) and cardiovascular diseases (+ 0,8%). By the other hand, data from various cities in Europe, the United States and Canada were analyzed, founding an increase of 10 μg m-3 in PM10 that corresponds to a 0.2-0.6% increase in mortality from all causes with similar effects for the United States and Europe and greater effects in Canada [3-5, 33, 46-51].

Furthermore, in Europe, higher effects of PM_{10} on daily mortality are observed in cities with upper temperatures and a greater contribution of traffic emissions to PM. Other studies carried out in Spain, France, Italy, Sweden, and Netherlands show a significant association between PM_{10} and $PM_{2.5}$, with daily mortality. The incidence of the characteristics of each PM was also demonstrated according to its emitting source; however, these studies do not provide convincing evidence that any specific PM source, component, or size class can be excluded as a possible contributing factor to PM toxicity. Furthermore, a large-scale study conducted in the USA [4] reported significant associations between PM (both fractions) and hospital admissions for ischemic heart disease, cerebrovascular disease, and heart failure [33, 46-51].

2.2. Exposure to Inorganic Gases: O_3, NO_x, SO_2

There are serious health risks not only from exposure to particulate matter, but also to ozone (O_3), nitrogen dioxide (NO_2), and sulfur dioxide (SO_2). As with particulate matter, the highest concentrations tend to be found in urban areas in low- and middle-income countries. Ozone is a major factor in asthma mortality and morbidity, while nitrogen dioxide and sulfur dioxide can influence asthma, bronchial symptoms, and respiratory failure.

Ozone

Ozone, the main compound ingredient in photochemical smog, poses a serious problem for air quality in many parts of the world. Ozone is formed in the atmosphere by photochemical reactions in the presence of sunlight and precursor pollutants, such as nitrogen oxides (NOx) and various volatile organic compounds (VOCs). It is destroyed in reactions with NO_2 and is deposited in the soil [5, 52]. These precursor pollutants are emitted from automobiles, power plants, industrial boilers, refineries, chemical plants, and other sources. Pollution occurs during the summer months when weather conditions are conducive to forming ozone at ground level: lots of sun and high temperatures. Even at low levels, ozone can cause a number of respiratory problems. Excess ozone in the air can have serious adverse effects on human health, as breathing problems, cause asthma, reduce lung function, and lead to lung disease. Several European studies have revealed that daily mortality due to heart disease increases by 0.3% and 0.4% with an increase of 10 $\mu g\ m^{-3}$ of O_3 [52, 53].

A new guideline level for ozone, 100 $\mu g\ m^{-3}$ (maximum daily average of eight hours), was proposed in the 2005 global update of the WHO air quality guidelines [53]. This decision is based on numerous epidemiological time-series studies that provided a considerable body of new evidence on health effects. These studies have shown that there are small but convincing positive associations between daily mortality and ozone levels that are independent of the effects of particulate matter for European and North American cities. As ozone concentrations rise above the guideline value, the effects on the health of the population are more and more numerous and serious. These effects can occur in places where concentrations are already high due to human activities or rise during episodes of very hot weather. Also, when the O_3 concentration for eight hours was higher than 240 $\mu g\ m^{-3}$, there is a probability of significant health effects. Based on time series evidence, exposure to ozone concentrations of this magnitude would lead to an increase in the number of attributable deaths of 5-9% relative to exposure at the estimated background level [5, 52, 53].

Nitrogen Oxides

As nitrogen oxides (NOx) includes nitrogen monoxide (NO) and nitrogen dioxide (NO_2). Both gases have an anthropogenic origin, the main sources being transport and any other process of combustion at high temperatures, such as heating and electricity generation. The toxicity of NOx is mainly due to NO_2, so this text describes the health effects with respect to NO_2. Most atmospheric NO_2 is emitted in the form of NO, which is rapidly oxidized to NO_2 by ozone. The latest update of the WHO Air Quality Guidance Levels indicates for NO_2: 40 μg m^{-3} as annual mean and 200 μg m^{-3}, as one-hour mean [53].

Experimental studies carried out with animals and people, in short-term, indicate that NO_2 above 200 μg m^{-3} is a toxic gas with significant health effects, causing significant inflammation of the respiratory tract. Regarding prolonged exposure, epidemiological studies have shown that the symptoms of bronchitis in asthmatic children increase in association with the annual concentration of NO_2, and that the lower increase in lung function in children is associated with high concentrations of NO_2 in communities already subjected to current levels in the urban environment of North America and Europe.

On the other hand, in several short-term experimental human toxicology studies, acute health effects have been reported after exposure to concentrations of more than 500 μg m^{-3} of NO_2 for one hour, however, studies on bronchial responsiveness in asthmatics indicates that it increases with levels greater than 200 μg m^{-3} [49, 52-54].

Sulfur Dioxide

The possible sources of SO_2 can be both natural and anthropogenic. Natural sources include the oxidation of hydrogen sulfide (H_2S) originated by the anaerobic degradation of organic matter from swamps, mudflats, oceans, terrestrial, and biological sources. Then the anthropogenic sources are fundamentally the combustion of fuels that contain sulfur, as example in thermoelectric plants, industrial combustion sources and domestic heating. The oxidation of SO_2, especially on the surface of particles and with the presence of catalytic metals, leads to the formation of sulfurous and sulfuric acids. Inhalation is the only route of exposure to SO_2 that is of concern about its health effects [52]. The latest update of the WHO Air Quality Guidance Levels indicates for SO_2: 20 μg m^{-3}, 24-hour mean, and 500 μg m^{-3}, 10-minute mean [53].

SO_2 can affect the respiratory system and lung functions, and cause eye irritation. Inflammation of the respiratory system causes coughing, mucous discharge, and exacerbation of asthma and chronic bronchitis; likewise, it increases the propensity of people to contract infections of the respiratory system. Hospital admissions for heart disease and mortality increase on days when SO_2 levels are higher. In combination with water, SO_2 turns into sulfuric acid, which is the main component of acid rain that causes deforestation.

During short-term exposures, controlled studies with exercising asthmatics indicate that some of them experienced changes in lung function and respiratory symptoms after periods of exposure to SO_2 of as little as 10 minutes. Based on these tests, it is recommended that a SO_2 level of 500 μg m^{-3} is not exceeded for periods with an average duration of 10 minutes. Since short-term exposure to SO_2 is highly dependent on the nature of local sources and prevailing meteorological conditions, it is not possible to apply a simple factor to this value to estimate the corresponding guideline values over longer periods of time, such as an hour.

The analysis of prolonged exposures (more than 24 hours) was necessarily based on epidemiological studies in which the population was normally exposed to a mixture of contaminants. In the second edition of the air quality guidelines for Europe [52] it was noted that subsequent epidemiological studies documented separate and independent adverse public health effects for PM and SO_2, because of which it was established for SO_2 125 μg m^{-3}, 24-hour mean.

The latest available evidence includes a study in Hong Kong, China, which achieved a significant reduction in the sulfur content of fuels over a very short period. This result has been associated with a substantial reduction in health effects (e.g., respiratory diseases in childhood and mortality at all ages). Recent time series studies of hospitalizations for heart disease in Hong Kong and London (England) did not obtain any evidence of a threshold for health effects with 24-hour SO_2 concentrations in the range 5-40 µg m^{-3}. If there was a threshold for effects in any of these studies, it would have to be very low. Considering the uncertainty about the causality of SO_2; the practical difficulty of obtaining levels that are certainly associated with the absence of effects, and if the reduction of SO_2 concentrations will achieve a decrease in exposure to a causal and correlated substance, the revision of the SO_2 guide for 24 hours downwards is based, adopting a value of 20 µg m^{-3} as a precautionary approach. An annual guide is not necessary, since if the 24-hour level is respected, low annual average levels are guaranteed. These recommended guide values for SO_2 are not linked to those for PM [52, 53].

3. Evaluation of Associated Effects: Epidemiological Surveys, Respiratory Function Tests, Metabolites in Blood and Urine. Studies in Ocular Alterations

3.1. Epidemiological Surveys

To collect information regarding exposure and socio-economic factors, questionnaires developed based on international recommendations and criteria are usually applied, and among which those developed by ISSAC[1] and GINA[2]. From them, information is obtained on the characteristics of

---

[1] *ISAAC*: International Study of Asthma and Allergies in Childhood. http://isaac.auckland.ac.nz/ International Study of Asthma and Allergies in Childhood is a unique program of worldwide epidemiological research established in 1991 to investigate asthma, rhinitis and eczema in children. Is the world's largest collaborative research project, involving more than 100 countries and nearly 2 million children, and its goal is to develop environmental measures and disease monitoring to form the basis for future interventions to reduce the burden of allergic and non-allergic diseases, especially in children in developing countries. New population-based studies are urgently needed to discover more about the underlying mechanisms of the non-allergic causes of asthma, rhinitis and eczema and the burden of these diseases. The ISAAC Program formally ended in December 2012. The Global Asthma Network (http://www. Globalasthmanetwork.org/index.php) was founded in 2012 as an extension of ISAAC's work.

[2] *GINA*: Global Initiative for Asthma. http://ginasthma.org/

the population, participant data, medical history (individual and family), characteristics of the area of residence, the dwelling, the demographic data of the family group and the corresponding authorizations (informed consent and assent) to carry out the survey, and the different tests, such as spirometry, and the study of biomarkers in tears with the approval of a Bioethics Committee [11, 56-60].

3.2. Respiratory Function Tests

Regarding the measures of respiratory functionality, one of the effects described by air pollution consists of obstructive pulmonary pathologies, including asthma and other conditions related to inflammation and hyperresponsiveness of the respiratory tree.

For this measure, the spirometric test is used, a technique that measures the respiratory flows and volumes useful for the diagnosis and monitoring of respiratory diseases, in accordance with international recommendations. (ATS[3], ERS[4], SEPAR[5]).

It was released in 1993 in collaboration with the National Heart, Lung, and Blood Institute, the United States National Institutes of Health, and the World Health Organization. The GINA program and its strategies are determined for asthma care, according to committees made up of leading asthma experts from around the world. It works with healthcare professionals and public health officials around the world to reduce asthma prevalence, morbidity, and mortality.

[3] *ATS*: American Thoracic Society. http://www.thoracic.org
Founded in 1905 to combat tuberculosis, the ATS has grown to address asthma, COPD, lung cancer, sepsis, acute respiratory distress, and sleep apnea, among other diseases. Today, the ATS has grown into an international society with more than 15,000 members, dedicated to improving our clinical and scientific understanding of lung diseases and sleep-related breathing disorders.

[4] *ERS*: European Respiratory Society. https://www.ersnet.org/.
ERS is an international organization that brings together physicians, healthcare professionals, scientists and other experts working in respiratory medicine, representing more than 140 countries. Our mission is to promote lung health in order to alleviate the suffering of disease and the leading standards of respiratory medicine globally. ERS participates in the promotion of scientific research and access to high quality educational resources. It also plays a key role in promoting lung disease awareness among the public and politicians.

[5] *SEPAR*: Spanish Society of Pulmonology and Thoracic Surgery. http://separcontenidos.es/site/
SEPAR is a Scientific Society that brings together more than 4,000 respiratory health professionals in Spain, that is, practically all pulmonologists, thoracic surgeons and respiratory nursing and physiotherapy, along with other national and foreign specialists, with common

Usually are considered the following spirometric parameters: *FVC* (Forced Vital Capacity), volume of air that the patient can expel in a maximum expiration, after filling her lungs to maximum capacity; *FEV1* (Forced Expiratory Volume) amount of air that the patient manages to forcefully expire in the first second; *FEV1/FVC,* relationship between the previous parameters; *FEF25-75%*, forced mid-expiratory flow between 25 and 75% of FVC; *FEF* (maximum forced expiratory flow), maximum flow reached during the forced expiratory maneuver [61-64].

Spirometry is indicated in the medical clinic for the diagnosis of patients with respiratory symptoms, the assessment of the response to certain drugs or diseases with pulmonary involvement, and for patients exposed to air pollution. In the case of suspected asthma or COPD, forced spirometry is the basic method for the study of lung function both in children and in adults. Before interpreting the results of spirometry, it is necessary to check whether it has been performed correctly, since this test presents, among other operational difficulties, the requirement for the active collaboration of the examined patient subject. Another requirement originates in the way of expressing the values of the spirometric parameters. The usual is the comparison of the values obtained with reference values established through epidemiological studies carried out by ATS or ERS in industrialized countries [11, 60-67]. However, due to the ethnic, age, social, geographical, and climatic conditions, and characteristics of the different regions, including air pollution and the socio-sanitary condition, it is recommended that each population have its own reference values, and in this way to achieve a better interpretation of the studies that are carried out [66-68].

Anyway, to verify the correct performance of the spirometric maneuver, the acceptability criteria proposed by the regulations for forced spirometry recommended by SEPAR, ERS or ATS are used, performing a minimum of three maneuvers and a maximum of eight when these are not considered adequate.

---

interests. Its objective is to work on scientific projects that advance pulmonology and thoracic surgery and carry out initiatives on respiratory health that have a positive impact on society.

As additional acceptance criterion is the repeatability of the maneuver, for which the difference between the two best FVC and the acceptable FEV1 must be less than 0.15 l. As an essential criterion, the two best expirations of the three best acceptable curves should not vary from each other by more than 200 ml of FVC or FEV1 [60-65].

The highest value of FVC and FEV1 of all the acceptable maneuvers and without artifacts will be selected for each individual, even if their values do not come from the same maneuver. The rest of the parameters will be obtained from that acceptable curve where the sum of the FVC and FEV1 values reaches its maximum value. Finally, a spirometry will be considered "normal" when it presents values around 80% of the theoretical value of FEV1, FVC and 0.7 for the FEV1/FVC ratio, for people under 65 years of age and whose height is not extreme. Likewise, SEPAR recommends using the limits themselves calculated from the reference equations for each region [65-68].

3.3. Metabolites in Blood and Urine (Biomarkers)

Epidemiological studies are extremely valuable, but their contribution can be supported and complemented by biomarker studies. A biomarker is defined as the variation induced by a contaminant at the level of the biochemical or cellular components of a process, structure or function, which can be measured in an individual. It is a complementary methodological approach, based on the measurement of biochemical, cellular, and physiological responses, which can be measured in individuals or at the population level, and which express a potential danger due to exposure to one or more pollutants. Each of these biomarkers represents an integrated signal of the level of exposure in a certain area and, therefore, constitutes an indicator of the level of toxicological risk to which a certain population may be subjected [69-78].

Its specific contribution lies in its ability to avoid the serious limitations of the classical approaches to environmental toxicology, such as the measurement of chemical residues in organisms or in the environment and relate the acute effects through toxicity tests. Biomarkers, especially physiological and non-specific, have been widely used in the laboratory to document and quantify both, the exposure, and the effects of environmental pollutants on organisms. For exposure monitoring, biomarkers have the advantage of reacting quickly to exposure and thus being able to quantify only biologically available contaminants, that is, to assess the internal or biologically effective dose. In turn, as effective monitors they can integrate the effects of multiple stressors [69, 76-78].

Biomarkers can improve understanding of potential mechanisms of action and pathophysiological pathways, and potentially assist in the identification of susceptibility and variations in response. However, few biomarkers have been identified as indicators of specific substance exposures. This is the case of the hydroxylated metabolites of benzene, toluene and xylenes (which are later exemplified in occupational groups). In the case of PM exposure, the most toxicologically relevant dosed metabolites generated by PAHs adsorbed in PM are hydroxy-PAH (OH-PAH) and amino-PAH (NH_2-PAH). Other biomarkers used frequently, but of greater no specificity, are DNA adducts, chromosomal aberrations, micronuclei, oxidative damage to nucleic bases, and changes in RNA methylation.

These biomarkers cover the entire spectrum of disease onset and progression from external exposure to tumor formation and some have also been suggested as indicators of future cancer risk, reinforcing the causal reasoning. In this sense, they are used in many studies carried out in workers (occupational exposure) and populations of different cities with high levels of vehicular traffic or proximity to prominent sources. Additionally, other highly useful markers of exposure derive from the associations between exposure to PM in the air and the inflammatory response. In this context, the most relevant blood markers are C-reactive protein (CRP), interleukin-6 (IL-6) and fibrinogen [69-79].

3.4. Cancer Biomarkers

As previously mentioned, the results of the different epidemiological studies carried out show a strong correlation between air pollution and the increased risk of developing lung cancer, particularly due to exposure to PM_{10} or $PM_{2.5}$ [2, 5-7]. Various biomarkers have been introduced in epidemiological studies of chronic diseases, mainly exposure biomarkers, among them DNA and protein adducts, oxidative modifications as a product of the inflammatory process; as well as early damage markers, including mutagenicity, other genotoxic effects, and molecular biomarkers of cancer, considering not only direct evidence related to biomarkers related to outdoor air pollution, but also evidence on other sources of compounds present in the air contaminated. In particular, tobacco smoke and emissions from fuel combustion (mainly coal) as sources of PAHs [69, 80-84].

PAH metabolites and, to a lesser extent, PAH-DNA adducts in lymphocytes correlated well at the group level with exposure to benzo[a]pyrene, even at low levels of air pollution, with higher values among subjects most exposed to air pollutants, including occupational groups, presenting significant differences with the least exposed. Interestingly, within 24 hours of exposure both DNA adducts, and oxidized lipids are expressed. On the other hand, the PAHs-protein adducts –as benzo[a]pyrene-hemoglobin- also maintain distinctive concentrations in exposed individuals compared to those not exposed. Besides, numerous publications show that various PAHs present in practically all complex mixtures related to combustion, and extracted from PM, constitute an important source of genotoxicity. In several studies, when evaluated with salmonella TA98 (microsomal test), the comet test and the micronucleus test, the extracts usually show genotoxic effects even when the $PM_{2.5}$ concentrations were lower than those recommended by the air quality standards. Other factors, such as particle size and chemical reactivity in the atmosphere, are also known to affect the genotoxicity of ambient air [59, 69, 80-84].

3.5. Occupational Groups

Exposures to occupational air pollution in the trucking industry, due regular emission of exhaust gases and fuel, is associated with risk of mortality from ischemic heart disease. Specific studies have shown associations between this type of work and PAH metabolites (1-hydroxypyrene), biomarkers of endothelial function and general markers of inflammation, such as the cytokines IL-1β, IL-6, IL-10, TNF -α, IFN-γ, and CRP, all associated with increased cardiovascular risk and predictors of acute coronary events. Furthermore, CRP has recently been linked to mortality from lung cancer. PAHs-DNA adducts are also increased [10, 11, 81, 82]. Moreover, the findings of several studies formed the basis that allowed the IARC to indicate that outdoor air pollution is a cause of lung cancer, and also to classify outdoor air pollution, and PM, including exhaust of diesel, as Group 1 carcinogens. The positive association between exposure to PM, in particular $PM_{2.5}$ or less, and cancer mortality was evidenced also in the miners. In fact, the incidence of lung cancer becomes 5 times higher in these cases [85-87].

In this context, the results found in studies in populations that work in coal mines are illustrative. Coal dust is mainly made up of hydrocarbons and metals. During coal mining, large amounts of coal dust particles are emitted, contributing to environmental pollution and miners are constantly exposed to coal dust and its derivatives. In these cases, it is very useful to evaluate the potential genotoxic effects of coal and oxidative stress in those who were exposed to coal as part of their occupation. The comet assay, the micronucleus assay (MN), and inflammation biomarkers are good candidates for evaluating these effects. Significant and positive correlations were observed between the MN frequencies in the lymphocytes and the buccal cells of the control and exposed individuals. Chronic inflammation leads to the generation of cytokines and other oxidative stress markers, all of them increased. The evidence that exposure to coal and its derivatives presents a genetic risk demonstrates the need for protective measures and educational programs for coal miners [69, 73, 84-87].

3.6. Studies in Eye Alterations

The damaging effects of air pollution on the eyes are not well documented, and it is of great importance, since the eyes could be vulnerable due to their constant exposure to the air.

Symptoms of dry eyes are commonly associated with air pollution. A common cause of eye irritation is exposure to cigarette smoke, although dryness and eye irritation has also been related to indoor air or indoor pollution, more specifically to sick building syndrome, representing a quarter of occupational diseases in the countries developed, as well as effects generated by photochemical contamination have also been found. Some research shows symptoms that appear due to exposure to polluted air, such as pain and itching, irritation, reduced vision, and increased sensitivity to light. In the case of contact lens wearers, intolerance to them could be manifested [88-94].

It is considered that dry eyes can cause instability of the tear film and the concentrations of its components can be compromised. The acidification of pH by the high levels of oxidants (NO, NO_2, and SO_2) in the air could generate these effects [95, 96]. Several studies show the relationships between CO, $PM_{2.5}$ or total petroleum hydrocarbon exposures in water, and the ocular symptoms and signs of the population living near sources of pollution, as well as the different risk factors in the prevalence of dry eye [97-100]. More recent studies have determined certain alterations in the anterior segment of the eye attributed to air pollution, such as high levels of alterations of the subclinical ocular surface (Tear breakup time invasive or with fluorescein "Butin" and Schirmer test) were found in people who travel in highly contaminated areas, as well as effects on the ocular mucosa, showing positive and significant associations between exposure to air pollution (NO_2) and goblet cell hyperplasia in the human conjunctiva [101, 102].

Decreased tear film stability and subclinical ocular surface disorders (Butin and Schirmer) and the incidence of eyelid conditions such as blepharitis were also seen [103-107]. It is suggested that the symptoms of ocular discomfort and tear break time could be used as convenient bioindicators of the adverse health effects generated by air pollution derived from traffic [104, 108].

On the other hand, the presence of different types of mucins on the ocular surface has been described: those secreted by Goblet cells (GC) (MUC5A, MUC19), those expressed by the lacrimal acinus glands (MUC7), and those associated to the corneal and conjunctival epithelia (MUC1, MUC4, MUC16). The production of some of them has been altered in certain ocular conditions such as "dry eye", but to date, there are no reports on the effects of environmental contamination on the expression of mucin on the ocular surface or in tear secretion [109, 110]. However, studies have been found that indicate abnormal protein profiles, or more specifically mucins that can be used as biomarkers of some ocular disorders, such as some subtype of dry eye [108-114]. Furthermore, C-reactive protein concentrations have been increased in populations exposed to high vehicular traffic, as well as a decrease in interleukin (IL) -2, IL-8, IL-10 and TNF-α compared to exposure to nitrogen compounds [115-116].

The exact mechanisms through which air pollutants interfere with the tear film, cornea and conjunctiva are not very clear to date, so further studies are required on the matter to also determine how the body compensates for the changes induced by chronic exposure to air pollution and how to identify susceptible individuals who require early treatment, in order to prevent chronic disorders, and promote eye health.

4. Tools for Evaluating Health Risk. Indexes and Maps

As we have seen previously, numerous studies warn about the effects of atmospheric pollutants on the health of the population. However, the perception of being in a risk situation is not a common factor for all social groups. Bruno Latour proposes that a part of risk is constructed in discourse [117] and, although we will not stop to explore the work in risk sociology, it is interesting to analyze the resources available to communicate, raise awareness and propose public policies that improve the quality of life of the population.

Postmodern societies maintain a model of economic development that constantly generates agents that are harmful to health [118]. Risk reduction management must ensure that existing levels are reduced and that the construction of new opportunities for production and settlement in the territory be promoted, under conditions of acceptable safety and sustainability. This is not an easy task, since it involves a complex process of social control that implies economic, political, and cultural changes, in a context of conflict of interest. To face this, social actors (citizens, politicians, social movements, scientists, etc.) must have descriptive and predictive methodologies that allow a detailed analysis of the situation. In turn, these must maintain a high quality of information, require a relative time of execution, and have a reasonable cost.

In this sense, and as a synthesis tool, indices are used that aim to obtain a synthesis of the relevant information which can be used as a basis for decision-making. As an advantage, the indexes facilitate the task of communicating, summarizing in a single number a long process of theoretical construction. On the other hand, the explanatory capacity of the phenomenon is resigned, making it difficult to visualize the causalities. Another way to present the information is through risk maps. Given that the risk has strong spatial components, it is possible to generate cartographies that allow analyzing the areas most affected by a contamination event.

4.1. Models for Managing Risk Reduction

Given the lack of a consensus in the scientific paradigms of risk studies, the terminology can be somewhat confusing without proper study of the subject. An example is the alternation of use between natural hazard and natural disaster. It was common in technical works from the mid-20th century to call risk to the estimation of the probability that a phenomenon will manifest itself (e.g., seismic risk, risk of storms), omitting the effects on the population or the environment that they could generate.

When this happens, the world conventions of experts are of great influence, where the most consecrated work teams in the academic field exhibit. In this case, the United Nations declaration of the 1990s defined a terminology as the "International Decade for Natural Disaster Reduction". But not everything is so simple when there is no dominant paradigm. A few years later, the UNISDR (United Nations Office for Risk Reduction) published a book which specifies that disasters are not natural but are the result of the omission and lack of prevention and planning in the face of the phenomena of nature [119].

This lack of agreement, where dialogue and the confrontation of theoretical frameworks prevail, is typical of a post-normal science [120]. Understanding its construction process requires observing the contributions that the different scientific disciplines have made. From the perspective of human ecology, Gilbert White's pioneering work "Human adjustment to floods" [121] was published in 1945, laying the foundations for geographic contributions in studies of risks from hazards of natural origin. Here, risk is interpreted as a process that arises from the relationship between two systems: natural and human. Based on this work, a series of public policies such as the National Flood Insurance Program [122] were promoted.

After this, Burton et al. [123] led an international group to explore the concepts and methods developed at the time, making great contributions in their work "The environment as hazard". This paper summarizes different works on the subject and proposes a risk analysis methodology. The main criticism of this work centers on its conclusions, where it was stated that overcoming the impacts of these risks would come when societies had economic growth from adopting the prevailing development model. In the 1980s, and particularly in the 1990s, political ecology was on the rise, reinforcing the social character of vulnerability in risk analysis. From this perspective, disasters are not extraordinary events that sporadically can affect human societies but are based on daily socio-environmental life.

Risk reduction, therefore, will only be achieved with an improvement in the population's living conditions. In turn, it postulates that exposed populations should not be treated as passive agents, but rather has capacities to organize and respond to risks (resilience). In other words, a disaster occurs only when the losses caused by an event exceed the population's capacity to bear them or when the effects prevent it from being easily recovered.

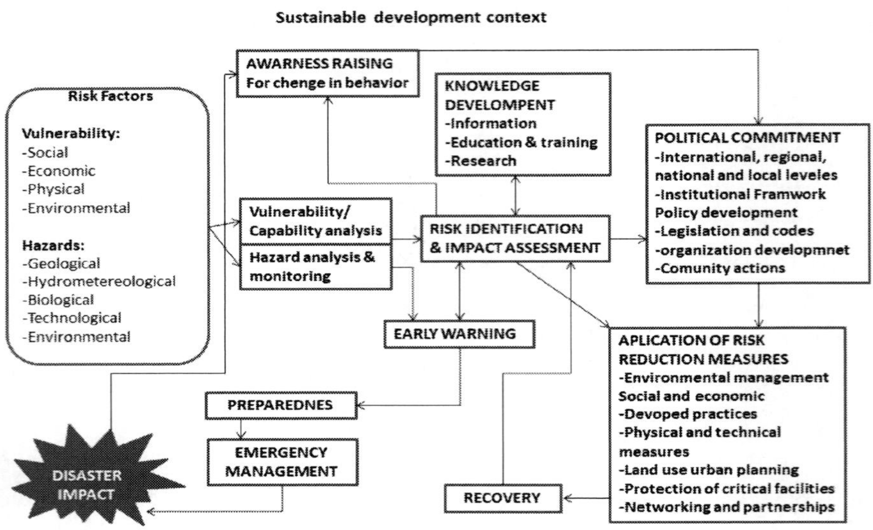

Figure 1. Conceptual diagram of the action framework model for disaster risk reduction.

Within the large number of theoretical frameworks for risk analysis, two works are worth highlighting for their different methodological approaches in risk analysis: The Framework for Action for Disaster Risk Reduction [124] with a strong systemic character and the Pressure-Response Model [125] with a more descriptive approach. The first case, outlined in Figure 1, is based on the equation Risk = Danger x Vulnerability. The methodology clearly shows the management phases for risk reduction: vulnerability analysis, hazard analysis, risk management, early warning, and response. However, it does not provide clarity to identify how reducing vulnerability contributes to reducing risk.

Figure 2 presents the Pressure-Response Model that focuses on vulnerability, looking at the causes of disasters. Therefore, it proposes to analyze disasters as the confluence of a hazard with a process of vulnerability, which has main causes, exerts dynamic pressures, and generates environmental conditions of insecurity. In other words, it advances in questioning the causes of disasters and not only their immediate effects [125]. It is important to emphasize that no theoretical model is better than another by itself, but that it will be based on our work objectives. In turn, regardless of the theoretical model that we propose, it is essential to consider that any simplification that we make of a complex system, both facilitates its understanding and loses descriptive capacity.

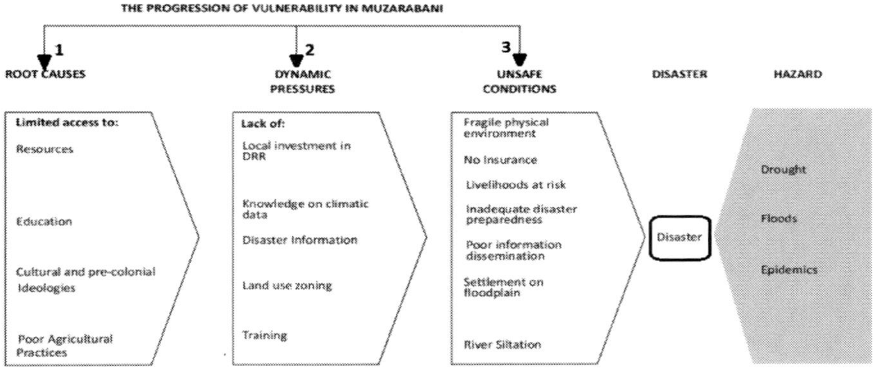

Figure 2. Conceptual diagram of the pressure-response model: the progression of vulnerability.

In reference to the vulnerability analysis of the exposed population, several components are recognized, modulated in turn by the characteristics of that population and a series of contributing factors, such as age, the presence of diseases or a poor state of health, gender, and socio-economic-environmental condition, with respect to three central components: Susceptibility, Context of the exposure and Response capacity. The context of the exposure, in the air pollution situation, remains defined for an inventory of emissions from fixed and mobile sources. Finally, the vulnerability analysis can be expressed through index or maps of vulnerability [126-128, 138].

4.2. Risk Index

To facilitate risk management, it is possible to generate indicators of the level of risk or indices that allow temporal or spatial control (Table 1). We can define an indicator as a parameter that provides information about the state of a phenomenon, with a significance that extends beyond that directly associated with the value of the parameter. The tropospheric ozone concentration is a parameter, while the ozone concentration relative to a base level is an indicator of air pollution. The indicators are always sectorial elements that tend to respond in a linear way to the increases or decreases of the parameters that compose them. The indicators can express rates, ratios, or percentages, but they are always the result of simple arithmetic calculations and linear responses.

The indices, on the other hand, are an aggregation of indicators using complex algorithms. They are usually transversal, integrate parameters of different nature and their construction responds to mathematical models. Indices are useful for monitoring over time and for comparison between different sites.

The selection of the indicator depends on the stressor studied, the availability of data, computing resources, skills, and the purpose of the study. The table below presents a set of indexes of interest in risk management for exposure to atmospheric pollutants [129].

We can mention three good reasons why it is convenient to work with indexes. First, because indexes transform complex data into general information, facilitating the development of public policies and the verification of compliance with regulations. Second, because it allows public communication, generating awareness and understanding of the situation. And finally, because it allows citizens to adopt corrective measures and adapt their behavior to benefit their quality of life. [144-147]. An example of this are the applications for cell phones that report the type of activities recommended to carry out, according to the levels of atmospheric pollution present, to avoid suffering from cardiovascular diseases (example, "CALIOPE: Air Quality" in Google Play).

Table 1. Most commonly used risk indexes

Index	Description	References
Excess lifetime cancer risk (LCR)	Estimates the probability of cancer growth in a population of individuals during a specific lifetime from projected lifetime exposure and response data against the specific dose of the contaminant. The WHO considers LCR values in the range $1x10^{-5}$ and $1x10^{-6}$ as acceptable, while the US EPA recommends LCR less than $1x10^{-6}$	59, 130-135.
Hazard Quotient (HQ)	The quotient between the concentration value of the non-carcinogenic compound in the environment under analysis and the value that is considered acceptable is made. HQ> 1 indicates risk. The values obtained are compared with those established by legislation (national or provincial), or with the values of the WHO.	132, 133, 136
Health Risk (HR)	Relates the exposure concentration and the lowest observed adverse effects (health) levels (LOAEL).	133, 136-138
Environmental burden of disease (EBD)	EBD analysis provides a standardized framework for integrating all available information on mortality, causes of death, individual health status, and condition-specific epidemiology to provide an overview of the levels of population health and the causes of loss of health from environmental risk factors.	139-143
Air pollution index (API)	Index based on the relative risk of daily mortality associated with short-term exposure to PM_{10}, $PM_{2.5}$, SO_2, O_3, NO_2 y CO.	144-146
Aggregate risk index (ARI)	It is based on the relative risk of daily mortality or morbidity. Allows to evaluate the additive effects of short-term exposure to PM_{10}, $PM_{2.5}$, SO_2, O_3, NO_2 y CO.	147-149

Finally, the data showed by risk index will then be cross-linked with the vulnerability index, through the application of risk matrices, and in this way the risk levels to which the population is exposed are obtained [128, 150-151].

4.3. Risk Maps

Maps are cartographic products that make it possible to perceive the levels of exposure or risk quickly and clearly in the territory, as showed in the Figure 3, for example. The visualization facilitates the identification of spatial patterns, allowing delimit the areas that require greater attention.

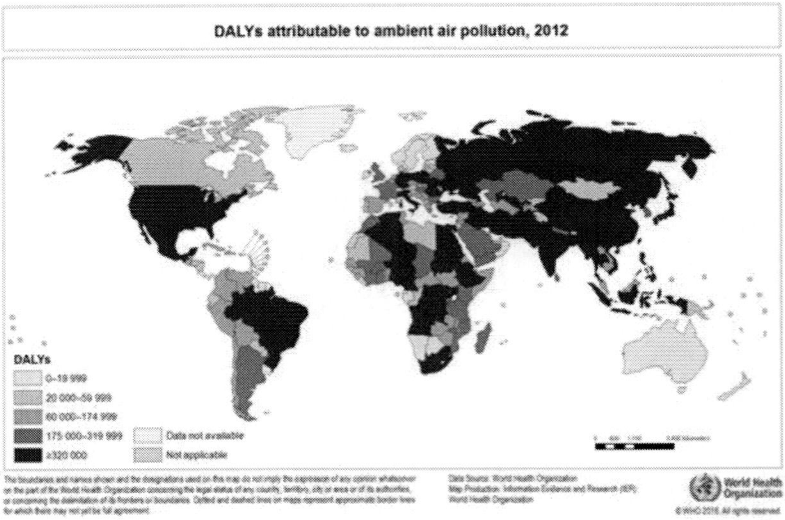

Figure 3. Map of DALYs attributable to ambient air pollution. (http://gamapserver.who.int/gho/interactive_charts/phe/aap_mbd/atlas.html)

In the maps, different variables that show spatial behavior can be visualized. For example, the *Air Pollution in World: Real-time Air Quality Index Visual Map*, available in: https://aqicn.org/map/world/; or the diverse maps of the World Health Organization (WHO) where you can see estimates of the impact of air pollution in the different countries of the world (Official site: http://gamapserver.who.int/ mapLibrary/).

For spatial calculation, there are specific computer tools that allow statistical and deterministic analysis of the variables of interest, being the geographic information systems (user, software, and hardware) central for the preparation and visualization of the final maps.

Risk maps make it possible to identify groups of vulnerability or risk exposed to pollutants that require greater preventive or health protection action.

This type of support tools for the management of urban air quality are used in other cities, for example AirWare[6], AirQUIS[7] y EnviMan[8] [152-155].

---

[6] *AirWare:* air quality management information system for urban and industrial applications (www.ess.co.at/AIRWARE)

References

[1] Karagulian F., C.A. Belis, C.F.C. Dora, A.M. Prüss-Ustün, S. Bonjour, H. Adair-Rohani, M. Amann, 2015. Contributions to cities' ambient particulate matter (PM): A systematic review of local source contributions at global level. *Atmospheric Environ.,* 120: 475-483.

[2] WHO, 2013. Review of evidence on health aspects of air pollution – REVIHAAP Project. *WHO European Centre for Environment and Health, Bonn, WHO Regional Office for Europe,* World Health Organization. 309 pp. www.euro.who.int/__data/assets/pdf_file/ 0004/ 193108/REVIHAAP-Final-technical-report-final-version.pdf.

[3] Pope 3rd C.A., Dockery D.W. 2006. Health effects of fine particulate air pollution: lines that connect. *Journal of the Air & Waste Management Association.* 56, 709–742.

[4] Dockery D.W., 2009. Health Effects of Particulate Air Pollution. *Annals Epidemiology,* 19(4): 257–263.

[5] Khafaie M.A., Yajnik C.S., Salvi S.S., Ojha A., 2016. Critical review of air pollution health effects with special concern on respiratory health. *Journal of Air Pollution and Health,* 1(2): 123-136.

[6] Guarnieri M., Balmes J.R., 2014. Outdoor air pollution and asthma. *Lancet,* 383(9928): 1581–1592.

[7] Newby D.E., Mannucci P.M., Tell G.S., Baccarelli A.A., Brook R.D., Donaldson K., Forastiere F., Franchini M., Franco O.H., Graham I., Hoek G., Hoffmann B., Hoylaerts M.F., Künzli N., Mills N., Pekkanen J., Peters A., Piepoli M.F., Rajagopalan S., Storey R.F., 2015. Expert position paper on air pollution and cardiovascular disease. *European Heart Journal,* 36(2): 83-93.

[8] WHO, 2014. WHO Indoor Air Quality Guidelines: Household Fuel Combustion. *Review 4: Health effects of household air pollution (HAP) exposure.* https://www.who.int/publications/m/item/who-indoor - air-quality-guidelines-household-fuel-combustion.

---

[7] *AirQUIS*: Air Quality Information System (www.airquis.com)

[8] *Opsis_EnviMan*: Air Quality Management Solutions. (www.opsis.se/Products/ProductsAQM/EnviManSoftware)

[9] *Public health, environmental and social determinantes of health (PHE).* Issue 63/March 2014. WHO, Department PHE. https://www.who.int/phe/eNews_63.pdf.

[10] IARC, 2013. *Outdoor air pollution a leading environmental cause of cancer deaths.* Press release 221, 17 October 2013. https://www.iarc.fr/wp-content/ uploads/ 2018/07/ pr221_E.pdf

[11] Wiwanitkit V., Suwansaksri J., Soogarun S., 2005. Cancer risk for Thai traffic police exposed to traffic benzene vapor. *Asian Pac. J. Cancer Prev.,* 6(2):219-20.

[12] Neophytou A.M., Hart J.E., Cavallari J.M., Smith T.J., Dockery D.W., Coull B.A., Garshick E., Laden F., 2013. Traffic-related exposures and biomarkers of systemic inflammation, endothelial activation and oxidative stress: a panel study in the US trucking industry. *Environmental Health.,* 12:105. doi: 10.1186/1476-069X-12-105.

[13] Brucker N., A.M. Moro, M.F. Charão, J. Durgante, F. Freitas, M. Baierle, S. Nascimento, B. Gauer, R.P. Bulcão, G.B. Bubols, P.D. Ferrari, F.V. Thiesen, A. Gioda, M.M.M.F. Duarte, I. de Castro, P.H. Saldiva, S.C. Garcia, 2013. Biomarkers of occupational exposure to air pollution, inflammation and oxidative damage in taxi drivers. *Science of the Total Environment,* 463–464, 884–893.

[14] Wichmann F., Busi L., Cianni N.F., Massolo L., Müller A., Porta A., 2009. Increased asthma and respiratory symptoms in children exposed to petrochemical pollution. *J. Allergy Clinical Immunol.,* 12: 632-38.

[15] Gorai A.K., P.B. Tchounwou, F. Tuluri, 2016. Association between Ambient Air Pollution and Asthma Prevalence in Different Population Groups Residing in Eastern Texas, USA. *International Journal of Environmental Research and Public Health,* 378: 1-17.

[16] Brunekreef B., Beelen R., Hoek G., Schouten L., Bausch-Goldbohm S., Fischer P., Armstrong B., Hughes E., Jerrett M., van den Brandt P., 2009. *Effects of Long-Term Exposure to Traffic-Related Air Pollution on Respiratory and Cardiovascular Mortality in the*

Netherlands: The NLCS-AIR Study. Health Effects Institute, Research Report 139, 1-93.

[17] Lu F., Xu D., Cheng Y., Dong S., Guo C., Jiang X., Zheng X., 2015. Systematic review and meta-analysis of the adverse health effects of ambient PM2.5 and PM10 pollution in the Chinese population. *Environ Research,* 136: 196-204.

[18] Romieu I., Gouveia N., Cifuentes L.A., de Leon A.P., Junger W., Vera J., Strappa V., Hurtado-Díaz M., Miranda-Soberanis V., Rojas-Bracho L., Carbajal-Arroyo L., 2012. *Multicity study of air pollution and mortality in Latin America* (ESCALA study). Research report, Health Effects Institute., 171: 1-120.

[19] Health Effects Institute, 2011. *Public health and air pollution in Asia (PAPA): coordinated studies of short term exposure to air pollution and daily mortality in two Indian cities.* Boston, Health Effects Institute, Research Report 157; http://pubs.healtheffects.org/getfile.php?u=623.

[20] Henschel S., Atkinson R., Zeka A., Le Tertre A., Analitis A., Katsouyanni K., Chanel O., Pascal M., Forsberg B., Medina S., Goodman P.G., 2012. Air pollution interventions and their impact on public health. *International Journal of Public Health,* 57(5):757–768.

[21] EEA (2011). *Impact of selected policy measures on Europe's air quality.* Copenhagen, European Environment Agency (http://www.eea.europa.eu/publications/impact-of-selected-policy-measures/at_download/file).

[22] Lin S., Jones R., Pantea C., Özkaynak H., Trivikrama Rao S., Hwang S., Garcia V.C. 2013. Impact of NO(x) emissions reduction policy on hospitalizations for respiratory disease in New York State. *Journal of Exposure Science & Environmental Epidemiology,* 23(1):73–80.

[23] Foster A, Kumar N. 2011. Health effects of air quality regulations in Delhi, India. *Atmospheric Environment,* 45(9):1675–1683.

[24] Yorifuji T., Kawachi I., Kaneda M., Takao S., Kashima S., Doi H. 2011. Diesel vehicle emission and death rates in Tokyo, Japan: a natural experiment. *Sci. Total Environment,* 409(19): 3620–3627.

[25] Breitner S., Stölzel M., Cyrys J., Pitz M., Wölke G., Kreyling W., Küchenhoff H., Heinrich J., Wichmann H., Peters A. 2009. Short-term mortality rates during a decade of improved air quality in Erfurt, Germany. *Environmental Health Perspectives,* 117(3):448–454.

[26] Peters A., Breitner S., Cyrys J., Stölzel M., Pitz M., Wölke G., Heinrich J., Kreyling W., Küchenhoff H., Wichmann H. 2009. *The influence of improved air quality on mortality risks in Erfurt, Germany.* Boston, Health Effects Institute. https://www.healtheffects.org/system/files/Peters137.pdf.

[27] Couto Gomes E., Florida-James G., 2014. Lung Inflammation, Oxidative Stress and Air Pollution. In: *"Lung Inflammation", Kian Chung Ong*, Ed. Chap. 1. ISBN 978-953-51-1373-7. http://www.Intechopen.com/books/statistics/lung-inflammation/lung-inflammation-oxidative-stress-and-air-pollution.

[28] Gauderman W.J., Avol E., Gilliland F., Vora H., Thomas D., Berhane K., McConnell R., Kuenzli N., Lurmann F., Rappaport E., Margolis H., Bates D., Peters J., 2004. The effect of air pollution on lung development from 10 to 18 years of age. *New England Journal of Medicine,* 351(11):1057-67.

[29] Raaschou-Nielsen O., Z.J. Andersen, R. Beelen, E. Samoli, M. Stafoggia, G. Weinmayr, B. Hoffmann, P. Fischer, M.J. Nieuwenhuijsen, B. Brunekreef, W.W. Xun, K. Katsouyanni, K. Dimakopoulou, J. Sommar, B. Forsberg, L. Modig, A. Oudin, B. Oftedal, P.E. Schwarze, P. Nafstad, U. De Faire, N.L. Pedersen, C. Östenson, L. Fratiglioni, J. Penell, M. Korek, G. Pershagen, K.T. Eriksen, M. Sørensen, A. Tjønneland, T. Ellermann, M. Eeftens, P.H. Peeters, K. Meliefste, M. Wang, B. Bueno-de-Mesquita, T.J. Key, K. de Hoogh, H. Concin, G. Nagel, A. Vilier, S. Grioni, V. Krogh, M. Tsai, F. Ricceri, C. Sacerdote, C. Galassi, E. Migliore, A. Ranzi, G. Cesaroni, Ch. Badaloni, F. Forastiere, I.Tamayo, P.

Amiano, M. Dorronsoro, A. Trichopoulou, C. Bamia, P. Vineis, G. Hoek, 2013. Air pollution and lung cancer incidence in 17 European cohorts: prospective analyses from the European Study of Cohorts for Air Pollution Effects (ESCAPE). *The Lancet Oncology,* 14: 813–22

[30] Ananya R., Gong J., Thomas D.C., Zhang J., Kipen H.M., Rich D.Q., Zhu T., Huang W., Hu M., Wang G., Wang Y., Zhu P., Lu S., Ohman-Strickland P., Diehl S.R., Eckel S.P., 2014. *The Cardiopulmonary Effects of Ambient Air Pollution and Mechanistic Pathways: A Comparative Hierarchical Pathway Analysis. PLoS ONE* 9(12): e114913. doi: 10.1371/ journal.pone.0114913

[31] Vivek D. Gandhi V.D., H. Vliagoftis, 2015. Airway epithelium interactions with aeroallergens: role of secreted cytokines and chemokines in innate immunity. Review article. *Frontiers in immunology,* April 2015, volumen 6, Article 147, 1-13. doi: 10.3389/ fimmu.2015.00147.

[32] Rückerl R., Schneider A., Breitner S., Cyrys J., Peters A. 2011. Health effects of particulate air pollution: a review of epidemiological evidence. *Inhalation Toxicology,* 23(10):555–592.

[33] Lippmann M., Chen L., Gordon T., Ito K., Thurston G.D. 2013. *National Particle Component Toxicity (NPACT) Initiative: Integrated Epidemiologic and Toxicologic Studies of the Health Effects of Particulate Matter Components.* Boston, Health Effects Institute, Research Report 177, 1-332. https://www.healtheffects.org/system/ files/RR177-Lippmann.pdf.

[34] Brook R.D., Sanjay Rajagopalan M.D., Arden Pope III C., Brook J.R., Bhatnagar A., Diez-Roux A.V., Holguin F., Hong Y., Luepker R.V., Mittleman M.A., Peters A., Siscovick D., Smith Jr S.C., Whitsel L., Kaufman J.D. 2010. Particulate matter air pollution and cardiovascular disease: an update to the scientific statement from the American Heart Association. *Circulation,* 121(21):2331–2378.

[35] Anderson H.R., Favarato G., Atkinson R.W. 2013. Long-term exposure to outdoor air pollution and the prevalence of asthma:

meta-analysis of multi-community prevalence studies. *Air Quality, Atmosphere and Health,* 6:57–68.

[36] Liu L., Poon R., Chen L., R. Poon, L. Chen, A.M. Frescura, P. Montuschi, G. Ciabattoni, A.Wheeler, R. Dales, 2009. Acute effects of air pollution on pulmonary function, airway inflammation, and oxidative stress in asthmatic children. *Environ Health Perspective;* 117:668–74.

[37] Rückerl R., Hampel R., Breitner S., Cyrys J., Kraus U., Carter J., Dailey L., Devlin R.B., Diaz-Sanchez D., Koenig W., Phipps R., Silbajoris R., Soentgen J., Soukup J., Peters A., Schneider A. 2014. Associations between ambient air pollution and blood markers of inflammation and coagulation/fibrinolysis in susceptible populations. *Environmental International* 70: 32–49.

[38] Grunig G., Marsh L.M., Esmaeil N., Jackson K., Gordon T., Reibman J., Kwapiszewska G., Park S., 2014. Perspective: ambient air pollution: inflammatory response and effects on the lung's vasculature. *Pulmonary Circulation,* 4(1):25-35.

[39] Pope 3rd C.A., Burnett R.T., Turner M.C., Cohen A., Krewski D., Jerrett M., Gapstur S.M., Thun M.J. 2011. Lung cancer and cardiovascular disease mortality associated with ambient air pollution and cigarette smoke: shape of the exposure–response relationships. *Environmental Health Perspectives,* 119(11):1616–1621.

[40] Hamra G.B., Guha N., Cohen A., Laden F., Raaschou-Nielsen O., Samet JM., Vineis P., Forastiere F., Saldiva P., Yorifuji T., Loomis D., 2014. Outdoor particulate matter exposure and lung cancer: a systematic review and meta-analysis. *Environmental Health Perspectives*, 122(9):906-11.

[41] Genc S., Zadeoglulari Z., Fuss S.H., Gen K. 2012. Review Article: The Adverse Effects of Air Pollution on the Nervous System. *Journal of Toxicology,* 782462: 1-23. doi: 10.1155/ 2012/782462.

[42] Eze I.C., Hemkens L.G., Bucher H.C., Hoffmann B., Schindler C., Künzli N., Schikowski T., Probst-Hensch N.M. 2015. Association between Ambient Air Pollution and Diabetes Mellitus in Europe and

North America: Systematic Review and Meta-Analysis. *Environmental Health Perspectives,* 123(5): 381-389.

[43] Proietti E., Röösli M., Frey U, Latzin P. 2013. Air pollution during pregnancy and neonatal outcome: a review. *J. Aerosol Medic. Pulmonary Drug Delivery,* 26(1):9–23

[44] Shah PS, Balkhair T. 2011. Air pollution and birth outcomes: a systematic review. *Environment International,* 37(2):498–516.

[45] Lamichhane D.K., Leem J., Lee J:, Kim H. 2015. A meta-analysis of exposure to particulate matter and adverse birth outcomes. *Environmental Health and Toxicology,* 30: e2015011, 1-19.

[46] Colais P., Faustini A., Stafoggia M., Berti G., Bisanti L., Cadum E., Cernigliaro A., Mallone S., Pacelli B., Serinelli M., Simonato L. 2012. Particulate air pollution and hospital admissions for cardiac diseases in potentially sensitive subgroups. *Epidemiology,* 23(3):473–481.

[47] Ostro B., Tobias A., Querol X., Alastuey A., Amato F., Pey J., Pérez N., Sunyer J. 2011. The effects of particulate matter sources on daily mortality: a case crossover study of Barcelona, Spain. *Environmental Health Perspectives,* 119(12):1781–1787.

[48] Meister K., Johansson C., Forsberg B., 2012. Estimated short-term effects of coarse particles on daily mortality in Stockholm, Sweden. *Environmental Health Perspectives,* 120(3):431–436.

[49] Chiusolo M., Cadum E., Stafoggia M., Galassi C., Berti G., Faustini A., Bisanti L., Vigotti M.A., Dessì M.P., Cernigliaro A., Mallone S., Pacelli B., Minerba S., Simonato L., Forastiere F. 2011. Short-Term Effects of Nitrogen Dioxide on Mortality and Susceptibility Factors in 10 Italian Cities: The EpiAir Study. *Environmental Health Perspectives,* 119(9): 1233-1238.

[50] Janssen N.A.H., Fischer P., Marra M., Ameling C., Cassee F.R. 2013. Short-term effects of PM2.5, PM10 and PM2.5–10 on daily mortality in the Netherlands. *Science of the Total Environment* 463–464: 20–26

[50] Bentayeb M., Wagner V., Stempfelet M., Zins M., Goldberg M., Pascal M., Larrieu S., Beaudeau P., Cassadou S., Eilstein D., Filleul

L., Le Tertre A., Medina S., Pascal L., Prouvost H., Quénel P., Zeghnoun A., Lefranc A. 2015. Association between long-term exposure to air pollution and mortality in France: A 25-year follow-up study. *Environmental International* 85: 5–14.

[51] WHO regional publications. 2000. *Air quality guidelines for Europe*. European series, No. 91, Copenhagen, second edition.

[53] WHO. 2006. *WHO Air quality guidelines for particulate matter, ozone, nitrogen, dioxide and sulfur dioxide.* Global update 2005https://www.who.int/airpollution/publications/ aqg2005/en/

[54] Roger, L.J., Horstman D.H., Mcdonnell W., Kehrl H., Ives P.J., Seal E., Chapman R., Massaro. 1990. Pulmonary function, airway responsiveness, and respiratory symptoms in asthmatics following exercise in NO2. *Toxicology and industrial health,* 6: 155–171.

[55] ISAAC Steering Committee. *ISAAC --- The International Study of Asthma and Allergies in Childhood.* Last updated 13 May 2011. http://isaac.auckland.ac.nz

[56] Weiland S.K., Bjorksten B., Brunekreef B., Cookson W.O.C., von Mutius E., Strachan D.P. 2004. Phase II of the International Study of Asthma and Allergies in Childhood (ISAAC II): rationale and methods. *European Respiratory Journal;* 24: 406–412

[57] Colman Lerner J.E., Morales A., Aguilar M., Giulani D., Ditondo J., Dodero V.I., Massolo L., Sánchez E.Y., Porta A. 2012. Air Pollution and Associated Effects on Children's Health. Comparative Study between La Plata and Bahía Blanca, Province of Buenos Aires. Environmental Science and Technology, an integrative approach. *Argentine Association for the Advancement of Science.* ISBN 978-987-28123-1-7 p 149-154.

[58] Massolo L.A. 2004. Ph.D. Thesis. *"Exposure to atmospheric pollutants and risk factors associated with air quality in La Plata and surroundings".* Faculty of Exact Sciences, National University of La Plata. Argentina. http://sedici.unlp.edu.ar/handle/10915/2267.

[59] Colman Lerner J.E., Morales A., Aguilar M., Orte M., Giulani D., Ditondo J., Dodero V.I., Massolo L., Sánchez E.Y., Matamoros N., Porta A. 2014. The effect of Air Pollution on Children's Health: a

Comparative Study between La Plata and Bahía Blanca, Buenos Aires Province, Argentina. In: *Enviromental Impact* II, Vol 181, 659-670.

[60] Chapman R.S., Hadden W.C., Perlin S.A. 2003. Influences of Asthma and Household Environment on Lung Function in Children and Adolescents. The Third National Health and Nutrition Examination Survey. *American Journal of Epidemiology,* 158:175–189.

[61] Díaz Vázquez C.A. Carvajal Urueña I., 1999. Prevalencia de asma diagnosticada en la población infantil en Asturias [Prevalence of diagnosed asthma in the child population in Asturias]. *Anales Españoles de Pediatría,* 51 Nº 5, 1999, 479-484.

[62] American Thoracic Society. 1995. Standardization of Spirometry, update. *American Journal of Epidemiology. Critical Care Medicine,* 152: 1107-36.

[63] Miller MR.., Hankinson J., Brusasco V., Burgos F., Casaburi R., Coates A., Crapo R., Enright P., van der Grinten CP, Gustafsson P., Jensen R., Johnson DC., MacIntyre N, McKay R, Navajas D., Pedersen OF., Pellegrino R., Viegi G., Wanger J. 2005. Standardization of spirometry. *European Respiratory J.,* 26: 319-338.

[64] Spirometry regulations. *Review 2013. SEPAR.* ISBN 63: 978-84-940708-0-8. https://issuu.com/separ/docs/normativa_1_espirometr__a__2013_

[65] Quanjer PH, Stanojevic S., Cole T.J. 2012. Multi-ethnic reference values for spirometry for the 3–95 year age range: the global lung function 2012 equations. *European Respiratory Journal,* 40(6): 1324-1343.

[66] Gutiérrez M., Valdivia G., Villarroel L., Contreras G., Cartagena C., Lisboa C. 2014. Proposal of new equations to calculate spirometric reference values in the Chilean adult population. *Revista médica de Chile,* 142: 143-152.

[67] Aguilar M.E., Porta A. 2020. *Regional reference values for spirometric parameters in young people in the province of Buenos*

Aires. Revista Argentina de Salud Pública, in press. http://www.rasp.msal.gov.ar/

[68] Demetriou C.A., Raaschou-Nielsen O., Loft S., Møller P., Vermeulen R., Palli D., Chadeau-Hyam M., Xun W., Vineis P. 2012. Biomarkers of ambient air pollution and lung cancer: a systematic review. *Occupational and Environmental Medicine,* 69: 619–627.

[69] Neophytou A.M., Hart J.E., Chang Y., Zhang J., Smith T.J., Garshick E., Laden F. 2014. Short-term traffic-related exposures and biomarkers of nitro-PAH exposure and oxidative DNA damage. *Toxics* 2: 377-390.

[70] Dadvand P., Nieuwenhuijsen M.J., Agustı A., de Batlle J., Benet M., Beelen R., Cirach M., Martinez D., Hoek G., Basagaña X., Ferrer A., Ferrer J., Rodriguez-Roisin R., Sauleda J., Guerra S., Anto J.M., Garcia-Aymerich J. 2014. Air pollution and biomarkers of systemic inflammation and tissue repair in COPD patients. *European Respiratory Journal,* 44: 603–613.

[71] Rückerl R., Greven S., Ljungman P., Aalto P., Antoniades C., Bellander T., Berglind N., Chrysohoou C., Forastiere F., Jacquemin B., von Klot S., Koenig W., Küchenhoff H., Lanki T., Pekkanen J., Perucci C.A., Schneider A., Sunyer J., Peters A. 2007. Air Pollution and Inflammation (Interleukin-6, C-Reactive Protein, Fibrinogen) in Myocardial Infarction Survivors. *Environmental Health Perspective,* 115: 1072–1080.

[72] Delfino R.J., Staimer N., Tjoa T., Arhami M., Polidori A., Gillen D.L., Kleinman M.T., Schauer J.J., Sioutas C. 2010. Association of Biomarkers of Systemic Inflammation with Organic Components and Source Tracers in Quasi-Ultrafine Particles. *Environ Health Perspective,* 118:756–762.

[73] Mostafavi N., Vlaanderen J., Chadeau-Hyam M., Beelen R., Modig, Palli D., Bergdahl I.A., Vineis P., Hoek G., Kyrtopoulos S.A., Vermeulen R. L. 2012. Inflammatory markers in relation to long-term air pollution. *Environment International,* 81: 1–7.

[74] Gil Hernández Fernando. 1998. *The role of biomarkers in Human Toxicology.* Department of Legal Medicine and Toxicology, Faculty

of Medicine, University of Granada, Spain. http://www.ugr.es/~fgil/biomarcadoresrevtoxicol.pdf.

[75] Delfino R.J., Staimer N., Vaziri N.D. 2011. Air pollution and circulating biomarkers of oxidative stress. *Air Quality, Atmosphere, and Health,* 4: 37–52.

[76] Faseeha Suhaimi N., Jalaludin J. 2015. Review Article Biomarker as a Research Tool in Linking Exposure to Air Particles and Respiratory Health. *BioMed. Research International,* 2015, Article ID 962853, 1-10.

[77] Thompson A.M.S., Zanobetti A., Silverman F., Schwartz J., Coull B., Urch B., Speck M., Brook J.R., Manno M., Gold D.R. 2010. Baseline Repeated Measures from Controlled Human Exposure Studies: Associations between Ambient Air Pollution Exposure and the Systemic Inflammatory Biomarkers IL-6 and Fibrinogen. *Environmental Health Perspectives,* 118: 120–124.

[78] Torres Lemos A., Torres de Lemos C., Negreiros Flores A., Ozorio Pantoja E., Vaz Rocha J.A., Ferrao Vargas V.M. 2016. Genotoxicity biomarkers for airborne particulate matter (PM2.5) in an area under petrochemical influence. *Chemosphere,* 159: 610-618.

[79] Vineis P., Husgafvel-Pursiainen K. 2005. Air pollution and cancer: biomarker studies in human populations. *Carcinogenesis,* 26 (11): 1846–1855.

[80] Castaño-Vinyals G., D'Errico A., Malats N., Kogevinas M. 2004. Biomarkers of exposure to polycyclic aromatic hydrocarbons from environmental air pollution. *Occupational and Environmental Medicine,* 61: e12.

[81] Vineis Paolo. 2013. Biomarkers of air pollution: DNA and protein adducts. In *Air Pollution and Cancer*, IARC Scientific Publication No. 161, Editors: K. Straif, A. Cohen, *J. Samet.* Chapter 12. 149-164.

[82] Møller P., Loft S. 2010. Oxidative Damage to DNA and Lipids as Biomarkers of Exposure to Air Pollution. Review. *Environmental Health Perspectives,* 118:1126–1136.

[83] Demetriou C.A., Vineis P. 2015. Carcinogenicity of ambient air pollution: use of biomarkers, lessons learnt and future directions. *Journal of Thoracic Disease,* 7(1): 67-95.

[84] Rohr P., Kvitko K., da Silva F.R., Menezes A.P., Porto C., Sarmento M., Decker N., Reyes J.M., Allgayer Mda C., Furtado T.C., Salvador M., Branco C., da Silva J., 2013. Genetic and oxidative damage of peripheral blood lymphocytes in workers with occupational exposure to coal. *Mutation Research,* 758(1-2): 23-28.

[85] Rainho C.R., Corrêa S.M., Aiub C.A.F., Felzenszwalb I. 2015. Biomonitoring of genotoxic risk of workers exposed to heavy air pollution. Air Pollution XXIII, 117-128. *WIT Transactions on Ecology and the Environment,* Vol 198. ISSN 1743-3541.

[86] Gharibvand L., Shavlik D., Ghamsary M., Lawrence Beeson W., Soret S., Knutsen R., Knutsen S.F. 2017. The Association between Ambient Fine Particulate Air Pollution and Lung Cancer Incidence: Results from the AHSMOG-2 Study. *Environ. Health Persp.,* 125: 378–38.

[87] Shephard R.J., Ponsford E., LaBarre R., Basu P.K. 1979. Effect of cigarette smoke on the eyes and airway. *International Archives of Occupational and Environmental Health,* 43(2):135–44.

[88] Franck C, Bach E, Skov P. 1993. Prevalence of objective eye manifestations in people working in office buildings with different prevalences of the sick building syndrome compared with the general population. Internat. Archives Occup. *Environ. Health,* 65(1): 65–9.

[89] Kok A., Yassi T., Kjellström T., Guidotti L. 2009. *Salud Ambiental Básica* [*Basic Environmental Health*]. Organización Mundial de la Salud. México. pp. 56-63.

[90] Altshuller A.P. 1977. Eye Irritation as an Effect of Photochemical Air Pollution. *J. Air Pollution Control Association,* 27(11):1125–6.

[91] McMonnies C.W. 1986. Key questions in a dry eye history. *Journal of the American Optometric Association,* 57(7):512–7.

[92] Lemp M.A. 1995. Report of the National Eye Institute/Industry workshop on Clinical Trials in Dry Eye. *Contact Lens Association of Ophthalmologists Journal,* 21(4):221–32.

[93] Korb D.R, Henriquez A.S. 1980. Meibomian gland dysfunction and contact lens intolerance. *Journal of the American Optometric Association,* 51(3): 243–51.

[94] Holly F.J., Esquivel E.D. 1985. Colloid osmotic pressure of artificial tears. *Journal of ocular pharmacology.* 1(4):327–36.

[95] Andrés S., García M.L., Espina M., Valero J., Valls O. 1988. Tear pH, air pollution, and contact lenses. *American Journal Optometry Physiological Optics,* 65(8): 627–31.

[96] Siddiqui A.R., Lee K., Gold E.B., Bhutta Z. 2005. Eye and respiratory symptoms among women exposed to wood smoke emitted from indoor cooking: a study from southern Pakistan. *Energy for Sustainable Development,* 9(3): 58–66.

[97] Ellegård A. 1997. Tears while Cooking: An Indicator of Indoor Air Pollution and Related Health Effects in Developing Countries. *Environmental Research,* 75(1): 12–22.

[98] San Sebastián M., Armstrong B., Stephens C. 2001. Health of women living near oil wells and stations in the Ecuadorian Amazon. *Revista Panamericana de Salud Pública,* 9: 375–83.

[99] Sahai A., Malik P. 2005. Dry eye: prevalence and attributable risk factors in a hospital-based population. *Indian Journal of Ophthalmology,* 53(2): 87–91.

[100] Saxena R., Srivastava S., Trivedi D., Anand E., Joshi S., Gupta S.K. 2003. Impact of environmental pollution on the eye. *Acta Ophthalmologica Scandinavica,* 81(5): 491–4.

[101] Novaes P., do Nascimento Saldiva P.H., Kara-Jose N., Macchione M., Matsuda M., Racca L. 2007. Ambient levels of air pollution induce goblet-cell hyperplasia in human conjunctival epithelium. *Environmental Health Perspectives,* 115(12): 1753–6.

[102] Gupta S.K., Gupta S.C., Agarwal R., Sushma S., Agrawal S.S., Saxena R. 2007. A multicentric case-control study on the impact of

air pollution on eyes in a metropolitan city of India. *Indian Journal of Occupational and Environmental Medicine,* 11: 37–40.

[103] Novaes P., Hilário do Nascimento Saldiva P., Matsuda M., Macchione M., Peres Rangel M., Kara-José N. 2010. The effects of chronic exposure to traffic derived air pollution on the ocular surface. *Environmental Research,* 110(4): 372–4.

[104] Moen B.E., Norbäck D., Wieslander G., Bakke J.V., Magerøy N., Granslo J.T. 2011. Can air pollution affect tear film stability? A cross-sectional study in the aftermath of an explosion accident. *BMC public Health,* 11:235.

[105] Torricelli A.A.M., Matsuda M., Novaes P., Braga A.L.F., Saldiva P.H.N., Alves M.R. 2014. Effects of ambient levels of traffic-derived air pollution on the ocular surface: Analysis of symptoms, conjunctival goblet cell count and mucin 5AC gene expression. *Environmental Research,* 131(2): 59–63.

[106] Gutiérrez M.A., Giuliani D.S., Porta A., Andrinolo D. 2019. Ocular surface alterations in relation with different aerial particulate matter concentrations. *Journal of Ophthalmic and Vision Research,* 14 (4): 419–27.

[107] Gipson I.K., Hori Y., Argüeso P. 2004. Character of ocular surface mucins and their alteration in dry eye disease. *The Ocular Surface,* 2(2): 131–48.

[108] Gutiérrez M.A. 2016. *Ocular alterations related to air pollution. Comparative study in populations of La Plata and Ensenada.* PhD thesis, School of Exact Sciences, UNLP. http://sedici.unlp.edu.ar/handle/10915/52404.

[109] Floyd A.M., Zhou X., Evans C., Rompala O.J., Zhu L., Wang M. 2012. Mucin deficiency causes functional and structural changes of the ocular surface. *PloS one,* 7(12): e50704.

[110] Ohashi Y., Oka A., Rodrigues-Pousada R., Possenti M., Ruberti I., Morelli G. 2003. Modulation of phospholipid signaling by GLABRA2 in root-hair pattern formation. *Science* 300(5624): 1427–30.

[111] Zhou L., Beuerman R.W., Chan C.M., Zhao S.Z., Li X.R., Yang H. 2009. Identification of Tear Fluid Biomarkers in Dry Eye Syndrome Using iTRAQ Quantitative Proteomics. *Journal of Proteome Research,* 8(11): 4889–905.

[112] Albietz J.M. 2000. Prevalence of dry eye subtypes in clinical optometry practice. *Optometry and Vision Science,* 77(7): 357–63.

[113] Gutiérrez M.A., Colman Lerner J.E., Giuliani D.S., Porta A., Andrinolo D. 2019. Comparative study of tear lipid composition in two human populations with different exposure to particulate matter in La Plata, Argentina. *Environmental Science and Pollution Research,* 26: 6948–56.

[114] Fuks K.B., Weinmayr G., Foraster M., Dratva J., Hampel R., Houthuijs D. 2014. Arterial Blood Pressure and Long-Term Exposure to Traffic-Related Air Pollution: An Analysis in the European Study of Cohorts for Air Pollution Effects. *Environmental Health Perspectives*, 122: 896–905.

[115] Mostafavi N., Vlaanderen J., Chadeau-Hyam M., Beelen R., Modig L., Palli D. 2015. Inflammatory markers in relation to long-term air pollution. *Environment International,* 81:1–7.

[116] García Díaz, P. 2011. Critical review of Latourian ontological politics. *Athenea digital: Journal of social thought and research,* 11(1), 0155-170.

[117] Beck, U. 1993. From the industrial society to the risk society.. *Revista occidente,* 1993, pp- 19-40.

[118] UNISDR. 2010. *"Natural hazards, unnatural disasters: the economics of effective prevention"* 2010, World Bank.

[119] Funtowicz, S.O., Ravetz, J.R. 1994. "The worth of a songbird: ecological economics as a post-normal science". *Ecological Economics,* 10, p. 197-207.

[120] White G.F. 1945. *Human adjustment to floods.*

[121] Anderson D.R. 1974. The national flood insurance program. Problems and potential. *Journal of Risk and Insurance,* 579-599.

[122] Burton I., Kates R. W., White G.F. 1978. The Environment as Hazard. *Annals of the Association of American Geographers,* Vol. 70, No. 2 (Jun., 1980), pp. 306-311.

[123] UNISDR. 2004. *Living with risk: a global review of disaster reduction initiatives.* ISBN/ISSN: 9211010640. 429 p. United Nations Office for Disaster Risk Reduction (UNISDR). http://www.unisdr.org/we/inform/publications/657.

[124] Blaikie P., Cannon T., Davis I., Wisner B. 1994. At Risk. *Natural Hazards, People's Vulnerability, and Disasters* (London, New York: Routledge). ISBN 0-4152-5215-6

[125] Ezzati M., Lopez A.D., Rodgers A., Murray C. J. 2004. Comparative quantification of health risks. *Global and regional burden of disease attributable to selected major risk factors.* Geneva: World Health Organization, 1987-1997.

[126] Boström C.E., Gerde P., Hanberg A., Jernström B., Johansson C., Kyrklund T., Westerholm R. 2002. Cancer risk assessment, indicators, and guidelines for polycyclic aromatic hydrocarbons in the ambient air. *Environ. Health Perspectives,* 110(Suppl 3), 451-89.

[127] Guo, H., Lee, S.C., Chan, L.Y., Li, W.M., 2004. Risk assessment of exposure to volatile organic compounds in different indoor environments. *Environmental Research,* 94: 57-66.

[128] Colman Lerner J.E., Sánchez E.Y., Sambeth J.E., Porta A. 2012. *Volatile organic compounds (VOCs) and particulate matter (PM) in occupational environments.* Characterization and health risk assessment. Lap Lambert Pub., Saarbrucken, Germany.

[130] Colman Lerner J.E., Kohajda T., Aguilar M., Massolo L., Sánchez E.Y., Porta A., Opitz P., Wichmann G., Herbarth O. y Mueller A., 2014. Improvement of health risk factors after reduction of VOC concentrations in industrial and urban areas. *Environmental Science and Pollution Research*, 21, 9676-9688.

[131] Colman Lerner J.E., Elordi M.L., Orte M.A., Giuliani D.S., Gutiérrez M.A., Sanchez E.Y., Sambeth J.E., Porta A. 2018. Exposure and risk analysis to particulate matter, metals and polycyclic aromatic hydrocarbon and at different workplaces in

Argentina. *Environmental Science and Pollution Research,* 25(9): 8487-8496, 2018.

[132] Colman Lerner J.E., Gutierrez M.A., Mellado D., Giuliani D., Massolo L., Sanchez E.Y., Porta A., 2018. Characterization and cancer risk assessment of VOCs in home and school environments in gran La Plata, Argentina. *Environmental Science and Pollution Research,* 25(10): 10039-10048.

[133] Payne-Sturges D.C., Burke T.A., Breysse P., Diener-West M., Buckley T.J. 2004. Personal exposure meets risk assessment: a comparison of measured and modeled exposures and risks in an urban community. *Environ. Health Perspectives,* 112 (5): 589-98.

[134] García Reynoso J., Grutter M., Cintora Juarez D. 2007. "Risk assessment for criteria pollutants and formaldehyde in Mexico City". *Revista internacional de contaminación Ambiental,* 23 (4), México.

[131] Ostro B. (2004). *Outdoor air pollution. Assessing the environmental burden of disease at national and local levels.* Environmental Burden of Diseases Series No 5. World Health Organization. Protection of the Human Environment. Geneva, 2004.

[135] Hänninen O., Knol A.B., Jantunen M., Lim T., Conrad A., Rappolder M., Carrer P., Fanetti A.C., Kim R., Buekers J., Torfs R., Iavarone I., Classen T. Hornberg C., Mekel O.C.L. 2014. Environmental Burden of Disease in Europe: Assessing Nine Risk Factors in Six Countries (review). *Environmental Health Perspectives,* 122(5): 439-446.

[136] Prüss-Ustün A., Wolf J., Corvalán C., Bos R., Neira M., 2016. *Preventing disease through healthy environments: a global assessment of the burden of disease from environmental risks.* WHO. www.who.int/quantifying_ehimpacts/publications/preventing-disease/en/

[137] Sosa B.S., Porta A., Colman Lerner J.E., Banda Noriega R., Massolo L. 2017. Human health risk due to variations in PM10-PM2.5 and associated PAHs levels. *Atmospheric Environment,* 160: 27-35.

[138] Giuliani D.S. 2020. Ph.D. Thesis. *"Speciation of compounds associated with particulate matter (PM10 and PM2.5) in La Plata*

and surroundings: metals, polycyclic aromatic hydrocarbons, simple and derivatives. Evaluation of the associated risk". Faculty of Exact Sciences, UNLP, Argentina. http://sedici.unlp.edu.ar/handle/10915/110949.

[139] Cairncross E.K., Juanette J., Zunckel M. 2007. "A novel air pollution index based on the relative risk of daily mortality associated with short-term exposure to common air pollutants", *Atmospheric Environment,* 41:38, p 8442-8454.

[140] Kanchan A., Kumar Gorai A., Goya P., 2015. A Review on Air Quality Indexing Systems. *Asian Journal Atmospheric Environment,* 9(2): 101-113.

[141] World Air Quality Index project. *Air Quality Programmatic APIs.* https://aqicn.org/api/

[142] Sicard, P., Talbot C., Lesne O., Mangin A., Alexandre N., Collomp R. 2012. The Aggregate Risk Index: An intuitive tool providing the health risks of air pollution to health care community and public. *Atmospheric Environ.*, 46. 11–16. 10.1016/j.atmosenv.2011.10.048.

[143] Tan X., Hana L., Zhanga X., Zhouab W., Li W., Qian Y. 2021. A review of current air quality indexes and improvements under the multi-contaminant air pollution exposure. *J. Environ. Management,* 279: 111681. https://doi.org/10.1016/j.jenvman.2020.111681.

[144] van den Elshout S., Léger K., Heich H. (2014). CAQI common air quality index—update with PM 2.5 and sensitivity analysis. *Sci. Total Environment,* 488: 461-46.

[145] Represa N.S., Mellado D., Bali L., Colman Lerner, J.E., Sanchez E.Y., Porta A., 2016. Application of geographic information systems (GIS) technologies for risk estimation due to exposure to volatile organic compounds (VOCs). *Proceedings of the III Congress of the Latin American Society for Risk Analysis,* S. Paulo, Brasil, pp. 1-6.

[146] Represa N.S., Sanchez E.Y., Porta A. Study of Social Vulnerability in Argentina Using GIS. Construction of a Locally Applicable Index. *Anuário do Instituto de Geociências – UFRJ,* 41(2): 351-357, 2018.

[147] Lahr J., Kooistra L., 2010. Environmental risk mapping of pollutants: State of the art and communication aspects. *Science of the Total Environment* 408: 3899–3907.

[148] Karimi H., Soffianian A., Mirghaffari N., Soltani S., 2016. Determining Air Pollution Potential Using Geographic Information Systems and Multi-criteria Evaluation: A Case Study in Isfahan Province in Iran. *Environmental Processes,* 3: 229–246.

In: Air Pollution: Effects and Dangers ISBN: 978-1-53619-544-6
Editor: J. Esteban Colman Lerner

Chapter 4

TOP-DOWN AND BOTTOM-UP APPROACHES TO ESTIMATE AIR EMISSIONS INVENTORIES (AEI): CASE OF WASTE COLLECTION SERVICE IN AN INTERMEDIATE CITY

***Beatríz S. Sosa*[1,*] *and Nicolás E. Cisneros Basualdo*[1,2]**
[1]Centro de Investigaciones y Estudios Ambientales (CINEA-UNICEN), Tandil, Argentina
[2]Comisión de Investigaciones Científicas (CIC-PBA), La Plata, Argentina

ABSTRACT

To ensure the prevention of air pollution and to avoid short-term health effects or long-term effects on the environment, it is necessary that AEIs are approached as a preventive rather than a reactive tool to work on existing issues. To assess air quality linked to the mobile source, it is first necessary to define the objective of the inventory, the scope and the particularities of the site. In this work, the results of the emissions

---

* Corresponding Author's Email: bsosa@fch.unicen.edu.ar.

estimation are compared with the Top-Down and Bottom-Up methodological approaches, with three methodologies in increasing order of complexity - IPCC 2006, COPERT 4 and IVE - that analyze a heavy transport fleet whose driving pattern belongs to the collection of urban solid waste (USW) in an intermediate city by 2016. The pollutants included in the Air Emission Inventory are CO_2, CH_4, N_2O, PM_{10}, CO, HC, SO_x, and NO_x. The main results obtained are quite similar for the IPCC and COPERT 4, while they show great differences when compared with the results of the IVE if the particularities of the place for fuel consumption are not considered. The most notable finding is the difference in CO_2 values estimated with IVE because the IPCC 2006 considers that it is not necessary to make estimates with high levels of information when it comes to this pollutant. The values of PM and SO_2 emitted during start-up are significant but do not contribute to the total emissions of pollutants. The work also concludes that the implementation of an AEI should be systematic and cyclical, and subject to improvements in planning and methodologies used; when the inventory is carried out periodically, it changes from being a diagnostic tool to one of prevention, being able to intervene with actions aimed at continuous improvement of the objectives of reducing air pollution.

1. INTRODUCTION

Emissions generated by anthropogenic activities cause impacts that extend beyond the geographic site where they were generated and can reach regional scales or even greater extensions. Historically, the analysis of environmental impacts has been based on specific experimental measures without being able to cover all emission sources and pollutants emitted. Most air pollution control systems combine technological and administrative tools in order to prevent high concentrations of air pollutants from reaching sensitive receptors. Air quality models, for example, as tools of the control system, contribute by providing knowledge about the emissions generated by the different sources and the concentrations expected at different distances from the emitting source. However, in order to carry out any modelling or impact studies, it is necessary to first characterize the emissions based on an Air Emission Inventory (AEI).

An AEI is a set of data that characterizes and consolidates by summation the emissions of air pollutants, according to the type of source and the type and amount of pollutants emitted, in a given geographical area and time interval [1]. The study of atmospheric emissions provides critical information for the following stages of air quality studies, although in the last decades not much progress has been made in its study in relation to the knowledge of atmospheric behavior both in terms of quality and quantity [2].

Depending on the purpose and scope, inventories can be developed from either a Bottom-Up or Top-Down approach. The methodologies with Bottom-Up approach are normally used for point sources, while Top-Down for area sources. Bottom-up approaches start from specific and detailed information on each of the emission sources in a given area, which when aggregated allow for general information on the required emissions. A Top-Down approach requires methodologies that estimate emissions from an activity in a general or aggregated form (e.g., at country or regional level) that, with the application of different analysis criteria, can be directly or indirectly related to emissions in specific study areas [1, 3, 4]. The European Monitoring and Evaluation Programme/European Environment Agency (EMEP/EEA) and the Intergovernmental Panel for Climate Change (IPCC) have a similar approach to calculating Greenhouse Gases (GHG) emissions, although they incorporate an intermediate level of complexity among their possibilities by proposing three levels. Tier 1 assumes a simple linear relationship between activity data and emission factors with activity information from statistical bases and emission factors from typical or averaged process conditions. Intermediate Tier 2 is similar to Tier 1 but with country-specific emission factors or more detailed activity statistics. Finally, Tier 3, the most complex in its information requirements, may include facility-level data and/or the use of models 2 and 3 are considered the most accurate and highest levels [5-7]. In addition, the methodological approach highlights four main elements: the pollutants emitted, the activities or sectors that generate these emissions, the temporal variation, and their geographical location – Dimensions of the Inventory, What, Why, When and Where [8].

Reviewing the availability of AEIs worldwide, there are databases with access to regional information with different histories and levels of development. In general, in each region, AEIs have emerged as a reactive tool to environmental pollution problems, starting in the sites with older and more economically developed societies, and spreading to other regions of the world as they also grew and the same environmental problems appeared. Among the IEAs available today are:

- The European Commission's *Emissions Database for Global Atmospheric Research* (EDGAR) estimates emissions from 1970 until today. This inventory contains emissions for particulate matter (PM_{10} and $PM_{2.5}$), mercury (Hg), direct greenhouse gases (GHG), ozone (O_3) and its precursors (CO, NO_x, NMVOC), acidifiers (NH_3, NO_x, SO_2) and black and organic carbon (BC, OC). This database is available to be downloaded as a georeferenced layer with a spatial resolution of 0.1° x 0.1° and a temporal resolution of one year. The information provided by EDGAR is based on activity levels obtained from international statistics and emission factors from the IPCC, the International Energy Agency (IEA), and the Food and Agriculture Organization of the United Nations (FAO) [9, 10], among others.

  In the European Union, efforts have been made to develop, improve, and update inventories since 1979, starting with the signing of the Convention on Long-Range Transboundary Air Pollution [11], which later led to the creation of the *Co-operative Programme for Monitoring and Evaluation of the Long-Range Transmission of Air Pollutants in Europe* (EMEP) [12] which includes inventories as one of its air quality control tools. EMEP emissions are provided by sector and geographically. Emissions of SO_x (SO_2, SO_3, H_2SO_4, H_2S), NO_x (NO_2, NO), NMVOC, NH_3, CO, PM ($PM_{2,5}$, PM_{10}) can be obtained on a grid with a spatial resolution of 0.1° x 0.1°, for the whole time series from 1990 to 2017, as well as Cd, Hg, Pb, PCDD/PCDF, Benzo(a)pyrene, Benzo(b)fluoranthene, Benzo(k)fluoranthene, Indene and HCB, and BC, TSP, and other metals (As, Cr, Cu, Ni, Se, Zn, and their compounds) at a resolution of 0.1° x 0.1° for the years 2015, 2016 and 2017 [13, 14].

In 1995, the Core Inventory of Air Emissions (CORINAR) program appeared which is being developed in parallel with the EMEP program that constitutes EMEP/CORINAIR and includes tasks of calculation, assessment, modeling and validation of atmospheric emissions from sources that have significant environmental impact. These inventories are currently being developed under the name EMEP/EEA [5].

- *Pollutant Release and Transfer Register* (PRTR) is a national or regional environmental database or inventory of hazardous chemical substances and pollutants released to air, water, and soil, and transferred off-site for treatment or disposal. Some PRTRs also include estimates of releases from diffuse sources, such as agriculture and transport, and from the end use of products. In 2007, 17 Organisation for Economic Co-operation and Development (OECD) countries had an operational PRTR and many more were in a process of developing such a system. OECD work on PRTRs is a part of the Environment, Health and Safety Program including: four inventories of available guidance or tools and best practices for better use and presentation of PRTR data. The four inventories are 1) available guidance on the use of PRTR data, 2) GIS mapping systems of PRTR data, 3) resources related to PRTR and risks including hazard information for PRTR data, and 4) activities related to pollution prevention/sustainability and PRTR data [15-17]. More information available at https://www.oecd.org/chemicalsafety/pollutant-release-transfer-register/publications-series-on-pollutant-release-and-transfer-registers.htm.
- In the United States, the first industrial emissions inventories were established in 1986 with the Emergency Planning and Community Right-to-Know Act (EPCRA), and in 1990 with the Pollution Prevention Act (PPA). These laws laid the foundation for the

development of the *Toxic Release Inventory* (TRI). Since 1993, emissions of substances that cause or are suspected of causing health effects have been collected from the most relevant sources; the National Toxics Inventory includes data on more than 21,000 facilities nationwide and covers 767 chemical substances and 33 chemical categories [18]. The Environmental Protection Agency (U.S. EPA) also has the *National Emissions Inventory* (NEI). The NEI is a national compilation of air emission estimates of criteria air pollutants (CAPs), the precursors of CAPs, and hazardous air pollutants (HAPs). The hazardous air pollutants that are included in the NEI are based on Section 112(b) of the Clean Air Act. State, local and tribal (S/L/T) air agencies submit emission estimates to EPA to be added to their emissions programs, such as the emissions trading program. Other products are available like Toxics Release Inventory (TRI), and data collected during rule development or compliance testing [19].

Greenhouse Gas Reporting Program (GHGRP). On October 30, 2009, the EPA published a rule for the mandatory reporting of GHG from sources that in general emit 25,000 metric tons or more of carbon dioxide equivalent per year. Approximately 8,000 facilities are required to report their emissions annually, and the reported data are made available to the public in October of each year [20]. More information available at https://www.epa.gov/air-emissions-inventories/national-emissions-inventory-nei and https://www.epa.gov/toxics-release-inventory-tri-program/tri-data-and-tools.

- For the area of South America, the only complete inventory that exists on a regional or national scale is that compiled in the global EDGAR database. Although there are cooperation initiatives between countries in the region to work with methodologies adapted to the area of study and which allow the aggregation and comparison of the results of AEI. An example of this is the project "Air Pollution Forecasting, Latin America and the Caribbean" financed by the European Union within the framework of the

Development and Innovation 2020 (PAPILA) project [21, 22]. More information available at https://edgar.jrc.ec.europa.eu/.

- In Asia, MIX is a new anthropogenic emission inventory for Asia for the years 2008 and 2010. The domain of MIX covers 30 countries and regions, stretching from Kazakhstan in the west to Far East Russia in the east, and from Indonesia in the south to Siberia in the north. Emissions are aggregated into five sectors: power, industry, residential, transportation, and agriculture. Ten chemical species are included in the MIX inventory, including both gaseous species and aerosol species: SO_2, NO_x, CO, NMVOC, NH_3, PM_{10}, $PM_{2.5}$, BC, OC, and CO_2. Monthly emissions are provided by sector at 0.25° x 0.25° resolution [23]. More information available at http://www.meicmodel.org/.
- In Africa, the study presented by Keita et al. [24] points out that there are not yet any inventories for the region, therefore global inventories are used as a basis for other air quality studies in the region. Publications of previous studies in Africa still have significant uncertainty in their results so air quality management policies have a major challenge ahead.
- Finally, the most well-known effort at world level is undoubtedly that carried out by the United Nations Framework Convention on Climate Change (UNFCCC). The UNFCCC has historically adopted various versions of these IPCC guidelines for their own reporting purposes. The IPCC leads an effort to develop and refine internationally agreed upon methodologies for calculating and reporting greenhouse gas emissions and removals. These IPCC national GHG inventory guidelines play an important role in fostering the incorporation of scientific evidence into national climate policy mechanisms. They are of particular interest now, as they were just "refined" in 2019, the first major overhaul since 2006 [6, 25]. At the 21st Conference held in Paris in 2015, the Parties to the UNFCCC reached a landmark agreement whose main objective is to strengthen the global response to the threat of climate change. In this century, the goal is to keep global

temperature rise below 2°C, compared to pre-industrial levels, and to continue efforts to further limit temperature rise to 1.5°C [26]. Currently, 197 countries have adopted the Paris Agreement.

The historical journey of the AEIs observed in different regions of the world shows a transition that begins with their use as a diagnostic tool or baseline for other studies and evolves towards their incorporation as a preventive tool for air pollution. There are still aspects to be improved, firstly, methodological aspects which facilitate the elaboration of the AEI as a tool of the Air Quality Management System in a greater number of regions and countries that are still lagging behind but with significant growth economies. Secondly, finding the mechanisms for the AEI to be used for the purpose of evaluating the risks to the health of the population aimed at sustainable development [17]. Thus, the bases to ensure the prevention of pollution and to avoid short term effects on health, or long term effects on the environment, require that AEIs are approached as a systematic and cyclical process, and subject to improvements in their planning and methodologies used; when the inventory is made periodically, it changes from being a diagnostic tool to one of prevention, being able to intervene with actions aimed at a continuous improvement of the objectives of reducing atmospheric pollution. The diversity of emission sources and the actors involved show the importance of having clear, updated information in sufficient quantity and quality (another critical aspect of the AEI); an accurate inventory will provide essential information for the understanding of the effects of air quality degradation. As a management tool, the AEIs are indispensable in the management of public relations between authorities and society because they demonstrate compliance with regulations, are an input for the estimation of impacts on air quality, allow the determination of baselines and collaborate with decision making in the management and control of the state of air quality [1, 27].

The importance of preserving air quality is considered, without a doubt, one of the main current challenges by international organizations that inform about climate change (as a global problem and its long-term effects) and health issues (local and urban problems, short and medium term) [28, 29, 30]. Management and control policies concerning anthropogenic activities that can impact the atmosphere are developed in four sectors: Energy; Industrial Processes and Product Use; Agriculture, Forestry and Other Land Use; and Waste. Emissions from Waste are estimated to contribute to 3-5% of total anthropogenic emissions [31], however, these figures do not include emissions from waste transportation and collection since those are included in the Energy sector. Therefore, research about urban solid waste (USW) management rarely considers collection emissions. In general, the study aims, and emission estimates are established considering only the management of USW (treatment and final disposal) [32].

The driving pattern of mobile sources has a significant influence on variations in vehicle emissions. In particular, heavy transport displays a considerable variety of driving cycles because it responds to different needs: public passenger transportation, merchandise transportation and distribution in urban areas, waste collection, etc. The appropriate method for estimating emissions should be selected according to the size of the fleet, the characteristics of the activity and the goal of the emission inventory. As mentioned, there are two contrasting approaches for estimating vehicle emissions, Top-Down and Bottom-Up. Each method has its own goals and requires considerably different input regarding the amount of detail of the information entered. This study aims to compare the results of different methodologies by estimating GHG emissions (CH_4, N_2O, and CO_2), criteria pollutant (PM_{10} and CO) and other pollutants (NO_2, SO_2, and hydrocarbons) for a case study of USW collection in Tandil (Argentina). This work starts from questioning the contribution of emissions from a waste collection truck fleet for a one-year period, 2016, when using different inventory methodologies that currently exist. It is expected that these findings contribute to understanding the need of appropriate methodologies according to vehicle activity and fleet characteristics when making an inventory.

2. Study Area

The city of Tandil, an example of an intermediate city, is a growing urban site with historical industrial activity within its urban area that has allowed local development. It is the capital city of the Municipality of Tandil. With a total population of 123,871 inhabitants (whole department), it comprises an approximate urban area of 50 km^2 and a considerably large vehicle fleet compared to its population size. It is located in the mountain system of Tandilia, which extends across the district in a NW-SE direction, and constitutes a tourist attraction in the city. The sustained growth of the city in recent decades has led to an urban sprawl pattern described as diffuse or dispersed. Territorial, connectivity, density or concentration patterns indicate (a priori) a type of growth that uses a lot of space [33]. These attributes mean a greater demand in transportation services and, consequently, in fuel and energy consumption for the urban system. At present, the municipal government is in charge of the USW collection, which covers an area of 47 km^2.

The Municipality of Tandil has a sanitary landfill where the USW from the city of Tandil and two towns, Vela and Gardey, are discharged. The collection has a daily frequency (Monday to Friday), for most of the city sectors, with collection services on Saturday for some sectors only. The number of daily trips to the landfill is higher on Monday (44), due to the accumulation of waste during the weekend, and decreases throughout the week (33 on Fridays).

Final waste disposal is carried out in the landfill site located in the NW area, 15 km away from the city center. There are two main access roads: National route 226 and Provincial route 30. Most of the USW treated in the sanitary landfill is domestic waste.

In reference to the local environmental conditions, Tandil has air pollution studies but not specifically with AEI [34]. As information of interest for this study it is important to mention that the topography of Tandil has heights that vary approximately between 168 and 317 meters above mean sea level (MAMSL), and the routes of the vehicles vary between 170 and 240 MAMSL. According to the data provided by the

National Meteorological Service for this study, in 2016 Tandil had 8.9°C as the average value of the coldest hour (6 AM) of the day and 19.2°C as the average value of the hottest hour (3 PM).

3. METHODS

The AEI for pollutants emitted during the management of USW were made for USW collection activities in Tandil. Stages of the collection activities are described below (3.1). Emissions from collection trucks were estimated using three methodologies (3.2) with increasing level of complexity. First, following IPCC level 1, where total kilometers traveled per vehicle (VKT) and suggested emission factors (Top-Down) were taken into account. Second, Copert 4 (Bottom-Up), here the vehicle category and associated emission factors were used for each pollutant according to fuel and average driving pattern speeds. Third, IVE (Bottom-Up), addressing multiple sources of information regarding the truck fleet, activity characteristics, and local environmental conditions. The inventoried pollutants and their estimations according to each model are presented in Table 1.

Table 1. Pollutants and models

Estimation model	Pollutants		References
	GHG	Other pollutants	
IPCC	CO_2 CH_4 N_2O	CO; NOx; SOx; HC (COVNM included)	N_2O y CH_4: IPCC 2006; for all other pollutants 2006 IPCC keeps 1996 IPCC values.
COPERT 4		CO; PM_{10}; NO_x^1; SO_x^2; HC (VOC, VOC evap; 1,3-butadiene; Acetaldehyde; Formaldehyde; NH_3; Benzene. Included)	[35]
IVE		CO; PM_{10}; NOx^1; SO_x^2; HC (VOC; VOC Evap; 1,3 butadiene; Acetaldehyde; Formaldehyde; NH_3; Benzene included)	ISSRC, 2008.

Note: When comparing between models $NO_x^1 = NO_2$ and $SO_x^2 = SO_2$.

Once GHGs emissions are calculated, CO_2, CH_4 and N_2O, are converted into CO_2 equivalents using the most recent Global Warming Potentials with a 100-year time horizon (GWP100) values from the IPCC Fifth Assessment Report (AR5), where CH_4 =28 and N_2O=265 [28, 29].

3.1. Characteristics of USW Collection in Tandil, Vehicle Activity and Fleet Description

Primary data on emission sources was obtained from the Municipality of Tandil. The database provided the number of trucks in activity, age of trucks (total vehicle kilometers travelled-VKT) and collection circuits in slots for each collection truck. A total of 11 trucks were used in each case. Emissions were calculated for a working day equivalent to 1020.24 km day^{-1}; results were extrapolated in order to obtain the one-year emissions inventory.

The collection service is mainly carried out with diesel compactor trucks. The total fleet consists of 19 trucks, while a maximum of 11 vehicles are used in the daily circuits. The surface covered by the circuits and the stages for the workday are shown in Figure 1.

A collection circuit of USW within a working day consists of three stages, with two sections each:

- **Trip.** *Urban trip*: departure of the collection truck from the municipal garage to the collection area and the transfer between collection zones in the urban area *Highway trip*: journey done by the truck on the main road.
- **Collection:** *Urban Collection*: door-to-door collection in the assigned urban area *Suburban Collection:* in some cases, with specific sites assigned as collection points.
- **Landfill**. *Landfill Access:* gravel access road. *Landfill*: activity of trucks inside the landfill.

Technical specifications of the fleet are provided in Table 2.

Table 2. Characterization of the waste collection truck fleet under study

Unit #	Year	GWR (lbs)	Make/Model	Air/Fuel Control
1	2011	7756.42	VW 17220 - CHASSIS	Bosch - P7100
3	2015	7615.81	IVECO 170E22 - CHASSIS	High pressure electronic injection system (Common Rail)
4	2010	3696.77	VW 9150 E - CHASSIS	Common Rail controlled by ECM (Engine control module)
12	1995	7257.47	FIAT 150N	Direct injection
100	2012	3696.77	VW 9150 E - CHASSIS	Common Rail controlled by ECM (Engine control module)
150	1997	6395.65	Ford 14000	Bosch Rotary pump (MWM 6.10 NA engine) Manual transmission.
156	2001	6395.65	Ford 14000	Bosch Rotary Pump (MWM 6.10 NA engine) Manual transmission
157	2001	6395.65	Ford 14000	Bosch Rotary Pump (MWM 6.10 NA engine) Manual transmission
160	2005	6350.29	IVECO 140 E 18	Direct injection. Bosch rotary with all-speed regulator, automatic advance feeder and LDA device
431	2010	7756.42	VW 17220 - CHASSIS	Bosch - P7100
432	2010	7756.42	VW 17220 - CHASSIS	Bosch - P7100

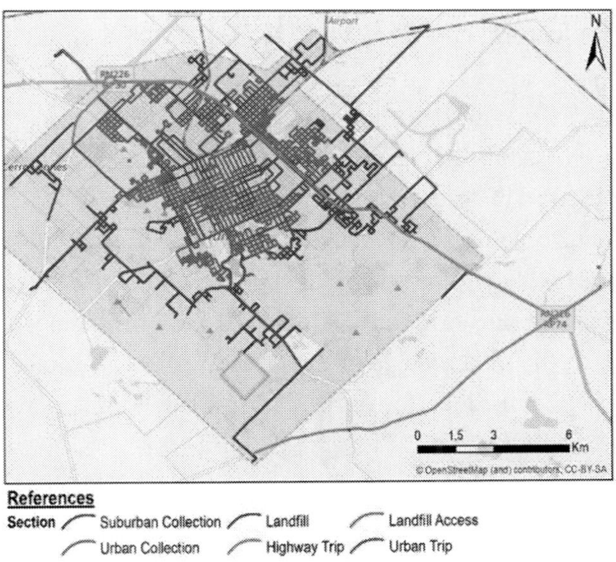

Figure 1. Collection circuits and sections.

3.2. Methodologies for Estimating Pollutant Emissions

3.2.1. 1996/2006 IPCC

The 2006 IPCC methodology used corresponds to Tier 1 equations of the Intergovernmental Panel on Climate Change. It uses data of vehicle activity based on fuel consumption or the number of l kilometers traveled (VKT). Estimations of GHG emissions were made from the number of kilometers traveled per day with a fuel consumption of 0.16 l km^{-1} according to the GHG inventory for Argentina in the year 2012 [36]. GHG emissions resulted from the sum of emissions generated by diesel fuel consumption of the waste collection service truck fleet classified as 1A3biii - Heavy duty vehicles (IPCC, 2006). The calculation was made taking into account the net calorific value of diesel fuel (NCV) of 0.043 TJ t^{-1}. The general equation (1) for estimating annual pollutant emissions (Pol.) is:

$$\text{Emission kg/Pol./year} = \sum [\text{Fuel } \alpha \text{ (TJ) x EF Pol.(kg/TJ) } \alpha] \text{ x } 365 \quad (1)$$

Where 2006 IPCC emission factors used for N_2O and CH_4 emissions; 1996 IPCC emission factors for NOx, CO, COVNM and SO_2 emissions.

3.2.2. COPERT 4

Copert 4 emission factors were used with the same fleet and its total distance traveled. The procedure was carried out as follows [35]: Geo-positioning data obtained from the GPS of each truck was gathered and recorded at thirty- second intervals for the entire fleet. It was analyzed and processed using the GIS Software, ArqGis 10.2. Georeferenced spatial data was adjusted to the city road map in order to reflect the real situation in Tandil. Start and end times for each section of the collection circuit were determined as well.

Due to significant differences found along the collection circuit during the analysis, average distances and speeds were calculated for different sections, as described in section 3.1.

Table 3. Average speed and distance for each section of the collection circuit and truck technology per day

Tech	Truck	Urban Trip		Highway Trip		Urban Collect		Suburban Collect		Landfill access		Landfill	
		km	km h^{-1}	km	km h^{-1}	km	km h^{-1}	km	km h^{-1}	km	km h^{-1}	km	kmh^{-1}
EI	01	62.1	26	48.2	51.2	31.6	5.75	19	6.74	8.64	36.4	1.83	6.94
EII	03, 04, 12 & 100	65	21.7	71.6	46.1	86	6.7	33.9	14.1	17.2	35	3.58	6.1
EIII	150, 156, 157 & 160	34.4	22.1	85.6	42.6	83.5	5.74	102	10.5	14.4	32.7	3.17	6.71
EV	431 & 432	38.6	21.2	70	49.3	104	7.41	NA	NA	20	36.6	4.21	6.77

Finally, distances and average speeds of each section were combined, according to the technology of each truck as determined in COPERT 4. This allowed to associate average speeds, distances, and technologies to an emission factor in order to calculate the total emissions.

Table 3 exposes the average speed and total distance traveled per day according for each technology and section of the collection circuit. Tech's column groups trucks with the same fuel quality (Euro standards). 11891 data points were used for the calculation, of which Euro I: 1455; Euro II: 3491; Euro III: 4366 and Euro IV: 2579.

3.2.3. International Vehicle Emissions (IVE)

The International Vehicle Emissions (IVE) model, version 2.0 provides a background for evaluation of control strategies and effectiveness of different transportation planning. IVE, as a bottom-up approach [37], can predict how different strategies will affect local emissions and measures progress in emission reductions over time (ISSRC, 2008). The IVE model has the ability to predict local air pollutants, greenhouse gas emissions and toxic pollutants from different mobile sources. Emission estimates are calculated using the following parameters: Vehicle emission rates, Vehicle activity, Vehicle fleet distribution and base adjustments with local environmental data such as temperature, humidity, altitude, and other settings such as fleet inspection/maintenance and fuel sulfur content. Six types of data are needed to characterize fleet technology: 1) Classification of Fleets, 2) Fuel type (Petrol, NG, Propane, Ethanol, Diesel, CNG/LPG and special), 3) Fuel delivery system (Carbureted, Single Point Fuel Injection, Multi Point Fuel Injection, Pre-Injection, Direct Injection, 2-cycle, and 4-cycle), 4) Vehicle type and feature, 5) Percentage distribution of the desired vehicle, and 6) Distribution of air conditioning system.

Vehicle activity is the distinctive element, since it observes driving behavior (speed and acceleration profiles) and start patterns (amount of ignitions of the engine). Speed data was obtained from vehicle activity monitors (VAM) in two collection trucks equipped with combined global satellite positioning units (CGPS). According to this methodology, the representativeness of the driving pattern is not given by the number of vehicles measured by VAM, but by the activity of the vehicular flow at different times of a day in a local representative area. The two circuits were selected by the researcher together with the municipal service (Figure 2).

Figure 2. Tandil topography and collection truck circuits observed with VAM unit.

With records of these driving patterns, truck emissions were observed as a function of vehicle specific power (VSP) and engine stress using SeedAnalisys software. VSP is the parameter used to characterize driving patterns in the specific area under study. The equation 2 for determining the instantaneous power of an on-road vehicle is:

$$VSP = Power/Mass \approx 1.1 \times v \times \alpha + 9.81 \times grade \times v + 0.213 \times v + 0.000305 \times (v + vw)^2 \times v \quad (2)$$

where, VSP is the vehicle specific power (kW t^{-1}) and variables v and vw are vehicle speed and headwind speed (m s^{-1}) respectively. Variable α is the acceleration in m s^{-2} and grade is the slope of the roadway (degrees). IVE emission factors (FE) (g km^{-1}) were also converted into g L^{-1} emission factors using the following equation 3 (Attach A ISSRC, 2008):

$$FE(g\,L^{-1}) = FE(g\,km^{-1}) \times F(km\,L^{-1}) \quad (3)$$

where, FE is the pollutant emission factor and F is the fuel economy.

Baseline running emission rates are equally defined in the IVE model and U.S. EPA MOBILE 6 model [39, 40]. IVE defines base start emissions as emissions produced by a cold start in grams [39]. Finally, base emission rates were modified taking into account the efficiency in fuel consumption according to U.S. EPA studies [40].

4. RESULTS

4.1. 1996/2006 IPCC

Fuel consumption figures for 2006 IPCC were taken by default from the national inventory made for the 3rd Argentina Communication for Climate Change, which sets a value of 0.16 $l\ km^{-1}$ for total heavy transport (all activities involving heavy transport use). The IPCC Tier 1 method used in this study considers that all the carbon is oxidised during combustion; the Guide refers to the fact that the carbon mass balance for gasoline light trucks shows that the solid carbon fraction value is insignificant. There is no specific reference to heavy-duty trucks and diesel fuel, consequently it is not possible to produce comparative results for PM_{10}. NO_x and SO_x values are related to the conversion of these compounds which are chemically bound to the fuel during combustion. The results obtained for each unit of the collection truck fleet are shown in Table 4.

4.2. COPERT 4

For the estimation of emissions in this AEI, the average speed for each section of the collection circuit were considered as indicated in the methodology. This section includes a comparison of the emissions per km traveled for each vehicle technology, obtained with COPERT 4 model emission factors. The results obtained with this methodology are shown in Table 5.

Table 4. Estimated pollutant emissions for the fleet according to 2006 IPCC methodology and a fuel consumption value of 0.16 l km^{-1}

			GHG			Other Pollutants			
Unit #	Kg unit^{-1} d^{-1}	PCI	CO_2 EF	CH_4 EF	N_2O EF	CO EF	NO_x EF	SO_2 EF	HC EF
		TJ t^{-1}	Kg CO_2 TJ^{-1}	KgCH_4 TJ^{-1}	KgN_2O TJ^{-1}	Kg CO TJ^{-1}	Kg NO_x TJ^{-1}	Kg SO_2 TJ^{-1}	Kg COVNM TJ^{-1}
		0.04	74100	3.9	3.9	900	1000	36.279	200
		GJ unit^{-1}	KgCO_2	Kg N_2O	Kg CH_4	Kg CO	Kg NOx	Kg SO_2	Kg COVNM
01	23.3	1.00	74.3923	0.0039	0.0039	1.0039	0.9035	0.2007	0.0364
03	2.67	0.11	8.51973	0.0004	0.0004	0.1149	0.1034	0.0230	0.0041
04	24.4	1.05	77.8461	0.0041	0.0041	1.0505	0.9455	0.2101	0.0381
12	8.52	0.37	27.1446	0.0014	0.0014	0.3663	0.3296	0.0732	0.0132
100	2.50	0.11	7.96653	0.0004	0.0004	0.1075	0.0967	0.0215	0.0039
150	18.8	0.81	60.1057	0.0031	0.0031	0.8111	0.7300	0.1622	0.0294
156	4.09	0.18	13.0390	0.0006	0.0006	0.1759	0.1583	0.0351	0.0063
157	3.41	0.15	10.8826	0.0005	0.0005	0.1468	0.1321	0.0293	0.0053
160	17.6	0.76	56.3314	0.0029	0.0029	0.7602	0.6841	0.1520	0.0278
431	15.7	0.68	50.0496	0.0026	0.0026	0.6754	0.6078	0.1350	0.0240
432	16.7	0.72	53.2281	0.0028	0.0028	0.7183	0.6465	0.1436	0.0266
Total kgd^{-1}			419.837	0.022	0.022	5.099	5.666	0.2060	1.133

Table 5. COPERT 4 estimated pollutant emissions from average speeds and emission factors (kg d^{-1})

Section	Exhaust Control	km/d	CO_2 kg/d	N_2O g/d	CH_4 g/d	SO_x kg/d	CO g/d	HC g/d	NO_x g/d	PM g/d
Urban trip	EuroI	62.08	42.65	0.002	10.8	1.36	171.4	66.55	699.8	36.38
	EuroII	65.04	44.66	0.001	11.3	1.42	156.9	52.45	846.8	17.65
	EuroIII	34.40	22.82	0.000	6.02	0.73	108.1	25.00	355.9	10.15
	EuroV	38.63	24.19	0.001	6.76	0.77	8.99	1.55	243.5	2.27
Highway trip	EuroI	48.20	31.10	0.001	8.44	0.99	79.48	28.62	398.2	16.40
	EuroII	71.56	45.72	0.001	12.5	1.46	107.2	29.62	645.3	11.41
	EuroIII	85.58	56.62	0.001	14.9	1.80	154.3	33.11	619.9	14.89
	EuroV	69.97	42.19	0.002	12.2	1.34	8.20	1.20	293.6	1.96
Urban collection	EuroI	31.58	38.79	0.001	5.53	1.24	265.4	107.6	699.4	62.44
	EuroII	85.95	86.04	0.001	15.0	2.74	449.9	171.9	1886	36.47
	EuroIII	83.46	105.8	0.001	14.6	3.37	723.6	176.2	2019	65.11
	EuroV	104.3	103.0	0.003	18.2	3.28	54.96	10.55	1089	15.76
Suburban collection	EuroI	18.99	17.25	0.001	3.32	0.55	142.9	59.57	379.2	33.73
	EuroII	33.93	22.68	0.000	5.94	0.72	110.3	40.46	531.7	11.49
	EuroIII	101.8	86.49	0.001	17.8	2.76	579.0	146.7	1617	54.56
	EuroV	0.00	0.00	0.000	0.00	0.00	0.00	0.00	0.00	0.00
Landfill access	EuroI	8.64	5.51	0.000	1.51	0.18	18.29	6.84	82.93	3.83
	EuroII	17.16	10.70	0.000	3.00	0.34	30.14	8.96	176.5	3.33
	EuroIII	14.35	9.11	0.000	2.51	0.29	32.26	7.09	119.6	3.08
	EuroV	19.97	11.83	0.001	3.49	0.38	2.91	0.45	96.09	0.72
Landfill	EuroI	1.83	1.53	0.000	0.32	0.05	13.49	5.65	35.91	3.18
	EuroII	3.58	3.39	0.000	0.63	0.11	19.77	7.51	82.87	1.55
	EuroIII	3.17	2.81	0.000	0.55	0,09	24.89	6.18	68.72	2.28
	EuroV	4.21	3.56	0.000	0.74	0.11	2.33	0.45	46.96	0.68
TOTAL [kg/d]			818.4	0.000	0.18	0.03	3.26	0.994	13.03	0.41

4.3. IVE

The characterization of the fleet performed for this model allowed the identification of index values, i.e., a single variable was identified containing other variables which, when combined, determined the emission factor used for each index and assessed pollutant (Table 6).

Table 6. Characterization of the heavy fleet using the IVE Model

Unit #	Category	Fuel	Air/Fuel Control	Exhaust Control	Evaporative Control	Age (km)	Index
01	Truck/Bus	Diesel	FI	EuroI	None	>161K	1124
03	Truck/Bus	Diesel	FI	EuroII	None	<79K	1131
04	Truck/Bus	Diesel	FI	EuroII	None	80-161K	1132
12	Truck/Bus	Diesel	FI	EuroII	None	>161K	1133
100	Truck/Bus	Diesel	FI	EuroII	None	>161K	1133
150	Truck/Bus	Diesel	FI	EuroIII	None	80-161K	1141
156	Truck/Bus	Diesel	FI	EuroIII	None	80-161K	1141
157	Truck/Bus	Diesel	FI	EuroIII	None	80-161K	1141
160	Truck/Bus	Diesel	FI	EuroIII	None	>161K	1142
431	Truck/Bus	Diesel	FI	EuroV	None	<79K	1158
432	Truck/Bus	Diesel	FI	EuroV	None	>161K	1160

Note: FI: fuel injected.

The driving pattern obtained with VAM and processed with SpeedAnalisys determined the power bins for different moments of engine stress. This information is the input to adjust default emission factors for each technology in the model. The driving pattern depends on the driver, the traffic flow, the role of the transport used and the slope. The selection of a truck whose circuit includes the diversity of slopes in the study area is important because it determines the engine stress. The representativeness of the driving pattern is not given by the number of vehicles measured by VAM, but by the activity of the vehicular flow in different hours of a day, in a characteristic local area. Calculations were made without determining an Allowed Idle (min) since it is typical of this heavy transport to work while compacting waste in the truck box, consequently this data was also used in the analysis.

Power bins (Figure 3) and concentrations of pollutants per day, which include start emissions and emissions per collection circuit are shown below (base emission rates for each technology are expressed as g/start or g/km for running).

Start emissions are estimated for the total number of starts at a specific time for each technology; the assigned value takes into account the temperature of the engine according to the previous period at rest. In this particular case start emissions are negligible in relation to total emissions.

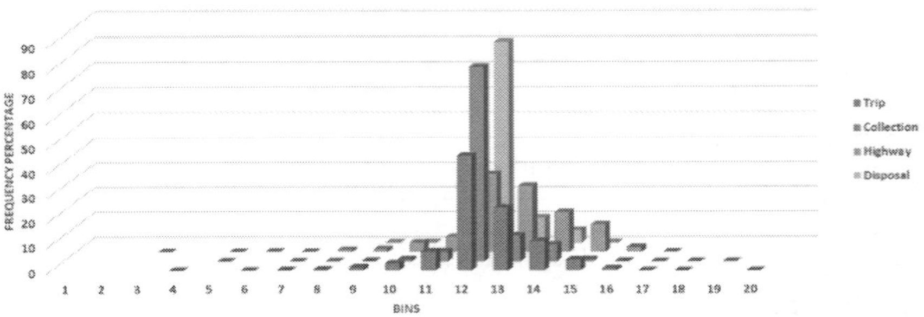

Figure 3. Low stress bins by section for two collection circuits.

Average stress values are caused by higher speeds along the highway section in the collection circuit in bins 36 to 38, although their values do not exceed the unit; no high stress is recorded in any case.

To complete the model's input requirement, humidity and hourly temperature information for one year (2016) and the particularity of Tandil's geography -data collected with VAM- were added. The results of the inventory are presented in Table 7.

Table 7. IVE estimated daily emissions for the truck fleet

	GHG			Other pollutants				
	CO_2	CH_4	NO_2	CO	PM	NO_X	SO_X	HC
start g d^{-1}	1,9660	0.00	1.18	22.07	28.01	22.07	0.02	1.71
running Kg d^{-1}	1,679	0.00	0.13	17.02	3.36	17.02	0.02	1,320
Total Kg d^{-1}	1,675	0.00	0.13	17.04	3.39	17.04	0.02	1,320
Start/Total (%)	0.12%	0.00%	0.90%	0.13%	0.83%	0.13%	0.12%	0.00%

5. Inventory, Emissions and Discussion

Annual results obtained with each of the models selected to compare methodologies for estimating pollutant emissions are presented in Table 8.

Table 8. Pollutant emissions for the truck fleet

Model	Pollutants (t $year^{-1}$)								
	GHG[1]				Other pollutants				
	CO_2	CH_4	N_2O	CO_2 eq[2]	CO	PM_{10}	NO_x	SO_x	HC
IPCC	160.419	0.008	0.008	160.419	1.95	---	2.164	0.078	0.432
IPCC[3]	300.786	0.015	0.015	300.786	3.653	---	4.059	0.147	0.812
COPERT	298.732	0.064	0.000	300.537	1.192	0.149	4.757	0.009	0.363
IVE	611.284	0.000	0.048	623.998	1.414	1.237	6.219	0.006	0.482

[1] It is not possible to produce significantly better results for CO_2 other than by using the existing Tier 2.
[2]GHGs emissions converted to CO_2 equivalents using the most recent Global Warming Potentials with a 100-year time horizon (GWP100) values from the IPCC Fifth Assessment Report (AR5).
[3]Results according to a fuel consumption value of 0.3 l km^{-1}., as reported in the study area.

The most significant differences among the GHGs can be observed in CO_2 and CO_2 eq. concentrations estimated with IVE are 3.8 times higher than those estimated with the 2006 IPCC Tier 1 emission factors. In the case of IPCC Guide, a fuel consumption value of 0.16 l/km was applied for the total heavy transport (all activities involving heavy transport). When the value is adjusted to consumption between 0.3 and 0.6 l/km - which are the values reported by the truck drivers in the study area - the total estimated emissions are close to the inventories for Copert and IVE. According to the 2006 IPCC Guide the most appropriate way to estimate CO_2 emissions is from the amount of fuel sold, or from the consumption per kilometer traveled, using proper data of the fuel carbon content. This is the reason why Tier 3 is not recommended to calculate estimates [41]. But, at the same time that is why it is so important the accuracy of the information provided even more when it is the only information required.

In this case study, the total fuel consumption for the inventory with IPCC is obtained from knowing the kilometers traveled by the trucks (and not from the fuel sold). When the consumption is adjusted to the particular characteristics of the area, results appear that correlate better with COPERT. This is because this inventory increases the level of complexity of the calculation considering technologies, sections of the circuit and speeds that affect fuel consumption.

The inventory carried out with IVE is the one that has estimated the highest emissions for the year under study, accentuating the increase in pollutant concentrations as the level of complexity used to carry out the study increases. The IVE model does not estimate the contribution of CH_4 to diesel fuel, so it is not possible to make comparisons.

There are significant differences between the results based on COPERT emission factors for average speeds and those obtained with IVE, mainly when taking into account stress engine. Start emissions do not represent a significant contribution, they only reach 0.07% of the total in the case of collection trucks.

NO_x and SO_x values during combustion are related to the conversion of these compounds, which are chemically linked to the fuel. The contribution of local information on fuel characteristics and vehicular activity explains the differences.

CONCLUSION

The work done reviews the importance of inventories and approaches to accomplish them according to the objectives and available resources. The results achieved allow the determination of a baseline on emissions from the waste collection service with some particularities according to the methodology adopted.

Methodologies are compared using a small fleet, which is representative of the heavy transport activity in service in intermediate cities; conclusions can be drawn in two directions.

The first and most important conclusion points out that both IPCC and COPERT4 produce similar results in the treatment of GHG and NOx. This is an advantage when data, time, and resources are scarce as only the fuel consumption data is needed - with good accuracy of this information - to achieve good knowledge of a fleet pollutant emissions, regardless of its activity and even if it is small, using IPCC methodology.

If we compare the Top-Down and Bottom-Up approaches (IPCC tier 1 and IVE) strictly respecting the fuel consumption per km traveled suggested by the IPCC (0.16 km^{-1}), the most striking finding is the difference in CO_2 values estimated with the IVE methodology, which tends to disappear when local reported consumption is taken into account. Here again the importance of considering the particularities of the study area in the input data for the inventory when economic and human resources allow it.

Second, the IVE model would be the most reliable option to obtain more detailed knowledge about other pollutants (such as those which are mainly related to health), because it requires a high information input, and the quality of the data is decisive to achieve reliable results. For example, PM and SO_2 have their greatest contribution at the time of vehicle start (which is not considered by COPERT4 when evaluating average speeds). Note that although start emissions represent a small proportion of total emissions for this particular heavy transport activity (waste collection with a minimum number of starts per day); these results can vary significantly as the characteristics of the activity change (another type of transport service). This can be explained, as emission rates from diesel-operated engines are strongly dependent on the load. However, in the operation of most engines, the load is a function of the task and it varies throughout the period of operation. For the same technology, the amount of time spent at each engine load is dependent upon the driver's behavior during the assigned USW collection circuit (i.e., how fast the driver performs the job). These extremely variable factors change from job to job. However, an attempt at trying to identify the average load cycle should be considered, as not accounting for this variability can significantly underestimate emissions.

Acknowledgments

The authors thank the researcher Dr Mauricio Osses (Center for Climate and Resilience Research (CR)2, Department of Geophysics of the University of Chile), and the Federico Santa Maria Technical University (Chile) for contribution of equipment and knowledge for this study.

References

[1] United States Environmental Protection Agency, US EPA. 1999. "*Handbook for criteria pollutant inventory development: A beginner's guide for point and area sources*." Office of Air Quality Planning and Standards. EPA-454/R-99-037.

[2] Dios Noceda, M. 2013. "*Estimation, analysis and evaluation of anthropogenic air emissions inventories at local, regional and continental scales*." PhD diss, National University of Santiago de Compostela.

[3] Nicholls D., Barnes, F., Acrea F., Chen C., Buluç L. Y., Parker M. 2015. "Top-down and bottom-up approaches to greenhouse gas inventory methods—a comparison between national- and forest-scale reporting methods." *Gen. Tech. Rep.* PNW-GTR-906. Portland, OR: U.S. Department of Agriculture, Forest Service, Pacific Northwest Research Station.

[4] Olaguer E. P., 2017. "Emission Inventories" In: *Atmospheric Impacts of the Oil and Gas Industry.* Part II: Tools for Crafting Solutions, Chapter 7, 67-77.

[5] European Monitoring and Evaluation Programme/European Environmet Agency, EMEP/EEA. 2019. "EMEP/EEA air pollutant emission inventory guidebook 2019." *Technical guidance to prepare national emission inventories.* EEA Report No 13/2019. doi: 10. 2800/293657.

[6] Intergovernmental Panel on Climate Change, IPCC. 2019. "*Refinement to the 2006 IPCC Guidelines for National Greenhouse Gas Inventories*." Calvo Buendia, E., Tanabe, K., Kranjc, A., Baasansuren, J., Fukuda, M., Ngarize S., Osako, A., Pyrozhenko, Y., Shermanau, P. and Federici, S. (eds.). Chapter 1: Introduction to National GHG Inventories. Chpter 2: Approaches to Data Collection.

[7] Pulles, T. 2018. "Twenty-five years of emission inventorying" *Carbon Management* 9.1, 1-5.

[8] Pulles, T., Heslinga D. 2007. "*The Art of Emission Inventorying*." TNO, the Netherlands. doi: 10.13140/RG.2.1.2082.8007.

[9] Muntean M., Guizzardi D., Schaaf E., Crippa M., Solazzo E., Olivier J.G.J., Vignati E. 2018. "*Fossil CO_2 emissions of all world countries - 2018 Report*." EUR 29433 EN, Publications Office of the European Union, Luxembourg. doi: 10.2760/30158, JRC113738.

[10] Crippa M., Guizzardi D., Muntean M., Schaaf E. 2019. "*EDGAR v5.0 Global Air Pollutant Emissions. European Commission, Joint Research Centre*." [Dataset] Accessed December 14 http://data.europa.eu/89h/377801af-b094-4943-8fdc-f79a7c0c2d19.

[11] United Nations Economic Commission for Europe, UNECE. 1979. "*Convention on Long-Range Transboundary Air Pollution*." Accessed December 10 http://www.unece.org/fileadmin//DAM/env/lrtap/full%20text/1979.CLRTAP.e.pdf.

[12] United Nations Economic Commission for Europe, UNECE. 1984. "*Protocol to the 1979 Convention on Long-Range Transboundary Air Pollution on Long-Term Financing of the Cooperative Programme for Monitoring and Evaluation of the Long-Range Transmission of Air Pollutants In Europe* (*EMEP*)." Accessed December 10 https://www.unece.org/fileadmin/DAM/env/lrtap/full%20text/1984.EMEP.e.pdf.

[13] *Centre on Emission Inventories and Projections, CEIP*. 2020. Accessed December 10 https://www.ceip.at/the-emep-grid.

[14] United Nations Economic Commission for Europe, UNECE. 2015. "*Guidelines for reporting emissions and projections data under the Convention on Long-range Transboundary Air Pollution*."

Accessed December 10 https://www.unece.org/fileadmin/DAM/env/documents/2015/AIR/EB/English.pdf.

[15] United Nations Economic Commission for Europe, UNECE. 1996. "*Recommendation of the Council on Implementing Pollutant Release and Transfer Registers.*" C(96)41/FINAL. Adopted by the Council at its 869th Session on 20 February 1996 [C/M(96)4/PROV]. General Distribution. Organisation de Coopération et de Développement Economiques, Organisation for Economic Co-operation and Development.

[16] United Nations Economic Commission for Europe, UNECE. 2003. "*Recommendation of the Council Amending The Recommendation of the Council on Implementing Pollutant Release and Transfer Registers (PRTRs)*" For Official Use C(2003)87Draft [C(96)41/FINAL] General Distribution. Organisation de Coopération et de Développement Economiques, Organisation for Economic Co-operation and Development.

[17] United Nations Economic Commission for Europe, UNECE. 2013. "*Environment Directorate Joint Meeting of the Chemicals Committee and The Working Party on Chemicals, Pesticides and Biotechnology. Application, Use and Presentation of Pollutant Release and Transfer Registers (PRTR).*" Unclassified ENV/JM/MONO (2013)1. Data Series on Pollutant Release and Transfer Registers No. 14. Organisation de Coopération et de Développement Economiques, Organisation for Economic Co-operation and Development.

[18] United States Environmental Protection Agency US EPA, 2020. "*Find Out What's Happening in Your Neighborhood. Using EPA's Toxcs Release Inventory (TRI).*" Accessed December 12 https://www.epa.gov/sites/production/files/2020-08/documents/tri_for_communities_fact_sheet_summer_2020.pdf.

[19] United States Environmental Protection Agency US EPA, 2020. "*2017 National Emissions Inventory Complete Release Technical Support.*" Office of Air Quality Planning and Standards Air Quality Assessment, Division Emissions Inventory and Analysis Group Research Triangle Park, North Carolina Document. Accessed

December 12 https://www.epa.gov/sites/production/files/2020-04/documents/nei2017_tsd_full_30apr2020.pdf.

[20] United States Environmental Protection Agency US EPA. 2013. "*Fact Sheet. Greenhouse Gases Reporting Program Implementation.*" 40 CFR Part 98.

[21] Center for Cimate and Resilience Research (CR)2, 2017. "*(CR)2 liderará proyecto internacional para predecir episodios de contaminación atmosférica en Latinoamérica.*" Nicole Tondreau-Michelle Ferrer. 7 de septiembre, 2017. ["*(CR)2 will lead an international project to predict episodes of air pollution in Latin America."* Nicole Tondreau- Michelle Ferrer. September 7th, 2017].

[22] Center for Cimate and Resilience Research (CR)2, 2018. "Mauricio Osses: "*Queremos juntar a todas las personas que trabajamos temas de emisiones en Latinoamérica y proponer un inventario mucho más actualizado.*" Michael Lieberherr Pacheco. 17 de diciembre, 2018. ["Mauricio Osses: "*We want to bring together all the people who work on emissions issues in Latin America and propose a much more updated inventory.*" Michael Lieberherr Pacheco. December 17, 2018].

[23] Li, M., Zhang, Q., Kurokawa, J. I., Woo, J. H., He, K., Lu, Z., Ohara, T., Song, Y., Streets, D. G., Carmichael, G. R., Cheng, Y., Hong, C., Huo, H., Jiang, X., Kang, S., Liu, F., Su, H., and Zheng, B. 2017. "MIX: a mosaic Asian anthropogenic emission inventory under the international collaboration framework of the MICS-Asia and HTAP" *Atmos. Chem. Phys.*, 17, 935-63.

[24] Keita, S., Liousse, C., Assamoi, E. M., Doumbia, T., Touré, N. E., Gnamien, S., Elguindi, N., Granier, C., and Yoboué, V. 2020. "African Anthropogenic Emissions Inventory for gases and particles from 1990 to 2015," *Earth System Science Data*, in review, 2020. https://doi.org/10.5194/essd-2020-328.

[25] Yona L., Cashore B., Jackson R. B., Ometto J., Bradford M. A., 2020. "Refining national greenhouse gas inventories." *Ambio* 49:1581-86.

[26] *UNFCC*, 2015. Paris Agreement.

[27] Day M., Pouliot G., Hunt S., Baker K., Beardsley M., Frost G., Mobley D., Simon H., Henderson B., Yelverton T., Rao V. 2019. "Reflexionando sobre el progreso desde el informe de inventario de emisiones de NARSTO de 2005, *Revista de la Asociación de Gestión de Residuos y Aire*" 69. 9, 1023-48. ["Reflecting on Progress since the 2005 NARSTO Emissions Inventory Report, *Waste and Air Management Association Journal*" 69. 9, 1023-48].

[28] Intergovernmental Panel on Climate Change, IPCC, 2014. "Climate Change 2014: Synthesis Report." *Contribution of Working Groups I, II and III to the Fifth Assessment Report of the Intergovernmental Panel on Climate Change*. [Core Writing Team, R. K. Pachauri and L. A. Meyer (eds.)]. IPCC, Geneva, Switzerland, 151 pp.

[29] Intergovernmental Panel on Climate Change, IPCC, 2013. "Climate Change 2013: The Physical Science Basis." *Contribution of Working Group I to the Fifth Assessment Report of the Intergovernmental Panel on Climate Change* [Stocker, T. F., D. Qin, G. K. Plattner, M. Tignor, S. K. Allen, J. Boschung, A. Nauels, Y. Xia, V. Bex and P. M. Midgley (eds.)]. Cambridge University Press, Cambridge, United Kingdom and New York, NY, USA, 1535 pp, doi: 10.1017/CBO9781107415324.

[30] World Health Organization, WHO, 2016. "*Ambient air pollution: A global assessment of exposure and burden of disease*." Available at https://apps.who.int/iris/handle/10665/250141.

[31] Bogner, J., M. Abdelrafie Ahmed, C. Diaz, A. Faaij, Q. Gao, S. Hashimoto, K. Mareckova, R. Pipatti, T. Zhang, 2007. "Waste Management, In Climate Change 2007: Mitigation." *Contribution of Working Group III to the Fourth Assessment Report of the Intergovernmental Panel on Climate Change* [B. Metz, O. R. Davidson, P. R. Bosch, R. Dave, L. A. Meyer (eds.)], Cambridge University Press, Cambridge, United Kingdom and New York, NY, USA.

[32] Institut für Energie- und Umweltforschung Heidelberg GmbH, IFEU, 2009. "*Calculadora MRS-GEI: herramienta de cálculo de gases de efecto invernadero (GEI) en el manejo de residuos sólidos*"

Frankfurt, Alemania [*MRS-GEI Calculator: greenhouse gas (GHG) calculation tool for solid waste management*] Accessed December 18 https://www.bivica.org/file/view/id/1081.

[33] Fernandez, G., Ramos, A., 2013, El Crecimiento Urbano de Tandil: ¿Modelo Territorial de la Ciudad Difusa? *Revista Geográfica Digital.* 10.20: 1-12. ISSN 1668-5180 Resistencia, Chaco. ["Tandil Urban Growth: Territorial Model of the Diffuse City?" *Digital Geographic Magazine* 10.20: 1-12. ISSN 1668-5180 Resistencia, Chaco].

[34] Sosa B., Porta A., Colman, E., Banda Noriega. R., Massolo, L., 2017. "Human health risk due to variations in PM_{10}-$PM_{2.5}$ and associated PAHs levels." *Atmospheric Environment* 160: 27-35.

[35] Gkatzoflias, D., Kouridis, C., Ntziachristos, L., Samaras, Z., 2012. "*COPERT 4 – Computer programme to calculate emissions from road transport.*" User manual version 9.0. Report for European Environment Agency, Emisia, Thessaloniki, Greece.

[36] Secretaría de Ambiente y Desarrollo Sustentable, SAyDS, 2015. "*Inventario de Gases de Efecto Invernadero de la República Argentina-Años 2010-2012. Energía.*" Volumen 1. ["*Greenhouse Gas Inventory of the Argentine Republic - Years 2010-2012. Energy.*" Volume 1] Accessed December 18 https://www.argentina.gob.ar/sites/default/files/1.-inventario-geis-energia.pdf.

[37] Wang H., Chen C., Huang C., Fu L., 2008. "On-road vehicle emission inventory and its uncertainty analysis for Shanghai, China." *Science of the Total Environment.* 398. 60-67.

[38] ISSRC, 2008. *International Sustainable Systems Research Center, IVE Model User Manual*, Version 2.0.

[39] EPA, 2001. *MOBILE6 Estimates of Exhaust Emissions for 1994-and-later Light Duty Diesel Cars and Trucks*, M6.EXH.011, EPA420-R-01-041.

[40] EPA, 2002. *Update of Heavy-Duty Engine Emission Conversion Factors for MOBILE6: Analysis of BSFCs and Calculation of Heavy-Duty Engine Emission Conversion Factors.* M6.HDE.004, EPA420-R-02-005.

[41] Intergovernmental Panel on Climate Change, IPCC, 2006. *Guidelines for National Greenhouse Gas Inventories,* Volume 2 Energy, Inventories Published by the Institute for Global Environmental Strategies (IGES) for the IPCC, ISBN 4-88788-032-4.

In: Air Pollution: Effects and Dangers
Editor: J. Esteban Colman Lerner
ISBN: 978-1-53619-544-6

Chapter 5

TRANSPORTATION RISK ASSESSMENT METHODOLOGY FOR HAZARDOUS MATERIALS TO IMPROVE URBAN DISASTER PREVENTION

Micaela B. Del Sole[1], Natacha S. Represa[1,2,*], Karina B. Balbi[1], Santiago Aguilar Ferraro[1] and Erica Y. Sanchez[1]

[1]Centro de Investigaciones del Medioambiente (CONICET-UNLP), La Plata, Argentina

[2]Geo-Environmental Cartography and Remote Sensing Group, Polytechnic University of Valencia, Valencia, Spain

ABSTRACT

The transport of hazardous materials presents a strong inherent risk, in particular when industrial buildings are located in urban areas. In this chapter, we presented a methodology for risk assessment in order to

* Corresponding Author's Email: solrepresa@quimica.unlp.edu.ar.

prevent disasters. In the first stage, we proposed the use of geographic information systems to detect conflict areas. In the identified hotspots on the hazardous materials transport route, we performed a simulation with ALOHA and analyzed the vulnerability by an index of social vulnerability. Finally, this information was integrated using a simple formula for risk estimation. As a specific case, we applied it to methanol transport in La Plata, Argentina. By implementing this strategy, we detected areas associated with higher risks, which may coincide either with both areas of high or low threat, understanding that vulnerability deserves a mitigation approach to prevent future disasters.

1. INTRODUCTION

Mainly in peripheral countries, industrial areas are integrated into the urban fabric, without being easily differentiated given the rapid development without urban planning [1, 2]. This situation of closeness and contemporality exposes the population to greater danger, potentially causing human casualties, property damage, and economic losses [2]. In this sense, industrial accidents are particularly interesting in the design of public policies to reduce disaster risk.

A particular case is that of accidents in the transport of hazardous materials (HAZMAT), especially road transport [3]. HAZMAT comprises explosives, flammables, oxidizing substances, poisonous gases, and radioactive materials. These materials can be extremely harmful to the environment and human health [4].

Therefore, a thorough risk analysis for good transport planning and management in order to prevent accidents and mitigate their effects is needed [5]. The most recent research incorporates social issues into risk studies and integrates them with the physical context to measure human vulnerability [6, 7]. By combining the hazard event and the exposed society, it is possible to estimate the degree of risk [8].

The literature contains a large number of software that serve to simulate threats, with varying degrees of complexity. However, the detail of the output model requires specific input data that are not known in all situations [9, 10]. Besides, other software only tells us about the hazard but they do not account for the risk. Areal Locations of Hazardous Atmospheres (ALOHA) software offers the versatility necessary to provide threat zones with readily available information [11, 12].

Similarly, the focus on modeling vulnerability has taken a simplistic form without affecting the representation of society. The development of indices for the study of social vulnerability has proven to be a robust instrument for the population description [13]. While physical vulnerability considers only the properties of the physical elements that determine their possible damage in the presence of a hazard event [14], social vulnerability focuses on the characteristics of society from a more comprehensive perspective [15].

With these components covered, the great challenge is the integration of hazard and vulnerability information in a transport accident context. The objective of this work has been to develop a comprehensive procedure for assessing and mapping the risk level of transporting hazardous materials in urban areas. This involved identifying the sites with the highest probability of an accident, delimiting the level of social vulnerability and the threat of the event, and evaluating the areas of risk in order to have the necessary information to plan actions to prevent a disaster.

2. METHODOLOGY

Accidents in the transport of HAZMAT are a potential source of urban disasters. Due to the distances traveled, trucks are often responsible for the highest number of HAZMAT events. To avoid these situations, a risk assessment is of paramount importance. We propose a simple methodology divided into 3 steps, as shown in Figure 1.

The first step was to identify the location of the main critical points of risk on the road according to a specific criterion detailed in Section 2.1.1. The second step was the prediction of the threat zones by ALOHA software and the calculation of the population's vulnerability. Finally, we compute the risk level for the areas of threat scope.

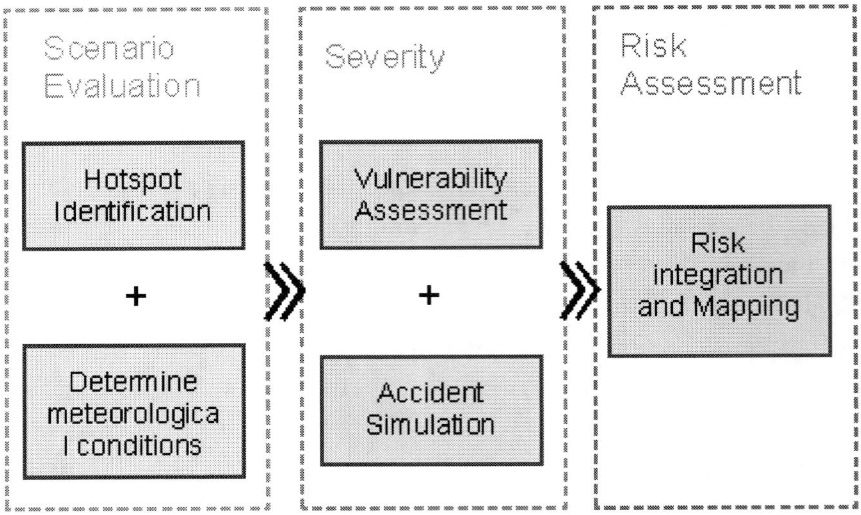

Figure 1. Diagram of the working methodology.

2.1. Scenario Evaluation

The evaluation of possible scenarios is the first step in simulating the conditions of an accident. We propose, on one hand, to identify the areas most likely to present a road accident, and on the other hand, to determine the most probable and critical weather conditions that may exist in the study area. In order to plan an appropriate response and ensure public safety, we evaluate the potential severity of these threats.

2.1.1. Hotspot Identification

The sites to simulate potential HAZMAT incidents must belong to one of the main HAZMAT transport routes. For the identification of the hotspots, we decided to use a multicriteria analysis, where it was considered:

- The occurrence of previous HAZMAT accidents.
- The occurrence of vehicular conflicts involving different types of vehicles. (light and heavy vehicles)

- High traffic of vehicles.
- The confluence of important access roads.
- Presence of sensitive receptors (high concentration of people).
- Presence of establishments with storage of explosive substances that may be at risk of suffering sequential events.

2.1.2. Meteorological Conditions

Meteorological factors directly affect the dispersion of pollutants, and consequently, the dimensions of the threat area. It is desirable to know the weather conditions that are most likely to be encountered in the presence of a disaster and also the most critical conditions. For this reason, we decided to work with the most likely (ML) and most conservative (MC) meteorological scenarios. Sanchez et al. [6] describe a scenario-building methodology using a decision tree that is very useful and easy to automate when working with hourly weather data.

2.2. Severity Determination

The severity of an accident is determined by the degree of vulnerability of the population and the size of the threat. In this step, we propose to evaluate the vulnerability by a local vulnerability index, and the threat through the simulation of the leakage by ALOHA software.

2.2.1. Vulnerability Assessment

The limitations of a community that affects its capacity to cope with the effects of hazards shape social vulnerability [16]. The latter contemplates a wide range of susceptibilities at the individual and the collective level, such as lack of access to resources, unmet needs, beliefs, and customs [17, 18]. The underlying causes of social vulnerability are economic, demographic, and political processes. Unlike physical vulnerability, social vulnerability does not rely on the type of hazard. A novel approach to assessing social vulnerability involves the development of multivariate models, based on easily accessible geolocated information such as census data [13]. The social vulnerability was categorized into 5 levels to be incorporated into the risk estimation.

2.2.2. Accident Simulation

The distance at which a physical magnitude reaches a threshold value corresponding to the onset of an undesirable effect delimits an area of threat. Appropriate software may estimate the extent of hazardous atmospheres (flammable, explosive, or toxic). The characteristics of the substance and the emission source define the emission type and determine its subsequent dispersion [19].

The widely adopted ALOHA software is employed to assess the hazard of a leak in a storage facility. The software is developed and supported by the National Oceanic and Atmospheric Administration (NOAA) in collaboration with the Office of Emergency Management of the Environmental Protection Agency (U.S. EPA) to provide emergency response personnel estimates of the spatial extent of chemical hazards [20].

This software was chosen because of its advanced gas dispersion modeling capabilities. It is one of the most accepted software for emergency management and is easily integrated with Geographic Information Systems (GIS). Compared to other programs, ALOHA is very easy to run and gives a quick estimate of the area impacted by a spill event [21].

2.3. Risk Assessment

Risk is defined as the result of threats of an accidental release and the vulnerability of the potentially affected population. The risk was evaluated by integrating both parts through the equation:

$$Risk_i = Threat\ Level_i + Vulnerability\ Level_i \tag{1}$$

where i represents each of the intersections of threat areas with areas of vulnerability.

For this, we made a classification of the threat in 3 levels (low, medium, high), while the vulnerability in 5 (very low, low, medium, high, and very high). Through integration in a GIS, we generate maps that highlight the areas of greatest risk that must be attended to as a priority in an accident.

3. Application Case

The Gran La Plata region covers the districts of La Plata, Berisso and Ensenada, on the west bank of the Río de La Plata, 50 km south of the city of Buenos Aires, Argentina (Figure 2). With a plain topography and a high building rate, the population exceeds 3,700 inhab/km^{-2} [22]. A large number of productive activities take place in the area. In particular, in the town of Ensenada, there is an important petrochemical complex, in which the major oil refinery of the country stands out with a processing capacity of about 35,000 m^3 day^{-1} of crude oil. Nearby, other subsidiary industries producing different chemical compounds are located, and the HAZMAT's transport is one support activity for regional industrial activity. In this context, the study region presents a very high exposure to incidents with hazardous materials compared to other types of threats [23, 24].

The Port of La Plata, located in the region of study, is one of the outlets for export methanol, an important HAZMAT [25]. Currently, the methanol trade has a high demand from biodiesel producing companies distributed throughout our country [26]. The Port of La Plata exported 154 kt per year of methanol, which enters the urban area through the route shown in Figure 2.

3.1. Scenario Planning

In order to know the hotspots of accidents, the traditional and new transport routes that could be developed in the region were taken into account. Information concerning accident rates, population, and infrastructure was not directly available and data had to be collected from different sources. For this work, we reviewed journalistic articles, in digital and printed versions, interviewed competent authorities in emergency response and planning, and conducted a visual inspection of methanol transportation routes. With the information obtained, we made a GIS to highlight the conflictive areas.

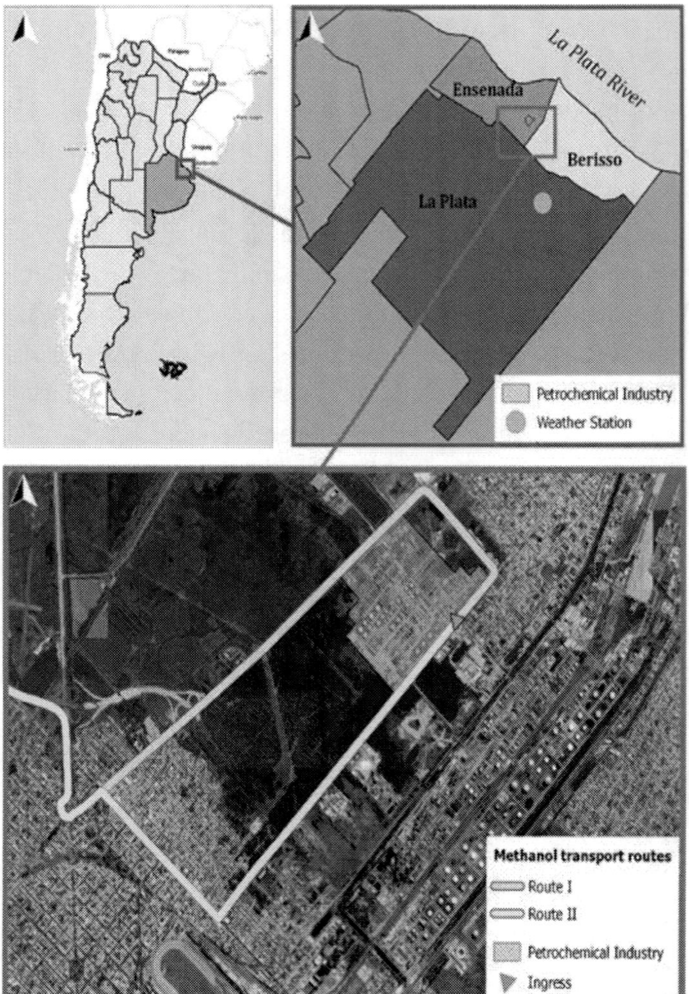

Figure 2. Methanol transport route from the petrochemical industry, La Plata, Argentina.

We used a meteorological database provided by the National Meteorological Service for the period 2014-2017 (Airport La Plata Station: 34° 58′ S, 57°54′ W). Since transport takes place at night (06:00 - 22:00), only that time range was analyzed to identify the ML and the MC meteorological scenarios.

3.2. Severity Calculation

To evaluate the vulnerability in this region, we used the index the Local Social Vulnerability Index (LSVI), obtained from sixteen characteristics available at the census area level taken from the 2010 Argentina Census [13]. The variables cover socioeconomic characteristics, such as the proportion of overcrowded households, gender, occupational class, employment status, and access to public services.

The most relevant information for representing the threat of HAZMAT transport is the size and the process condition of the tank (temperature, chemical status, percentage of filling by volume, type of tank failure). In the modeling with ALOHA, we defined a rectangular opening with a size of 60 x 4 cm, with the generation of a toxic evaporation pool. This simplification is intended to simulate a split in the tank that represents HAZMAT losses through the joints, a situation that has been frequently observed according to the information gathered.

As regards the ground roughness of the surroundings of the emission site, we considered a value of ground roughness (*Z0*) of 20 cm [27]. The surroundings of the emission source on the leeward directions present low houses with an average height of no more than 10 m. The time slot in which the most probable meteorological conditions exist defines the time of the potential event. Also, we considered three types of incidents: leaking tank where the chemical is not burning and forms an evaporating pool; leaking tank where the chemical is burning and forms a pool fire; and a boiling liquid expanding vapor explosion (BLEVE) where the tank explodes and chemical burns in a fireball.

4. Results and Discussion

After evaluating the area and the different conditions surrounding HAZMAT's transport route, we identified 2 conflictive areas with a high potential for damage (Figure 3). Area A is characterized by the main access to the city, where there are one gas station and 4 sensitive receptors less than 500 meters away. Area B is a roundabout with a high influx of vehicles, where there are 3 gas stations and 2 sensitive receptors. Vehicle accidents have already been recorded in both areas.

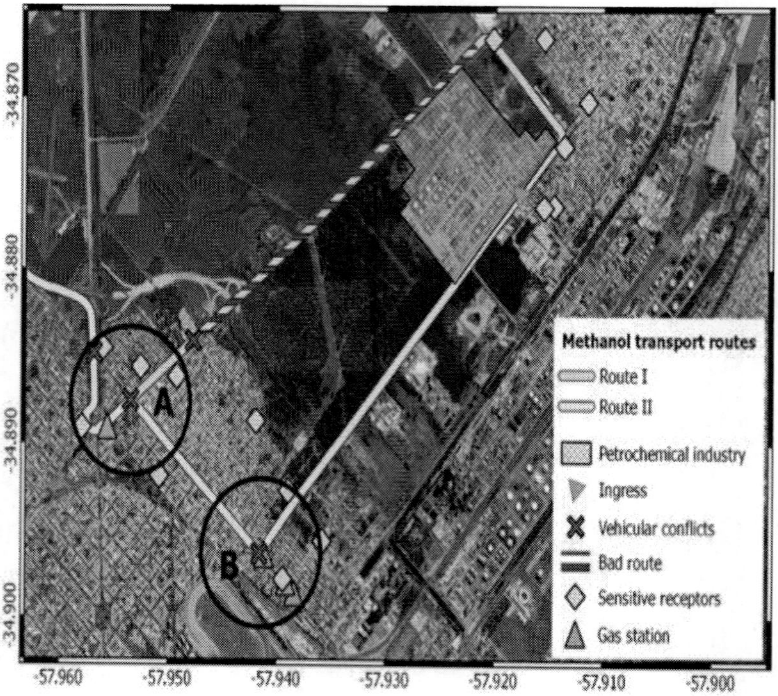

Figure 3. Identification of hotspots on the HAZMAT transport route.
A and B are the most conflictive areas.

Analysis of the weather service data revealed that the wind direction frequencies (%) were 23.5, 21.5, 17.3, 10.0, 10.2, 6.6, 5.9, and 4.8 for N, E, S, W, NE, SE, SW, and NW winds, respectively. We identified a total of 17 meteorological scenarios (12 ML and 5 MC). Then, we simulated with

ALOHA software the 17 scenarios under the different types of incidents described in Section 2.3. We obtained a total of 51 threat maps for the hotspot. Among them, we selected those with the longest distance to leeward of dangerous atmospheres for each type of incident. In that selection process, maps that simulated the burning of chemicals and that formed a pool of fire were discarded. A summary of the parameters considered for the selected scenarios is in Table 1.

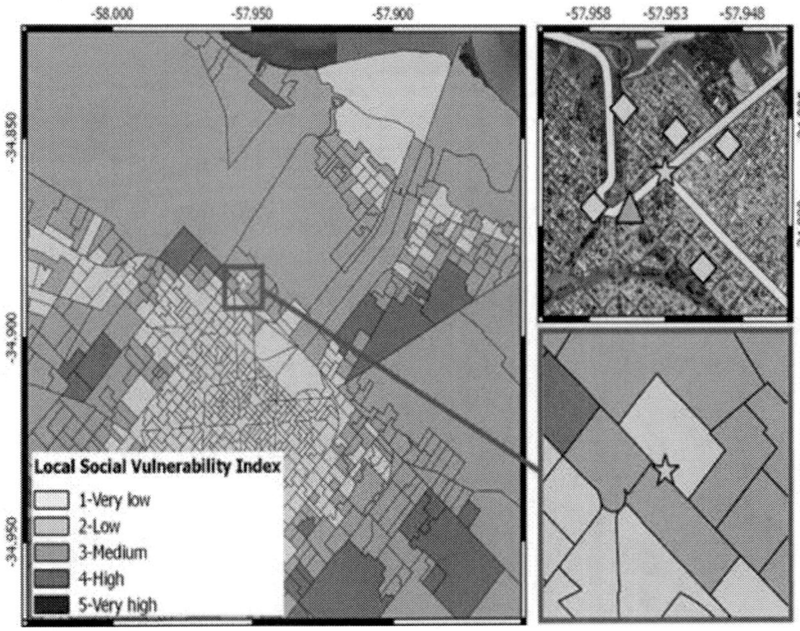

Figure 4. LSVI for the study area. Yellow star: hotspot (34° 53'20" S; 57° 57'21" W), orange triangle: gas station, cyan diamond: sensitive receptor.

The maps of the 3 selected threat scenarios are shown in Figure 5. The threat scenario A is defined according to the thermal radiation limits as suggested by the National Fire Protection Association for BLEVE simulations, and for scenarios B and C, it is based on the acute exposure guideline level for the toxic cloud formation [29, 30]. With respect to the final three scenarios, threat scenario A is the most geographically spread, while threat scenario C is the one furthest from the source.

Table 1. Modeling parameters for the selected scenarios

Scenario			A	B	C
Threat			BLEVE	Toxic Area of Vapor Cloud	Toxic Area of Vapor Cloud
Scenario Type			ML	ML	MC
Meteorological data	Wind	Speed (m s^{-1})	5.5	3.4	1
		Direction	N	N	E
	Humidity (%)		100	60	60
	Cloud Cover (10^{-1})		0	0	0
	Ambient temperature (ºC)		8	25	19
	Pasquill Stability Class		C	C	F
Emission	Source		Horizontal Cylindrical Tank (2 x 11.78 meters)		
	Hole		-	60 x 4 cm	60 x 4 cm
	Ground Type*		-	Concrete	Concrete
	Temperature (ºC)*		-	40	40

ML: the most likely. MC: the most conservative.
(*) This information is requested when modeling the evaporation from puddles caused by spills.

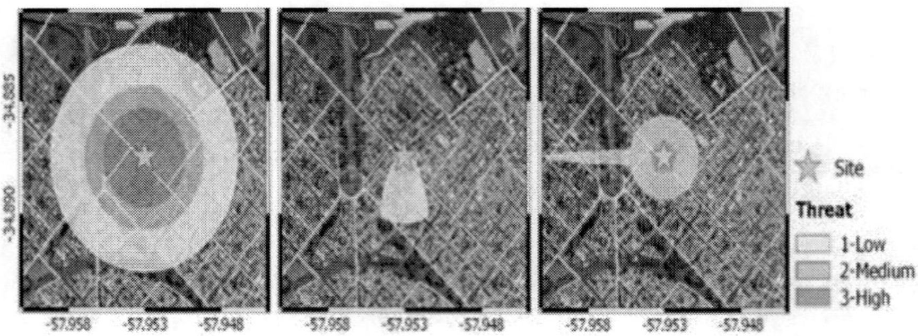

Figure 5. Threat zones in a- BLEVE scenario, b- ML toxic cloud scenario, c- MC toxic cloud scenario for the hotspot. In white are the census area polygons.

The zone with the highest threat level covered 23.22 ha and reached four census areas in scenario A, while it extended to only three census areas (2.37 ha) in scenario C. Scenario B did not present a high threat zone. The areas with a medium threat level covered six (25.15 ha), two (2.20 ha), and three (14.03 ha) census areas for scenarios A, B, and C, respectively. For the low threat areas, the threat reached 10 census areas for scenario A, while it remained at three for B and C.

As the last step in our methodology, we computed the risk level by the census area using equation 1. As shown in Figure 6, there are no high-risk areas on this hotspot. The highest level identified was 6, which is a moderate level. However, the areas at greatest risk are not above the incident point, the risk does not decrease equally as we move away. Some areas in scenario A, exposed to a low level of threat showed the same level of risk (R = 5) as other areas exposed to a high level of threat, governed by the different levels of social vulnerability presented by the areas reached by the thermal radiation. The highest risk level was in scenarios A and C (R = 6), covering 11.25 ha and 0.69 ha, respectively. These areas are smaller than the highest threat areas in each scenario (23.22 ha and 2.37 ha), allowing intervening actors to focus their resources on smaller, but more needy areas.

On the other hand, Figure 6 shows the area with the highest threat level in scenarios B and C. These only reach two census areas with different levels of vulnerability (V = 2 and V = 3). Although the risk levels that present the high threat zones are different (R = 4 and R = 5 for scenario B, and R = 5 and R = 6 for scenario C), they are the highest levels presented respectively in each case.

Based on the scenario planning paradigm, the methodology focuses on determining the critical points for risk assessment. In general, the process of risk scenario analysis includes the development of hypothetical scenarios, parameter adjustments, and the evaluation of simulation results in order to finally decide with the least possible impact on the population [31]. The bottom-up approach considers the risks along a specified route by a specified mode of transport, and then adds these to obtain the risk of the total hazardous materials traffic in the area of interest, as done in [32]. However here the Top-Down approach is used, where historical data were used to estimate particular situations. Through this methodology, it is possible to reduce the number of scenarios to be modeled for the transport route in a few hotspots.

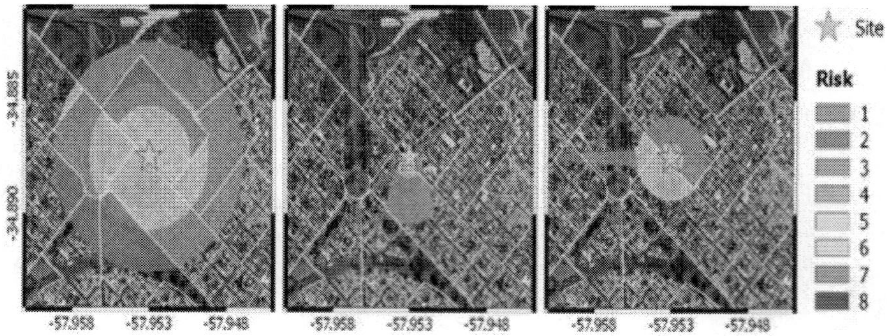

Figure 6. Zones are classified by risk levels. a- BLEVE scenario, b- ML toxic cloud scenario, c- MC toxic cloud scenario.

Conclusion

In this work, we present a 3-step methodology for assessing the risk of transporting hazardous materials. The first step focused on identifying hotspots, the second on modeling more likely/conservative scenarios and assessing vulnerability at those sites, and the last on estimating risk. Finally, we demonstrated the methodology in a concrete application case.

Regarding the determination of hotspots, the analysis of previous accidents on the transport route and the identification of vulnerable receptors were successful steps in the identification of critical areas. In this sense, it was useful to make use of GIS to visualize all these situations on a map. We consider it necessary to highlight places that house a large number of people (schools and clubs) and sites that can generate a concatenation of events (gas stations, HAZMAT stores).

In view of the results obtained, it is important and necessary to have risk maps that cross information on potential threats and intrinsic vulnerabilities, since a threat map may not be descriptive enough with regards to the associated risks. Combining territorial information on vulnerability with disaster simulation in a GIS, useful support is also provided for emergency management and urban planning.

In this sense, we understand that effective mitigation requires understanding the local risks, addressing difficult decisions, and investing in long-term community well-being policies. Thus, the methodology implemented in this work is useful for risk management by optimizing the direction of available resources to the areas of higher risk. Its application is viable for a wide variety of HAZMAT and different countries. It is only necessary to have reliable information on the socio-demographic characteristics of the studied population, and on regional weather, as well as, to know the characteristics of HAZMAT transport, such as transportation routes or quantity, and the emission source.

Finally, this work offers new perspectives to improve the future management of HAZMAT's transport logistics. The methodology developed simplifies planning, providing sufficient information to manage risk.

Acknowledgments

This work was supported by the National Agency for Scientific and Technological Promotion (PICT 2015-0618 and PICT 2017- 1670). We thank the support of the National Scientific and Technical Research Council (CONICET) and the National University of La Plata (UNLP). We also acknowledge the National Meteorological Service (SMN) belonging to the Argentine Ministry of Defense, and YPF S.A., for supplying information. Finally, we would like to highlight the advice provided by the Directorate of Ecological Prevention and Hazardous Materials of the Ministry of Security (Province of Buenos Aires, Argentina).

References

[1] Zhao, M. and Liu, X. 2018. "Development of decision support tool for optimizing urban emergency rescue facility locations to improve humanitarian logistics management." *Safety Science*, 102, pp.110-117.

[2] Zhao, M., Sun, Z. and Zeng, Y. 2020. "Exploring urban risk reduction strategy based on spatial statistics and scenario planning." *Journal of Cleaner Production,* p.121668.

[3] Cordeiro, F. G., Bezerra, B. S., Peixoto, A. S. P. and Ramos, R. A. R. 2016. "Methodological aspects for modeling the environmental risk of transporting hazardous materials by road." *Transportation research part D: transport and environment*, 44, pp.105-121.

[4] Erkut, E., Tjandra, S. A., & Verter, V. 2007. "Chapter 9 Hazardous Materials Transportation." En *Handbooks in Operations Research and Management Science* (Vol. 14, pp. 539-621). Elsevier. https://doi.org/10.1016/S0927-0507(06)14009-8.

[5] Torretta, V., Rada, E. C., Schiavon, M., & Viotti, P. 2017. "Decision support systems for assessing risks involved in transporting hazardous materials: A review." *Safety Science,* 92, 1-9.

[6] Sanchez, E. Y., Represa, S., Mellado, D., Balbi, K. B., Acquesta, A. D., Lerner, J. C., & Porta, A. A. 2018. "Risk analysis of technological hazards: Simulation of scenarios and application of a local vulnerability index." *Journal of hazardous materials*, 352, 101-110.

[7] Ardalan, A., Fatemi, F., Aguirre, B., Mansouri, N. and Mohammdfam, I. 2019. "Assessing human vulnerability in industrial chemical accidents: a qualitative and quantitative methodological approach." *Environmental monitoring and assessment,* 191(8), p.506.

[8] Birkmann, J. 2007. "Risk and vulnerability indicators at different scales: applicability, usefulness and policy implications." *Environ Hazards* 7:20–31.

[9] Cozzani, V., Bonvicini, S., Spadoni, G., & Zanelli, S. 2007. "Hazmat transport: A methodological framework for the risk analysis of marshaling yards." *Journal of Hazardous Materials,* 147(1-2), 412-423. https://doi.org/10.1016/j.jhazmat.2007.01.065.

[10] Landucci, G., Antonioni, G., Tugnoli, A., Bonvicini, S., Molag, M., & Cozzani, V. 2017. "HazMat transportation risk assessment: A revisitation in the perspective of the Viareggio LPG accident."

Journal of Loss Prevention in the Process Industries, 49, 36-46. https://doi.org/10.1016/j.jlp.2016.08.009.

[11] Inanloo, B., & Tansel, B. 2015. "Explosion impacts during transport of hazardous cargo: GIS-based characterization of overpressure impacts and delineation of flammable zones for ammonia." *Journal of Environmental Management*, 156, 1-9. https://doi.org/10.1016/j.jenvman.2015.02.044.

[12] Vianello, C., Milazzo, M. F., Guerrini, L., Mura, A., & Maschio, G. 2016. "A risk-based tool to support the inspection management in chemical plants." *Journal of Loss Prevention in the Process Industries,* 41, 154-168. https://doi.org/10.1016/j.jlp.2016.03.005.

[13] Represa, N. S., Sanchez, E. Y., & Porta, A. A. 2018. "Estudio de la vulnerabilidad social en Argentina mediante el uso de SIG. Construcción de un índice de aplicación local [Study of social vulnerability in Argentina through the use of GIS. Building a local application index]." *Anuario do Instituto de Geociencias*. 41(2):351-357.

[14] Ebert A., Kerle N., Stein A. 2009. "Urban social vulnerability assessment with physical proxies and spatial metrics derived from air- and spaceborne imagery and GIS data." *Nat Hazards* 48(2):275–294.

[15] Diaz-Sarachaga, J. M. and Jato-Espino, D. 2019. "Analysis of vulnerability assessment frameworks and methodologies in urban areas." *Natural Hazards*, pp.1-21.

[16] Hummell, B. M., Cutter, S. L., Emrich, C. T. 2016. "Social vulnerability to natural hazards in Brazil." *Int J Disaster Risk Sci* 7:111–122.

[17] de Oliveira Mendes, J. M. 2009. "Social vulnerability indexes as planning tools: beyond the preparedness paradigm." *Journal of Risk Research*, 12(1), 43-58.

[18] Martins, V. N., e Silva, D. S., & Cabral, P. 2012. "Social vulnerability assessment to seismic risk using multicriteria analysis: the case study of Vila Franca do Campo (São Miguel Island, Azores, Portugal)." *Natural hazards*, 62(2), 385-404.

[19] Orozco, J. L., Van Caneghem, J., Hens, L., González, L., Lugo, R., Díaz, S. and Pedroso, I. 2019. "Assessment of an ammonia incident in the industrial area of Matanzas." *Journal of Cleaner Production,* 222, pp.934-941.

[20] Robert J., William L., Debra S. and Michael R. 2013. *ALOHA* 5.4.4. Technical documentation.

[21] Anjana, N. S., Amarnath, A. and Nair, M.H. 2018. "Toxic hazards of ammonia release and population vulnerability assessment using geographical information system." *Journal of environmental management,* 210, pp.201-209.

[22] INDEC, Instituto Nacional de Estadísticas y Censos. 2011. Censo Nacional de Población, Hogares y Viviendas 2010 [National Institute of Statistics and Censuses. 2011. *National Census of Population, Households and Housing 2010*]. http://www.censo2010.indec.gov.ar/. last accessed July 2020.

[23] Barbier, J. L., Respighi, E., Etchichury, L., & Moscardini, O. 2012. "Documento País 2012 [Country Document 2012]." *Riesgo de Desastres en la Argentina*.

[24] Renda, E., Garay, M. R., Moscardini, O., & Torchia, N. P. 2017. "*Manual para la elaboración de mapas de riesgo* [*Manual for the preparation of risk maps*]." Ministerio de Seguridad de la Nación.

[25] INV, Instituto Nacional de Vitivinicultura. 2012. *Informe estadístico importación y exportación mundial de metanol: año 2009-2011* [*World methanol import and export statistical report: year 2009-2011*]. https://www.argentina.gob.ar/sites/default/files/informe_especial_metanol_2009_al_2011.pdf, last accessed July 2020.

[26] INV, Instituto Nacional de Vitivinicultura. 2019. *Estadísticas de alcohol etílico y metanol: año 2018* [*Ethyl alcohol and methanol statistics: year 2018*]. https://www.argentina.gob.ar/sites/default/files/alcoholes_informe_2018.pdf, last accessed July 2020.

[27] Fernández Long, H., Reimundín, J. C., Cudmani, R., Agosti, C. 1994. *Acción del viento sobre las construcciones* [*Wind action on buildings*]. INTI.

[28] Bubbico, R., Di Cave, S. and Mazzarotta, B. 2004. "Risk analysis for road and rail transport of hazardous materials: a GIS approach." *Journal of Loss prevention in the Process Industries*, 17(6), pp.483-488.

[29] Jones, R., Lehr, W., Simecek-Beatty, D., & Reynolds, M. 2013. *ALOHA® (Areal Locations of Hazardous Atmospheres) 5.4. 4: Technical Documentation*. https://response.restoration.noaa.gov/sites/default/files/ALOHA_Tech_Doc.pdf, last accessed February 2020.

[30] U.S.EPA. *Acute Exposure Guideline Levels*. https://www.epa.gov/aegl, last accessed February 2020.

[31] Panwhar, S. T., Pitt, R. and Anderson, M. D. 2000. "Development of a GIS-based hazardous materials transportation management system." *UTCA report*, 99244.

[32] Van Raemdonck, K., Macharis, C. and Mairesse, O. 2013. "Risk analysis system for the transport of hazardous materials." *Journal of safety research*, 45, pp.55-63.

In: Air Pollution: Effects and Dangers ISBN: 978-1-53619-544-6
Editor: J. Esteban Colman Lerner

Chapter 6

Health Impact Assessment of Benzene Levels Emitted by Gas Stations in the City of La Plata, Argentina

Maria L. Lefevre*[1,*]*, Santiago Aguilar Ferraro*[1]*,
***Jorge E. Colman Lerner*[2]* and Erica Y. Sanchez*[1]**
[1]Centro de Investigaciones del Medioambiente (CONICET-UNLP), La Plata, Argentina
[2]Centro de Investigación y Desarrollo en Ciencias Aplicadas (CONICET-UNLP-CICPBA), La Plata, Argentina

Abstract

Activities carried out in gas stations –mainly filling processes– expose the surrounding population to high levels of contamination due to the emissions originated in their normal operations. This work analyzes the impact of these emissions as diffused sources in the city of La Plata (Argentina) and assesses the risk of contracting cancer (LCR) from exposure to the benzene estimated concentrations. To achieve this goal,

* Corresponding Author's Email: marialauralefevre@gmail.com.

the AERMOD modeling system was used as a dispersion assessment tool. The results show that, although the estimated concentrations do not represent a risk for the surrounding population, they reach 7.2-8.5% of the limit concentration recommended by the United States Environmental Protection Agency (corresponding to a LCR of 1 $x10^{-6}$), regarding chronic exposure to benzene in the adult population. However, the levels may be alarming when considering other significant sources in the region that have not been included in the modeling, such as the petrochemical industrial area and the emission of gases coming from motor vehicles.

1. INTRODUCTION

On a global scale, increasing amounts of potentially harmful gases and particles are emitted that impact human health. According to the World Health Organization (WHO), more than two million deaths annually are attributed to the effects of air pollution in urban open spaces (outdoor) and in closed spaces (indoor). Developing countries are those with the highest mortality rate [1-3].

There is much evidence that correlates chronic exposure to pollutants related to vehicular traffic and chemical and petrochemical industry, such as particulate matter, metals, polycyclic aromatic hydrocarbons (PAHs), volatile organic compounds (VOCs) and inorganic gases (SO_2 and NO_x), with adverse effects on lung development and function. Other effects to be considered are the increase in mortality, mainly in children, and the increase in respiratory morbidity. The latter is frequently expressed in terms of the diagnosis of asthma or chronic obstructive pulmonary disease. Vulnerability to respiratory problems has been recognized as one of the main adverse effects of air pollution [4, 5]. Likewise, it has been shown that air pollution generates alterations in blood pressure, vascular tone, coagulation, etc. [6]. These small changes in blood pressure can cause a significant impact on strokes (cerebrovascular accidents), heart disease, transient ischemic attack and mortality [7, 8].

The number of respiratory diseases and allergies associated with environmental pollution has increased worldwide in the last two decades, with a strong consensus that exposure to air pollutants is an important health risk factor for the population [9-11].

An important source of VOCs emissions is represented by gas stations, which carry out retail activities for liquid fuels derived from petroleum. These activities, added to the high evaporation rate of fuels and the evaporative losses from vehicle emissions, are the main sources of change in air quality in urban areas [11].

Gasoline is one of the most important petroleum fuels and is chemically composed of hydrocarbons containing 6 to 10 carbon atoms, which mainly belong to the classes of paraffins (normal and branched), olefins, naphthenes and aromatics, while diesel fuels are mainly composed of hydrocarbons containing 10 to 28 carbon atoms [12, 13]. Within the aromatic hydrocarbons present in liquid fuels are benzene, toluene, ethylbenzene and xylenes (ortho, meta and para isomers), commonly known as BTEX. Among the BTEX compounds, benzene is considered the most dangerous and is classified as carcinogenic to humans (group 1) by the International Agency for Research on Cancer "IARC" [14].

The health effects of exposure to BTEX compounds depend on the dose and duration of exposure as acute or chronic effects. Chronic exposure primarily to benzene can cause neurological, endocrine, immunological, and haematological disorders, such as aplastic anaemia and myeloid leukaemia [15-19].

The levels of BTEX compounds in the atmosphere can be estimated through modeling systems, which provide information about their spatial and temporal distribution. There are several software that models the pollutants dispersion in the atmosphere, however, AERMOD is one of the favourites due to the compromise between the level of description and its ease of use. The modeling system has been efficiently tested in several recent studies of risk assessment of carcinogenic and non-carcinogenic compounds [20-22].

The study region includes the districts of La Plata, Berisso and Ensenada (Buenos Aires, Argentina), it has about 700,000 inhabitants and two relevant sources of pollutants. On the one hand, an important petrochemical industrial complex (Ensenada), which contains the most important oil refinery in the country (38,000m^3 day^{-1} of crude oil processed) and other subsidiary industries producing aromatic compounds (benzene, toluene, xylenes), aliphatics (pentane, hexane, heptane) and petroleum coke, among others. On the other hand, the urban area of La Plata has vehicular traffic of about 450,000 registered cars and 80 gas stations.

Previous studies in the same region show high levels of BTEX in the vicinity of the Pole and within the urban area, as well as adverse health effects, such as respiratory function determined by spirometry, ocular alterations and other complementary epidemiological studies [23-29].

The goal of this study is to estimate the contribution of BTEX emissions from gas stations to air quality levels in the region and the risk in population health.

2. Methodology

2.1. Emissions Estimation

The emissions of the 55 main gas stations from the study region were estimated using emission factors. These factors, to BTEX compounds, were calculated from the method proposed by AP-42 of the U.S. EPA [30] with the monthly fuel informed sales of each gas station in the region. To this end, there were assumed constant emissions and continue operations of all sources throughout each month.

2.2. BTEX Dispersion Modeling and Monitoring

The AERMOD View™ modeling system [31] was used to simulate the dispersion and transport of BTEX emissions coming from the main gas stations in the study region.

First, we compare the concentrations of benzene, ethylbenzene, toluene and xylenes estimated by AERMOD in a discrete receptor site, with the concentrations observed (monitoring) on the site mentioned. The aim was to analyze the representation of the emissions estimated by the emission factors in the study scenario (Section 2.1). Second, the model was used to estimate benzene concentrations contributed by the main gas service stations.

For the first case, the BTEX dispersion from the emission of an area source (264m^2 with center in 34° 54'24.71 "S, 57° 55'37.36" W) was simulated for a period of 8 hours (9am - 5pm), corresponding to the time of BTEX sampling, for two specific days (September 2 and 23, 2016) with prevailing south and southwest winds.

These choices made it possible to ensure the transport of BTEX compounds from the simulated source to the receptor site of interest (34° 54'17.95" S, 57° 55'34.51" W). The location and environment of the modeled gas station (area source), along with the monitoring site (discrete receptor), and the wind roses for the periods and dates simulated are shown in Figure 1.

BTEX concentrations at the receptor site were obtained by active monitoring with activated carbon tubes on the mentioned dates and the samples were analyzed by gas chromatography coupled to mass spectrometry according to the protocol developed by Mellado [32].

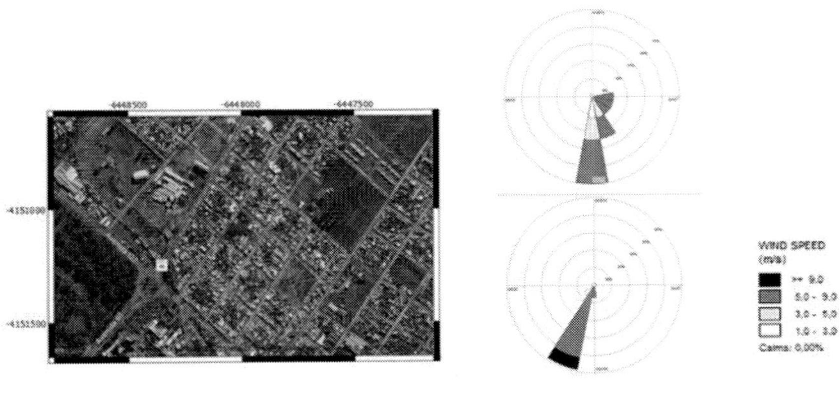

Figure 1. On the left, an indicative map of the area source (gas station) is shown represented with a red polygon, and the monitoring site represented with an orange star with UTM 21 S coordinates. On the right, the Wind roses correspond to September 2 and 23, 2016, from 9am to 5pm (up and down).

As a second step, and once the representativeness of the emissions estimated in the previous paragraph had been analyzed, the spatial distribution of benzene concentrations (monthly average) was estimated from the contribution of the 55 main gas stations, for July and December 2016.

The temporal selection mentioned is justified by the following two reasons, one associated with the differences in meteorological conditions in those months, and the other, with the characteristics of the information used to estimate the emissions (monthly sales of fuels in m^3 reported by companies, information published by the Secretary of Energy of Argentina).

For both steps, a meteorological database from 2016 was used, with hourly resolution, provided by the National Meteorological Service (SMN) for La Plata AERO station (ID 87593).

Modeled species were treated as inert tracer species, since typical photochemical losses are much slower than out-of-domain transport. The domain used in the first stage was represented by a 2km (N-S) x 1.5km (E-W) grid, with 50m resolution (1201 grid receptors).

For the second stage, the domain was represented by a grid of 8.85km (N-S) x 11.05km (E-W), with 50m resolution (39516 gridded receptors). The grid receptors heights were specified to match the height of the ground above sea level.

The flat topography of the study region was taken into account to evaluate the impact in the area surrounding of the gas stations.

2.3. Risk Analysis

There is a very useful and simple formulation for the cancer risk calculation by exposure to pollutants, based on data provided by recognized entities, such as the WHO, the U.S. EPA and the IARC. Knowledge of the carcinogenic potential of the specific compound and the degree of exposure of the target group is required. Thus, the lifetime cancer risk (LCR) associated with benzene is calculated (equation 1) by multiplying the daily chronic intake (CDI) by the power factor (FP, equivalent to the slope factor estimated at from the exposure/population curve and tabulated in the IRIS System) and by the absorption factor [23, 33-35].

Chronic daily intake is calculated from equation 2.

$$LCR = FP \; x \; CDI \quad (1)$$

$$CDI = \frac{CC \; x \; IR \; x \; ED \; x \; EF \; x \; LE}{BW \; x \; ATL \; x \; NY} \quad (2)$$

Where the variables for the calculation of the CDI were determined in previous studies for the region by Colman Lerner et al. [23] and are shown in Table 1, with the power factor for benzene being 0.029 $(mg/kg\ day)^{-1}$ and an absorption factor of 90%.

Table 1. Parameters used to estimate the CDI

PARAMETERS	ADULTS
IR	0.833
ED	42
EF	48
LE	40
BW	70
ATL	70
NY	365

Where each term is:

CC: pollutant concentration ($mg\ m^{-3}$).

IR: inhalation flow ($m^3\ hour^{-1}$).

ED: duration of exposure ($hours\ week^{-1}$).

EF: frequency of exposure ($weeks\ year^{-1}$).

LE: length of exposure (years).

BW: body weight (kg).

ATL: mean time (average of the period during which one will be exposed).

NY: number of days per year (365 days).

Table 2. Emission (g $s^{-1}m^{-2}$) calculated from the emission factors (EF) for the months of July, September, and December for each service station (B: Benzene, T: Toluene, E: Ethylbenzene, X: Xylenes)

Source	July				September				December			
	B	T	E	X	B	T	E	X	B	T	E	X
1	1.4E-5	3.6E-5	2.3E-6	9.2E-6	1.6E-5	4.0E-5	2.5E-6	1.0E-5	1.8E-5	4.4E-5	2.9E-6	1.1E-5
2	1.7E-5	4.2E-5	2.7E-6	1.1E-5	1.8E-5	4.6E-5	3.0E-6	1.2E-5	1.9E-5	4.9E-5	3.1E-6	1.3E-5
3	8.3E-6	2.1E-5	1.3E-6	5.3E-6	9.3E-6	2.3E-5	1.5E-6	6.0E-6	1.0E-5	2.6E-5	1.6E-6	6.5E-6
4	1.3E-5	3.3E-5	2.1E-6	8.5E-6	1.4E-5	3.6E-5	2.3E-6	9.2E-6	1.9E-5	4.8E-5	3.1E-6	1.2E-5
5	1.2E-5	2.9E-5	1.9E-6	7.5E-6	1.1E-5	2.8E-5	1.8E-6	7.2E-6	1.1E-5	2.7E-5	1.7E-6	6.9E-6
6	3.3E-5	8.3E-5	5.3E-6	2.1E-5	3.7E-5	9.2E-5	5.9E-6	2.4E-5	4.1E-5	1.0E-4	6.6E-6	2.6E-5
7	2.3E-5	5.9E-5	3.8E-6	1.5E-5	2.4E-5	6.0E-5	3.8E-6	1.5E-5	2.7E-5	6.7E-5	4.3E-6	1.7E-5
8	3.4E-6	8.6E-6	5.5E-7	2.2E-6	4.0E-6	1.0E-5	6.4E-7	2.6E-6	5.2E-6	1.3E-5	8.3E-7	3.3E-6
9	1.6E-5	4.0E-5	2.5E-6	1.0E-5	1.6E-5	4.0E-5	2.6E-6	1.0E-5	1.6E-5	4.1E-5	2.6E-6	1.0E-5
10	2.9E-5	7.2E-5	4.6E-6	1.8E-5	3.0E-5	7.5E-5	4.8E-6	1.9E-5	2.8E-5	7.1E-5	4.5E-6	1.8E-5
12	3.3E-6	8.3E-6	5.3E-7	2.1E-6	4.2E-6	1.0E-5	6.7E-7	2.7E-6	4.1E-6	1.0E-5	6.6E-7	2.6E-6
13	2.7E-6	6.7E-6	4.3E-7	1.7E-6	2.9E-6	7.3E-6	4.7E-7	1.9E-6	3.2E-6	8.1E-6	5.2E-7	2.1E-6
14	1.4E-5	3.6E-5	2.3E-6	9.2E-6	1.7E-5	4.2E-5	2.7E-6	1.1E-5	1.5E-5	3.9E-5	2.5E-6	1.0E-5
16	5.1E-6	1.3E-5	8.3E-7	3.3E-6	5.1E-6	1.3E-5	8.3E-7	3.3E-6	6.0E-6	1.5E-5	9.7E-7	3.9E-6
17	2.8E-5	7.1E-5	4.5E-6	1.8E-5	2.8E-5	7.1E-5	4.6E-6	1.8E-5	2.8E-5	6.9E-5	4.4E-6	1.8E-5
20	8.4E-6	2.1E-5	1.4E-6	5.4E-6	9.3E-6	2.3E-5	1.5E-6	6.0E-6	9.4E-6	2.4E-5	1.5E-6	6.0E-6
21	1.7E-5	4.3E-5	2.8E-6	1.1E-5	1.6E-5	3.9E-5	2.5E-6	1.0E-5	1.6E-5	4.0E-5	2.6E-6	1.0E-5
22	1.7E-5	4.3E-5	2.8E-6	1.1E-5	1.5E-5	3.8E-5	2.4E-6	9.8E-6	2.3E-5	5.7E-5	3.7E-6	1.5E-5

Source	July				September				December			
	B	T	E	X	B	T	E	X	B	T	E	X
23	1.4E-5	3.5E-5	2.2E-6	9.0E-6	2.7E-5	6.9E-5	4.4E-6	1.8E-5	1.5E-5	3.8E-5	2.4E-6	9.7E-6
24	4.0E-6	1.0E-5	6.5E-7	2.6E-6	5.0E-6	1.2E-5	8.0E-7	3.2E-6	5.5E-6	1.4E-5	8.8E-7	3.5E-6
25	3.8E-5	9.5E-5	6.1E-6	2.4E-5	3.4E-5	8.6E-5	5.5E-6	2.2E-5	3.9E-5	9.8E-5	6.3E-6	2.5E-5
26	2.5E-6	6.3E-6	4.0E-7	1.6E-6	6.0E-6	1.5E-5	9.7E-7	3.9E-6	6.7E-6	1.7E-5	1.1E-6	4.3E-6
29	2.9E-5	7.3E-5	4.7E-6	1.9E-5	2.8E-5	6.9E-5	4.4E-6	1.8E-5	2.9E-5	7.2E-5	4.6E-6	1.8E-5
30	2.8E-5	7.1E-5	4.5E-6	1.8E-5	2.6E-5	6.5E-5	4.2E-6	1.7E-5	2.9E-5	7.4E-5	4.7E-6	1.9E-5
31	3.3E-5	8.4E-5	5.4E-6	2.2E-5	3.9E-5	9.9E-5	6.4E-6	2.5E-5	4.4E-5	1.1E-4	7.1E-6	2.8E-5
36	5.4E-5	1.4E-4	8.7E-6	3.5E-5	5.3E-5	1.3E-4	8.5E-6	3.4E-5	5.4E-5	1.3E-4	8.6E-6	3.5E-5
37	2.0E-5	5.1E-5	3.3E-6	1.3E-5	2.2E-5	5.4E-5	3.5E-6	1.4E-5	2.2E-5	5.7E-5	3.6E-6	1.5E-5
39	1.1E-5	2.7E-5	1.7E-6	6.9E-6	1.1E-5	2.7E-5	1.7E-6	6.8E-6	1.2E-5	3.1E-5	2.0E-6	8.0E-6
43	7.3E-6	1.8E-5	1.2E-6	4.7E-6	7.8E-6	2.0E-5	1.3E-6	5.0E-6	8.1E-6	2.0E-5	1.3E-6	5.2E-6
44	3.7E-5	9.3E-5	6.0E-6	2.4E-5	4.1E-5	1.0E-4	6.6E-6	2.6E-5	4.6E-5	1.2E-4	7.4E-6	3.0E-5
45	4.3E-5	1.1E-4	7.0E-6	2.8E-5	4.2E-5	1.1E-4	6.9E-6	2.7E-5	4.3E-5	1.1E-4	6.9E-6	2.8E-5
46	1.1E-5	2.9E-5	1.8E-6	7.4E-6	1.2E-5	2.9E-5	1.9E-6	7.4E-6	1.1E-5	2.9E-5	1.8E-6	7.4E-6
47	1.6E-5	4.1E-5	2.6E-6	1.1E-5	2.0E-5	5.1E-5	3.3E-6	1.3E-5	2.1E-5	5.3E-5	3.4E-6	1.4E-5
48	3.0E-5	7.5E-5	4.8E-6	1.9E-5	2.8E-5	6.9E-5	4.4E-6	1.8E-5	2.9E-5	7.3E-5	4.7E-6	1.9E-5
50	6.2E-6	1.6E-5	1.0E-6	4.0E-6	7.1E-6	1.8E-5	1.1E-6	4.6E-6	6.6E-6	1.7E-5	1.1E-6	4.2E-6
51	1.8E-6	4.4E-6	2.8E-7	1.1E-6	3.4E-6	8.4E-6	5.4E-7	2.2E-6	8.1E-7	2.0E-6	1.3E-7	5.2E-7
52	1.8E-5	4.5E-5	2.9E-6	1.2E-5	2.1E-5	5.2E-5	3.3E-6	1.3E-5	2.2E-5	5.6E-5	3.6E-6	1.4E-5
56	3.6E-5	9.0E-5	5.8E-6	2.3E-5	3.7E-5	9.3E-5	6.0E-6	2.4E-5	3.8E-5	9.5E-5	6.1E-6	2.4E-5
57	2.2E-5	5.6E-5	3.6E-6	1.4E-5	2.6E-5	6.5E-5	4.2E-6	1.7E-5	2.4E-5	6.0E-5	3.9E-6	1.5E-5

Table 2. (Continued)

Source	July				September				December			
	B	T	E	X	B	T	E	X	B	T	E	X
58	2.8E-5	6.9E-5	4.5E-6	1.8E-5	3.0E-5	7.5E-5	4.8E-6	1.9E-5	3.1E-5	7.7E-5	4.9E-6	2.0E-5
59	4.8E-5	1.2E-4	7.8E-6	3.1E-5	4.9E-5	1.2E-4	7.9E-6	3.2E-5	4.7E-5	1.2E-4	7.6E-6	3.0E-5
60	1.1E-5	2.9E-5	1.9E-6	7.4E-6	1.2E-5	3.1E-5	2.0E-6	7.8E-6	1.2E-5	3.0E-5	1.9E-6	7.7E-6
61	5.2E-6	1.3E-5	8.3E-7	3.3E-6	6.0E-6	1.5E-5	9.7E-7	3.9E-6	5.2E-6	1.3E-5	8.4E-7	3.4E-6
62	2.0E-6	5.0E-6	3.2E-7	1.3E-6	1.9E-6	4.8E-6	3.1E-7	1.2E-6	2.1E-6	5.3E-6	3.4E-7	1.4E-6
63	3.9E-6	9.9E-6	6.3E-7	2.5E-6	3.9E-6	9.8E-6	6.3E-7	2.5E-6	4.6E-6	1.2E-5	7.4E-7	3.0E-6
64	1.3E-5	3.3E-5	2.1E-6	8.5E-6	1.3E-5	3.2E-5	2.1E-6	8.3E-6	1.4E-5	3.4E-5	2.2E-6	8.8E-6
66	1.6E-5	4.1E-5	2.7E-6	1.1E-5	1.9E-5	4.8E-5	3.1E-6	1.2E-5	2.0E-5	4.9E-5	3.2E-6	1.3E-5
68	6.6E-6	1.7E-5	1.1E-6	4.2E-6	7.5E-6	1.9E-5	1.2E-6	4.9E-6	7.0E-6	1.8E-5	1.1E-6	4.5E-6
70	1.7E-6	4.2E-6	2.7E-7	1.1E-6	4.6E-6	1.2E-5	7.4E-7	3.0E-6	4.6E-6	1.2E-5	7.5E-7	3.0E-6
73	3.0E-6	7.7E-6	4.9E-7	2.0E-6	1.6E-6	4.1E-6	2.6E-7	1.1E-6	8.8E-7	2.2E-6	1.4E-7	5.7E-7
74	4.2E-6	1.0E-5	6.7E-7	2.7E-6	7.9E-6	2.0E-5	1.3E-6	5.1E-6	1.9E-6	4.8E-6	3.1E-7	1.2E-6
75	4.1E-6	1.0E-5	6.7E-7	2.7E-6	3.3E-6	8.2E-6	5.3E-7	2.1E-6	2.2E-6	5.4E-6	3.5E-7	1.4E-6
77	3.8E-6	9.5E-6	6.1E-7	2.4E-6	4.3E-6	1.1E-5	6.9E-7	2.8E-6	5.2E-6	1.3E-5	8.3E-7	3.3E-6

3. RESULTS

3.1. BTEX Emissions

The BTEX compounds emissions from the gas stations were estimated for the months of July, September, and December with an emission factor [29] of 14.15 kg VOCs m^{-3} of fuel, from which 0.31% is Benzene, 0.78% is Toluene, 0.05% is Ethylbenzene and 0.20% are Xylenes (all isomers) with the monthly fuel sales values for those periods (without discriminating between gasoline and diesel), showing the results obtained for the 55 area sources in Table 2.

3.2. BTEX Dispersion Modeling and Monitoring

The BTEX values obtained through active monitoring for the dates 09/02/2016 and 09/23/2016, together with the concentrations estimated with AERMOD in the discrete receptor site, coinciding with the monitoring site, are shown in Table 3, except for benzene and toluene (09/02/2016) which results are in the same order of magnitude, the modeled values are lower than the monitoring values. Consequently, the possible contribution of BTEX compounds from other sources present in the region is expected [23, 27, 34].

Table 3. Concentrations (µg m^{-3}) obtained by active monitoring (M) and estimated with the AERMOD View modeling system at the discrete receptor site (A)

Date	Benzene		Toluene		Ethylbenzene		Xylenes	
	A	M	A	M	A	M	A	M
09/02/2016	0.095	0.082	0.238	0.164	0.015	0.037	0.061	0.861
09/23/2016	0.119	1.200	0.299	3.970	0.019	0.200	0.077	3.510

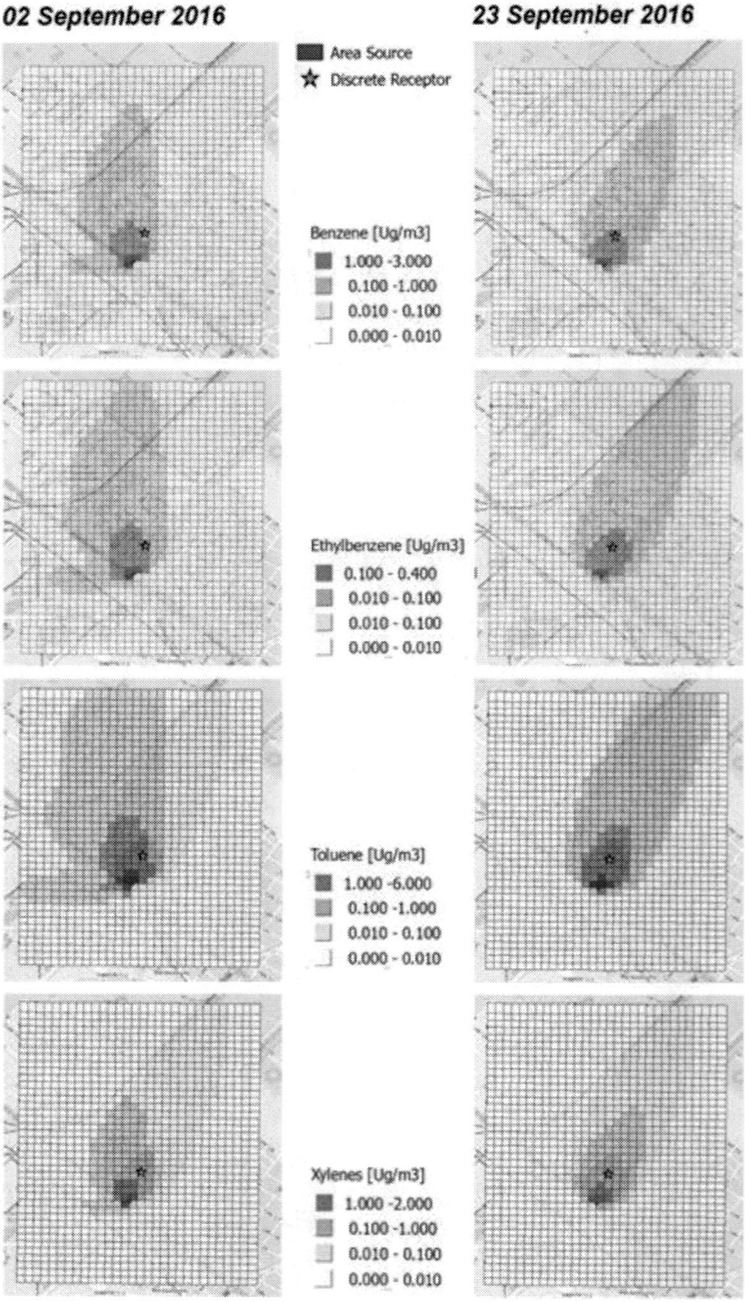

Figure 2. Results of the simulation using AERMOD for benzene, ethylbenzene, toluene, and xylenes from the area source closest to the monitoring site.

Figure 2 shows the simulations obtained by AERMOD for benzene, ethylbenzene, toluene and xylenes, for the gas station closest to the monitoring site, located to the northeast (34° 54'14"S, 57° 55'34" W).

For any point on the grid, the BTEX levels are the result of the dispersion and transport of these emissions from their sources to the receptors.

Modeling of the dispersion and transport of BTEX from gas stations to the discrete receptor site showed a correspondence between the estimated (modeling) and the observed (monitoring) levels. The differences may be attributable to the contribution to the discrete receptor site of other sources not considered in the modeling, such as emissions of gases coming from motor vehicles and the petrochemical pole, according to earlier studies [23]. This pole is located leeward of the source studied and the discrete receptor site.

Finally, to estimate the benzene concentrations in the study region by the contribution of the main gas stations, a multi-source simulation of the 55 service stations was carried out.

The spatial and temporal patterns of BTEX compounds in ambient air (Figure 3) show concentrations lower than 0.71μg m^{-3} in most of the study region, these values are exceeded in regions close to each gas station (radius of between 200 and 300 meters).

There are hotspots of concentration higher than average in the vicinity of service stations.

The values of the means and medians obtained for the region are shown in Table 4 for the months of July and December, these results are lower than those obtained by Colman Lerner et al. [23] in the same region between 2007 and 2010 where benzene values were between 1.8 and 2.0μgm^{-3}, toluene 4.3 and 5.7μg m^{-3}, ethylbenzene 0.9 and 1.1μg m^{-3} and xylenes 4.1 and 5.4μg m^{-3}, considering that these reported values include contributions from other sources present in the region, such as emissions coming from vehicular traffic and the petrochemical pole.

It was observed that the BTEX contribution, taking into account the median, represents 4% of the regional BTEX levels obtained in earlier studies.

3.3. Risk Analysis

To estimate Lifetime Cancer Risk (LCR) of the population from exposure to benzene it was taken into account the median values that were obtained from the simulation of all sources (gas stations) in the study region (Table 2).

They were in a range 8.02×10^{-8} to 9.44×10^{-8}. The World Health Organization (WHO) considers LCR values "acceptable" when they are in the range between 1×10^{-5} and 1×10^{-6}, while the U.S. EPA recommends a lower LCR (1×10^{-6}) [23].

When comparing the results obtained with the monthly medians, it is observed that the guide values for risk of both the U.S. EPA and the WHO are met. But it must be taken into account that the risk is only being calculated from the values obtained by the emissions of the gas stations and not with the actual environmental levels of benzene in the region, which is a pollutant that can be emitted by various sources, including vehicular traffic and the petrochemical pole present.

Taking into account this reality, the concentration of benzene necessary for the acceptable risk limit (LCR) for the WHO and U.S. EPA can be calculated in the opposite direction, with benzene values of 0.71 $\mu g\ m^{-3}$ and 7.1 $\mu g\ m^{-3}$ respectively.

Considering these limit values of benzene concentration for an unacceptable risk and the median values obtained when simulating the emissions of gas stations using AERMOD, it is observed that their contribution to benzene levels, necessary for it to be a health risk, are between 7.2 and 8.5%.

However, if it is considered the average values obtained in the modeling the contribution to health risk is doubled.

From what was observed from the multi-source simulation with AERMOD for benzene, if the risk in the areas surrounding the emission sources is calculated (Figure 3), the risk values obtained would exceed the guide values recommended by both the U.S. EPA and the WHO.

Previous results in the region [23, 27, 34] on the estimation of the risk of Lifetime Cancer Risk by monitoring with passive 3M 3500 monitors, were 1.8 and 2.0μg m^{-3} of benzene which correspond to LCR values between 1.9 x 10^{-6} and 2.1 x 10^{-6}, where the U.S. EPA guide values for acceptable risk are not fulfill.

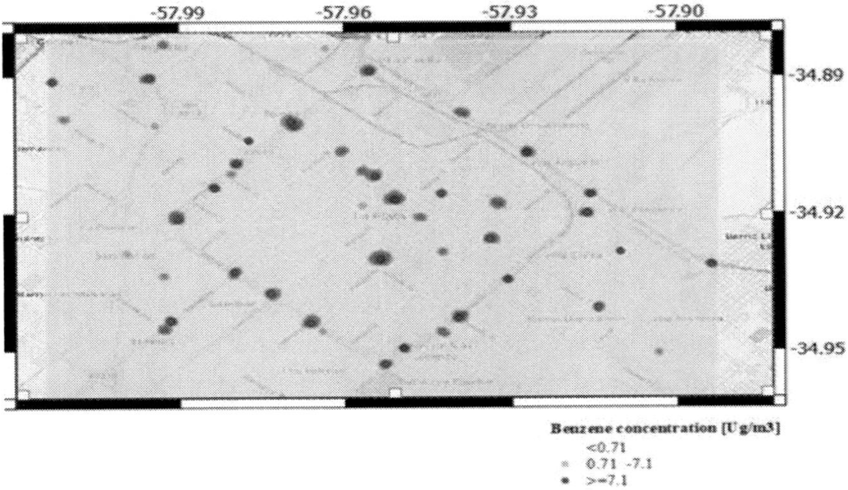

(a)

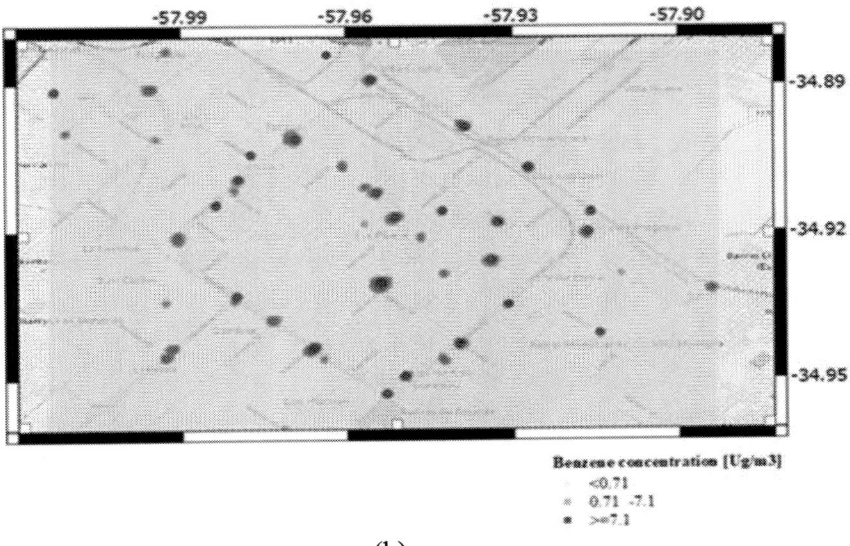

(b)

Figure 3. Multi-source simulation with AERMOD, for benzene in the study region: (a) July 2016 and (b) December 2016 with coordinates in decimal degrees.

Table 4. Monthly concentrations (μg m^{-3}) obtained from the simulations using AERMOD

Month	Benzene		Toluene		Ethylbenzene		Xylenes	
	mean	median	mean	median	mean	median	mean	median
July	0.111	0.060	0.893	0.386	0.018	0.010	0.136	0.067
December	0.099	0.051	0.097	0.093	0.064	0.036	0.066	0.037

CONCLUSION

In this work, it was possible to analyze the contribution of Benzene, Toluene, Ethylbenzene and Xylene (BTEX) emissions from gas stations (55) in an Urban-Industrial region (Gran La Plata, Buenos Aires, Argentina) by simulating the dispersion using AERMOD and comparing the results with monitoring values in the region, obtaining satisfactory results. It was observed that the BTEX contribution, considering the median, represents a contribution of 4% of the regional BTEX levels obtained in previous studies.

Taking into account benzene in particular (type I carcinogen according to IARC), it was observed that in most of the region the risk (LCR) is acceptable according to the WHO and U.S. EPA parameters. The benzene concentrations obtained were lower than 0.71μg m^{-3}, with the contribution of the emission sources (gas stations) to the benzene levels necessary for it to be a health risk of between 7.2 and 8.5%, doubling the contribution if the mean values obtained were taken into account.

Performing an analysis in nearby areas (200 – 300 meter radius) to the modeled emission sources (gas stations) the risk becomes unacceptable (concentrations higher than 0.71μg m^{-3}) according to the WHO and U.S. EPA recommendations.

It should be noted that for the risk calculation the other relevant emission sources in the region were not taken into account, such as traffic and the petrochemical pole, showing that the emissions from gas stations (in relation to BTEX) have a not negligible contribution to the average levels in the region.

ACKNOWLEDGMENT

The authors would like to thank the National Agency for Scientific and Technical Promotion (ANPCyT), the National Scientific and Technical Research Council (CONICET), the National University of La Plata which have financed this work. The National Meteorological Service (SMN) is also gratefully acknowledged for providing the meteorological information.

REFERENCES

[1] Payne-Sturges, D., Gee, G. C. National environmental health measures for minority and low-income populations: Tracking social disparities in environmental health. *Environmental Research,* 2006; 102:154 - 71. doi: 10.1016/j.envres.2006.05.014.

[2] WHO. *Environmental Health Criteria 237.* (2006). Principles for Evaluating Health Risks in Children Associated with Exposure to Chemicals. Published under the Joint Sponsorship of the United Nations Environment Programme, the International Labour Organization and the World Health Organization Geneva.

[3] WHO. *Environmental Health Criteria 239.* (2009). Principles for Modelling Dose–Response for the Risk Assessment of Chemicals. Published under the Joint Sponsorship of the United Nations Environment Programme, the International Labour Organization and the World Health Organization Geneva.

[4] Adar, S. D., Kaufman, J. D. (2007). Cardiovascular Disease and Air Pollutants: Evaluating and Improving Epidemiological Data

Implicating Traffic Exposure. *Inhalation Toxicology*, 19:135 - 49. doi: 10.1080/08958370701496012.

[5] Pope III, C. A., Dockery, D. W. (2006). Health Effects of Fine Particulate Air Pollution : Lines that Connect. *Journal of the Air and Waste Management Association,* 56:709 - 42. doi: 10.1080/ 10473289.2006.10464485.

[6] Fuks, K. B., Weinmayr, G., Foraster, M., Dratva, J., Hampel, R., Houthuijs, D. et al. (2014). Arterial blood pressure and long-term exposure to traffic-related air pollution: An analysis in the European Study of Cohorts for Air Pollution Effects (ESCAPE). *Environmental Health Perspectives*, 122:896 - 905. doi: 10.1289/ehp.1307725.

[7] Whelton, P. K., He, J., Appel, L. J. et al. (2002). Primary prevention of hypertension: Clinical and public health advisory from the national high blood pressure education program. *JAMA,* 288:1882 - 8.

[8] Cook, N. R., Cohen, J., Hebert, P. R., Taylor, J. O., Hennekens, C. H. (1995). Implications of small reductions in diastolic blood pressure for primary prevention. *Archives of Internal Medicine*, 155:701 - 9.

[9] Massolo, L., Müller, A., Herbarth, O., Ronco, A., Porta, A. (2008). Air pollution and children's health in urban and industrial areas of La Plata, Argentina. *Acta Bioquím Clínica Latinoamericana*, 42: 567 - 74.

[10] Wichmann, F. A., Müller, A., Busi, L. E., Cianni, N., Massolo, L., Schlink, U. et al. (2009). Increased asthma and respiratory symptoms in children exposed to petrochemical pollution. *The Journal of Allergy and Clinical Immunology,* 123:632 - 8. doi: 10.1016/j.jaci. 2008.09.052.

[11] Cianni, N., Aguilar, M., Massolo, L., Colman, E., Müller, A., Matamoros, N. et al. (2009). *Air quality in urban and industrial areas and respiratory pathology in children.* Round Table at the I International Congress on Childhood and Adolescent Toxicology XVI Argentine Congress of Toxicology 24 Al 27 de Junio de 2009, Puerto Madryn, Patagonia Argentina.

[12] Periago, J. F. and Prado, C. (2005). Evolution of Occupational Exposure to Environmental Levels of Aromatic Hydrocarbons in Service Stations. *Air Quality, Atmosphere and Health*, 49, 233 - 240.

[13] Agency for Toxic Substances and Disease Registry. (2004). *Interaction Profile for: Benzene, Toluene, Ethylbenzene, and Xylenes (BTEX).* Department of Public Health and Human Services, Public Health Service, Atlanta.

[14] U.S. EPA. *SW846 method 8015D revision 4*. (2003). https:// www.epa.gov/sites/production/files/2015-12/documents/8015d_r4. pdf.

[15] International Agency for Research on Cancer (IARC). (1987). Monographs on the Evaluation of Carcinogenic Risks to Humans. *Overall Evaluations of Carcinogenicity*, 1 - 42, Supplement 7.

[16] Agency for Toxic Substances and Disease Registry. (2004). *Interaction Profile for: Benzene, Toluene, Ethylbenzene, and Xylenes (BTEX).* Department of Public Health and Human Services, Public Health Service, Atlanta.

[17] World Health Organization (WHO). (2000). *Air Quality Guidelines for Europe.* 2nd Edition, WHO Regional Publications, European Series, No. 91, Copenhagen.

[18] US Environmental Protection Agency. (1998). *Carcinogenic Effects of Benzene: An Update* (Final). Washington DC.

[19] Smith, M. T. (2010). Advances in Understanding Benzene Health Effects and Susceptibility. *Annual Review of Public Health,* 31, 133 - 148.

[20] Snyder, R. (2012). Leukemia and Benzene. *International Journal of Environmental Research and Public Health,* 9, 2875 - 2893.

[21] Mirrezaei, M. A. and Orkomi, A. A. (2020). *Gas flares contribution in total health risk assessment of BTEX in Asalouyeh*, Iran. Process Safety and Environmental Protection.

[22] Sarkhosh, M., Shamsipour, A., Yaghmaeian, K., Nabizadeh, R., Naddafi, K., Mohseni, S. M. 2017. Dispersion modeling and health risk assessment of VOCs emissionsfrom municipal solid waste transfer station in Tehran, Iran. *J. Environ. Health Sci. Eng.,* 15, 4.

[23] Truong, S. C., Lee, M.-I., Kim, G., Kim, D., Park, J.-H., Choi, S.-D., Cho, G.-H. 2016. Acci-dental benzene release risk assessment in an urban area using an atmospheric dispersion model. *Atmospheric Environment,* 144, 146 - 159.

[24] Colman Lerner, Jorge E. et al. (2014). Improvement of health risk factors after reduction of VOC concentrations in industrial and urban areas. *Environmental Science and Pollution Research*, vol. 21, no 16, p. 9676 - 9688.

[25] Müller, Andrea et al. (2006). Cytotoxicity and oxidative stress caused by chemicals adsorbed on particulate matter. *Environmental Toxicology: An International Journal,* vol. 21, no 5, p. 457 - 463.

[26] Wichmann, Gunnar et al. (2009). Different immunomodulatory effects associated with sub-micrometer particles in ambient air from rural, urban and industrial areas. *Toxicology,* vol. 257, no 3, p. 127 - 136.

[27] Wichmann, Fernando A. et al. (2009). Increased asthma and respiratory symptoms in children exposed to petrochemical pollution. *Journal of Allergy and Clinical Immunology*, vol. 123, no 3, p. 632 - 638.

[28] Mueller, A. et al. (2009). Risk assessment of airborne particles and volatile organic compounds from industrial areas. *Industrial pollution including oil spills.* Nova Science Publishers, New York, p. 101 - 144. ISBN: 978-1-60456-917-9.

[29] Gutiérrez, M. D. L. A., Palmieri, M. A., Giuliani, D. S., Colman Lerner, J. E., Maglione, G., Andrinolo, D. and Tasat, D. R. (2020). Monitoring human genotoxicity risk associated to urban and industrial Buenos Aires air pollution exposure. *Environmental Science and Pollution Research,* 1 - 12.

[30] Gutierrez, M. D. L. A., Colman Lerner, J. E., Giuliani, D. S., Porta, A. A. and Andrinolo, D. (2019). Comparative study of tear lipid composition in two human populations with different exposure to particulate matter in La Plata, Argentina. *Environ. Sci. Pollut. Res. Int.*, 6948 - 6956.

[31] US. EPA. *AP-42: Compilation of Air Emissions Factors*. Available https://www.epa.gov/air-emissions-factors-and-quantification/ap-42-compilation-air-emissions-factors.

[32] Lakes Environmental. *AERMOD View*. Available: https://www.weblakes.com/products/aermod/index.html.

[33] Mellado, Daniela. (2020). "*Estudio de zonas críticas de emisión con modelos de receptores en regiones urbanas con entornos industrializados*". PhD diss. Universidad Nacional de La Plata. ["*Study of critical emission zones using receptor models in urban regions with industrialized environments*". PhD diss., National University of La Plata].

[34] Guo, H., Lee, S. C., Chan, L. Y. and Li, W. M. (2004). Risk Assessment of Exposure to Volatile Organic Compounds in Different Indoor Environments. *Environmental Research*, 94, 57 - 66.

[35] Massolo, L., Rehwagen, M., Porta, A., Ronco, A., Herbarth, O. and Mueller, A. (2010). Indoor-outdoor distribution and risk assessment of volatile organic compounds in the atmosphere of industrial and urban areas. *Environmental Toxicology*, 25(4), 339 - 349.

[36] Sosa, B. S., Porta, A., Colman Lerner, J. E., Banda Noriega, R. and Massolo, L. (2017). Human health risk due to variations in PM 10 - PM 2.5 and associated PAHs levels. *Atmospheric Environment*, 160, 27 - 35.

In: Air Pollution: Effects and Dangers ISBN: 978-1-53619-544-6
Editor: J. Esteban Colman Lerner

Chapter 7

Source Apportionment of Ambient Volatile Organic Compounds (VOCs) in Gran La Plata, Argentina

Daniela Mellado*[1,*]*, Daniela Giuliani*[1]*,
Jorge E. Colman Lerner*[2]*, Pablo M. Demetrio*[1]
***and Erica Y. Sanchez*[1]**
[1]Centro de Investigaciones del Medioambiente (CONICET-UNLP), La Plata, Argentina
[2]Centro de Investigación y Desarrollo en Ciencias Aplicadas (CONICET-UNLP-CICPBA), La Plata, Argentina

Abstract

In many countries, mostly - but not exclusively - developing countries, only small and/or medium databases obtained by discrete monitoring are available to analyze some characteristics of atmospheric pollution. Ambient Volatile Organic Compounds are pollutants that affect public health, mainly in urban regions close to a petrochemical complex.

* Corresponding Author's Email: dmellado@quimica.unlp.edu.ar.

In this context, the source apportionment is addressed across to different statistical tools oriented to receptors, applied to small and medium databases obtained by discrete monitoring, so that valuable information is acquired in spite of the technical constraint above mentioned. Through this application, it is possible to characterize the relevant emission sources, estimating the profiles of the sources and their contributions by principal component analysis with multiple linear regression (PCA – MRL). Finally, the most probable areas where the pollutants under study originated are estimated from the back trajectories of the contributions of the present sources by means of a Nonparametric Trajectory Analysis (NTA). As a case study, the analysis of ambient Volatile Organic Compounds (VOCs) levels in Gran La Plata (Argentina) is addressed. This region lacks continuous monitoring and is also characterized by the presence of a petrochemical area and high traffic. This study shows that the highest concentrations are near the petrochemical area but that the data collected at the receptor sites can be explained with two main sources, emission associated to the industry (70.6%) and the traffic (28.0%). Then, the integration of PCA-MRL and NTA allowed access to graphic and numerical information of simple interpretation, useful for stakeholders. Furthermore, the estimated lifetime cancer risk (LCR) in Gran La Plata shows that conglomerate inhabitants are not probably exposed to this risk.

1. INTRODUCTION

The receptor models are chosen by stakeholders in environmental air quality management as essential tools for the generation of mitigation policies and the development of public health plans. These models allow us to interpret which of the present sources are the most significant in the area and to design mitigation strategies. The choice of the model to implement will depend on the available information, such as source types and profiles and emissions inventory, among others. For a correct performance, it is required that both the receptor site and the characterized species are representative of the region. Meteorological information and updated source profiles are considered relevant in this instance [1-5].

The receptor models can be classified into classic or hybrid, depending on whether or not they use meteorological information. Within the classic models, most of the studies approach source apportionment using CMB

(Chemical Mass Balance), PMF (Positive Matrix Factorization) and PCA (Principal component analysis) [4-6]. The first two were offered by the United States Environmental Protection Agency [7]. These models seek to solve the mass balance by component and to estimate the source profiles, except CMB, because in this model the source profiles are needed as input data. Some applications of classical models can be found in the literature [8-15]. To reduce the uncertainty of the solution, characteristic of multivariate models, a large volume of datasets and a background of the current sources are needed. This technique allows to obtain qualitative information about the composition (profiles) of the sources and their relative weight for the measurements observed (contribution) in a receptor site. Such approximate models are useful, but they require a detailed characterization of the scenario of study, mainly emissions inventories and extensive monitoring databases [16] and, also, they do not account for directions of origin of the pollutant. A complementary strategy to overcome some of these limitations is to apply hybrid models, which make use of trajectory and wind analysis. The main objective is georeference the potential emission sources and its impact in greater urban centers. These hybrid modelling use statistical procedures to identify the area with the highest probability of emission of pollutants, through the concentration data affected by the local meteorological information, in a given monitoring site. Some examples of hybrid models are HYSPLIT, Conditional Probability Function (CPF), Nonparametric Wind Regression (NWR), Trajectory Sector Analysis (TSA), Potential Source Contribution Function (PSCF), and Nonparametric Trajectory Analysis (NTA) [17-29].

Different authors have applied many of the above-mentioned models as air management tools, mainly with large-volume databases obtained from continuous monitoring. They are usually applied individually in air quality studies; however, each method separately offers an incomplete view of the scenario of study, which is often saved by access to supplementary information, such as emission inventories [30-34].

On the other hand, the integration of receptor models to estimate pollutants origin is a relatively unexplored field with small or medium databases resulting from discrete monitoring. This scenario often occurs in developing countries, where access to continuous monitoring systems is limited [35, 36]. This chapter seeks to overcome limitations associated with the lack of continuous monitoring stations, capitalizing the results of different receptor models as a joint analysis of the scenario of study. In this context, we present an integration of different statistical tools to analyze small or medium databases of contaminants of interest. As a result, the main sources of emissions that contribute to increase pollution levels in the monitoring sites are identified and characterized. In particular, this chapter will address the impact of the possible sources of emission of ambient Volatile Organic Compounds (VOCs) that contribute to the pollution levels of Gran La Plata, Argentina. Exposure to high concentrations of VOCs can have significant acute and chronic effects on the health of the population, particularly for people living in regions with high emission levels, such as vehicular traffic or the petrochemical industry [37-39].

2. Methodology

2.1. Description of Region under Study

Gran La Plata, formed by the counties of La Plata, Ensenada and Berisso, is the sixth most populated region of Argentina with 799,523 inhabitants [40]. The large urban town coexists with one of the most important petrochemical complexes in the country. This petrochemical complex is one of the most important in South America, with a refining capacity of 189,000 barrels per day [41] and other subsidiary industries producing aromatic compounds, aliphatic and petroleum coke, among others. Petroleum refineries and petrochemical plants are VOCs contributors, with emissions originated in production processes as well as during storage and in waste areas [42, 43]. Likewise, the region has an estimated vehicle fleet of 310,000 vehicles (represents 19% of the country's automotive fleet), and it is shown as another prominent source of VOCs to be studied [44-46].

Given the complexity of the urban center in coexistence with a significant industrial activity, and the lack of official communication of the state of air quality, highlight the need to study this region for the purposes of environmental risk [47, 48].

Previous studies carried out in the region showed higher levels of benzene, toluene, xylenes and ethylbenzene (BTEX) in the Industrial zone (I) compared to Urban (U) and Residential (R) areas [49] for the period 2000-2002. These levels were associated with vehicular emissions and the industrial sector. Furthermore, Colman Lerner et al. [42] compared the results obtained for the period 2007-2010 with those obtained in the period 2000-2002, showing in I significant decreases in the industrial emissions: 85% decrease in the emission of benzene, between 30-50% for the rest of the aromatics and 80% in the emission of alkanes, after a program, that began in 2007, to improve the control and reduction of leaks and diffuse emission, increasing the influence of vehicle emissions in urban and residential areas, due also to the increase in the number of vehicles. Regarding the levels of VOCs for the same period of time in the remaining sites, the reductions of the concentrations in general were lower with respect to I, observing a decrease of around 40-50% for toluene and benzene in U and an increase of approximately 50% for ethylbenzene and xylenes in R. These results, together with the preliminary emissions inventory obtained by the research center, allow us to point out that there are two types of relevant sources: a fixed source comprised of emissions from the processes associated with the petrochemical area, and a diffuse source associated with emissions from the vehicles.

2.2. Sample Collection and Chemical Analysis

The ambiental VOCs monitoring campaign was carried out in three defined sites in Gran La Plata. One of them, the industrial site (I) located in Ensenada (34°54′14″S, 57°55′34″W), near the petrochemical complex, provincial routes, and the encirclement of La Plata city. Another one, the urban site (U) situated in La Plata (34°56′42″S, 57°57′29″W), around the main administrative and commercial center. The third, the residential site (R), placed in Gonnet (34°53′23″S, 58°1′26″W), relatively remote from industrial influences, with low buildings and close to three highways (Figure 1). The monitoring was carried out during working days from 8 am to 4 pm, simultaneously in the three sites, in the period August-November 2016. As residential monitoring points are located on private properties, access to them was not necessarily unrestricted, and therefore the sampling period for the residential site was larger than stipulated; anyway, taking into account the complexity of the campaign and the difficulties previously mentioned, the data obtained were used in the analysis.

The availability of the number and type of species when approaching the study using a receptor model depends on the objectives of the study, the characteristics of the region and the human, technical and financial resources [4]. Then, out of all detected VOC compounds, nine species were selected for discussion with the method detection limit for individual species criteria. In this context, the species to analyze are the alkanes (hexane, heptane, nonane, undecane and dodecane) and the aromatic (benzene, toluene, xylenes and ethylbenzene).

Figure 1. Location of the Gran La Plata monitoring sites: Residential (R), Industrial (I) and Urban (U). The purple polygon is the petrochemical area.

During the monitoring, the air passage was forced at a constant flow rate of 0.2 L min^{-1} through Gillian sampling pumps (GilAir 3) over a period of 8 hours. The VOCs were retained in coconut charcoal tubes (Anasorb 226-17-1A). The samples were desorbed with dichloromethane (1 ml) by a 30-minute shake. Subsequently, they were analyzed by gas chromatography coupled to mass spectrometry Thermo Trace 1310 ISQ, in single ion mode (SIM). The method detection limit (*MDL*) was calculated by 5 control tubes, using as value the standard deviation multiplied by 3. Then, the obtained values of MDL are 0.3 ng m^{-3} for BTEX and 0.25 ng m^{-3} for alkanes. The recovery factor (r) for each analyte was determined in triplicate, the values being 0.984, 0.963, 0.910, 0.941, 0.922, 0.916, 0.902, 0.912 and 0.925 for benzene, toluene, ethylbenzene, xylenes, hexane, heptane, nonane, undecane and dodecane, respectively [50].

2.3. Source Apportionment

The concept of Source Apportionment (SA) is approached from the assumptions of both types of receptor models when coupling them. As a result, the relevant sources in a region are identified through their emission profiles (SA in classical model) and the regions of greatest impact (SA in hybrid model) of the contributions for these sources are georeferenced by means of a back-trajectory analysis.

2.3.1. Identification of Emission Sources

Principal component analysis (PCA) is a statistical technique frequently applied to classical receptor models due to its ability to decompose a multivariate data [51-53]. By the used the PCA with multiple linear regression (PCA–MLR), we seek to estimate the profiles and contributions of each source [52-54]. Reducing the dimension number of the data set through the correlation matrix, allows finding a small number of orthogonal factors that can explain the variance of the data [55, 56].

The PCA technique decomposes the data set (*X*) into two matrices: factor loadings (*F*) and factor scores (*G*). The relationship between the components explained by the PCA and the species is observed through the factor loadings (*F*). To maximize their interpretation in the analysis of air quality data, a varimax rotation is applied [56]. The rotated factorial loads include those related between the factorial components and the studied species. The higher the load value, the greater the relationship of the species with the factor. In accordance with previous works, it is established that values greater than 0.5 are indicators of a good relationship between the species and the factor studied [53, 57, 58].

On the other hand, the absolute scores of the factor obtained by this technique are used as independent variables for the application of MLR, using the total concentration of VOCs as the dependent variable. In this way, the regression coefficients obtained by MLR were used to estimate the mass contributions of each source identified according to the authors Thurston [56], de Paula et al. [42] and Viana et al. [59].

2.3.2. Georeferencing of Emission Sources

The most probable areas of provenance of each source (*p*) are estimated through a Nonparametric Trajectory Analysis (NTA). To visualize the zones of greatest probability of contribution of VOCs levels to a specific monitoring site, spatial non-parametric regression is applied. This model [60] consists of approximating the back-trajectories to the local scale through the kinematic model of uniform rectilinear motion. The advantage of the method consists in the association of each point of the trajectory with a spatial coordinate. The use of spatial coordinates is a modification of the original model by the authors of this chapter.

Given the proximity of the potential sources to the receptor sites, a short period of time is established in the estimation of the back-trajectories. Specifically, we considered 8 minutes of trajectory with a temporal resolution of 30 seconds. As the temporal resolution is less than the periodicity of the meteorological data (hourly), the wind components are approached by linear interpolation. Measurements associated with calm winds (winds less than 1 m s^{-1}) were excluded and the maximum velocity observed during the monitoring campaign was 7.3 m s^{-1}.

The back-trajectories are estimated for the contributions of the samples for each source (g'_p) from the receptor site, according to the equation 1:

$$(x, y) = \begin{cases} x_k(t_j) = \sum_{i=0}^{k} v_x(t_{j-i})\,\Delta t \\ y_k(t_j) = \sum_{i=0}^{k} v_y(t_{j-i})\,\Delta t\ ,\ k = 1, \ldots, N. \end{cases} \tag{1}$$

where the coordinates (x,y) correspond to the point on back trajectory at time j, Δt is the time step (temporal resolution), N is the number of steps taken backwards in time, v_x and v_y are the decomposition of wind in Cartesian coordinates estimated from geometric expressions for the *k-th* time. Each point of trajectory is associated to the contribution of factor (g'_{pk}) perceived at the receptor site.

Then, each point (x_i, y_i) generated in each back-trajectory, hereinafter referred to as coordinate z_i, is assembled to the contributions associated with the source p summarizing the information in a new set (z_i, g'_{pi}). The resulting trajectory (z_i) expressed in the Cartesian system with longitudinal units (km), was approximated in its corresponding geographic coordinates. A square Cartesian 20 km x 20 km grid with center in the geographic center of city (34°55′13″ S, 57°57′12″ W) was implemented, with a spatial resolution of 0.1 km. Then, for any point (z) defined in this grid, the contribution of the factor (p) to the measurement matrix $\widehat{G'}_p(z)$ (formed by the contribution of each factor p to the obtained samples) is estimated using a nonparametric regression of previously generated back-trajectories, using the Epanechnikov kernel [60, 61]. The expected value of $\widehat{G'}_p$ of the air has passed over grid point z is a result of a kernel regression by the equation 2:

$$\widehat{G'}_p\,(z) = \frac{\sum_i K(\|z - z_i\|/h)\, g'_{pi}}{\sum_i K(\|z - z_i\|/h)} \tag{2}$$

where h is the smoothing parameter obtained by cross validation. It should be noted that this estimation is influenced by the proximity to the receptor site and the parameters used in the regression. Finally, the location of the major sources contributing to concentrations of the pollutant at the monitoring site is spatially visualized.

2.4. Risk Analysis

VOCs contribute to the most serious air pollution problems; first, have proved active in the formation of smog and ground level ozone production. Secondly, there are several VOCs to which they can be classified as carcinogenic. Among the compounds studied, benzene is one of the compounds of interest because it is considered carcinogenic to humans (Group 1) by the International Agency for Research on Cancer (IARC) [62]. In order to assess the risk of chronic benzene exposure, the index called Lifetime Cancer Risk (LCR) is used (equation 3). This index considers the chronic daily intake (CDI), affected by the potential factor of the pollutant. The potential factor (*FP*) value for benzene is 0.029 $(mg\ kg^{-1}\ day)^{-1}$ [63, 64]. The CDI calculation is carried out from the equation 4:

$$LCR = FP \ x \ CDI \tag{3}$$

$$CDI = \frac{CC \ x \ IR \ x \ ED \ x \ EF \ x \ LE}{BW \ x \ ATL \ x \ NY} \tag{4}$$

where it is taken into account: average concentration of contaminant (CC [$mg\ m^{-3}$]), daily inhalation rate (IR [$m^3\ hours^{-1}$]), exposure duration (ED [$hours\ week^{-1}$]), exposure frequency (EF [$week\ year^{-1}$]), the highest duration of exposure (LE [year]), arithmetic mean in body weight (BW [Kg]), the average lifetime for carcinogens (ATL [day]) and the number of days per year exposed (NY [$day\ year^{-1}$]). Furthermore, the VOCs absorption factor for human was supposed to be 90%. The parameters (Table 1) of the equations are differentiated according to age group, classifying them into adults and children, taken from previous work done in Gran La Plata region [42].

Table 1. Parameters used to estimate the CDI

Parameter	Adults	Children
IR	0.833	0.417
ED	42	42
EF	48	48
LE	40	10
BW	70	38
ATL	70	70
NY	365	365

3. RESULTS

3.1. Data Analysis

The data analysis was approached in two sections; from individual species through their statistics and evaluating the relationships between them.

3.1.1. VOCs Levels

The medians observed of benzene (B), toluene (T), ethylbenzene (E) and xylenes (X) (m+p-xylen and o-xylene) in the industrial (I), urban (U), and residential (R) monitoring sites are 7.1 µg m^{-3}, 3.2 µg m^{-3} and 2.5 µg m^{-3}, respectively (Table 2). The same pattern observed for BTEX is evidenced for alkanes, being the highest levels in the industrial zone. The highest concentrations of VOCs were observed at the I site, followed by the U site and finally the R site. The results are similar to those observed in previous studies of the region [42, 49]. In addition, it is observed that alkanes provide the highest contributions to the concentration of VOCs agreeing with similar studies [65, 66].

Table 2. Statistics (µg m^{-3}) of contaminants during monitoring campaign: minimum (Min), median ($\tilde{X}$), mean ($\underline{X}$) and maximum (Max)

Species	R Site (N = 22)				U Site (N = 25)				I Site (N = 22)			
	Min	$\tilde{X}$	$\underline{X}$	Max	Min	$\tilde{X}$	$\underline{X}$	Max	Min	$\tilde{X}$	$\underline{X}$	Max
hexane	0.13	0.75	1.41	4.65	0.14	1.12	2.56	9.26	0.90	5.03	7.72	27.6
heptane	0.01	0.08	0.16	0.49	0.03	0.09	0.20	0.77	0.04	0.18	0.46	3.30
nonane	0.01	0.34	0.41	1.39	0.05	0.47	0.64	1.98	0.11	1.01	2.59	22.9
undecane	0.30	1.18	1.62	6.99	0.84	1.97	3.36	12.7	1.28	3.55	7.63	26.5
dodecane	0.13	1.04	1.82	7.97	0.93	2.14	3.29	10.6	1.38	4.90	7.33	29.2
B	4E-3	0.12	0.14	0.39	0.01	0.12	0.20	0.87	0.05	0.28	0.56	3.3
T	2E-3	0.65	1.36	5.24	0.05	1.19	1.72	7.96	0.16	2.49	3.30	14.4
E	0.02	0.07	0.09	0.32	0.01	0.10	0.13	0.34	0.02	0.23	0.38	3.3
X	0.05	1.07	1.37	5.45	0.21	1.42	2.00	6.24	0.30	3.23	6.66	44.2
BTEX	0.08	2.45	2.96	6.97	0.79	3.16	4.04	10.4	1.08	7.10	10.9	58.7
alkanes	0.66	2.14	3.28	9.99	1.57	4.61	6.66	19.9	4.18	9.34	16.3	47.6
BTEX + alkanes	1.33	4.57	6.24	15.2	2.50	9.82	10.7	21.4	5.81	17.6	27.5	105.

Table 3. Levels of BTEX species in Latin American countries and in a global context

Region	B	T	E	m+p-X	o- X	References
Gran La Plata (I+U+R)	0.2	1.3	0.1	0.7	0.7	*
La Plata, (I)	0.3	2.5	0.2	1.4	1.9	*
Medellin, Colombia	3.0-3.8	12.9-16.4	3.0-3.4	8.4-9.6	2.2-2.8	[70]
Mexico City, Mexico	5.9	37.9	5.0	14.9	5.9	[71]
Tlaxcala, Mexico	1.1-5.3	2.3-14.0	0.4-2.2	1.4-8.0	0.5-3.2	[67]
Curitiba, Brazil	0.6-3.0	2.8-5.9	0.2-1.6	1.3-3.5	0.3-1.3	[68]
Review of studies in the world	1.5-7.0	7.2-26.9	0.6-2.1	3.1-13.3	0.9-4.2	[69]

(*) median values obtained in this chapter.
(**) averaged values.

From the bibliographic search of international mean values of BTEX concentrations, the benzene levels for La Plata are lower than those observed in other countries (Table 3). With respect to the rest of the species, the levels observed near the petrochemical site (I) are similar to those observed near the industrial area of Tlaxcala, Mexico [67], and the petrochemical area of Curitiba, Brazil [68]. In particular, the levels of the species T, E, m+p-X, are close to the minimum values found in these cities, while the o-X presents higher values comparable to those found globally [69]. Furthermore, we can observe that the values found for Medellin [70] and Mexico City [71] present higher mean values for all species. Finally, the species that represents the largest mass contribution to BTEX is toluene.

3.1.2. VOCs Relations

A preliminary analysis of the emission sources of VOCs can be approached through species-specific relationships. The ratios of toluene (T) and benzene (B) species are widely studied and allow to account for the types of emissions involved [66, 72-74]. The literature provides enough information on these ratios to identify mainly the emissions from the sources such as vehicle exhaust, industrial use, and fuel combustion. In an urban environment, T/B ratios close to 2 indicate that pollutants are being emitted mainly by vehicle exhausts [75]. Other authors establish ratios associated with these urban emissions at values between 1.5-3.0 [76-79]. Values above this range suggest that emissions of VOCs, emissions come from industrial activity as the main source, without excluding the contribution of traffic in the region [32, 74]. On the other hand, low ratios of these species may come from emissions from coal combustion and biomass burning, as the main sources [80].

The complexity of the BTEX analysis lies in the vast abundance of them in industrial productions. In particular, benzene is present in a great variety of industrial processes, while in the urban environment, it is a component emitted from fuel combustion and vehicular exhaust. With respect to the rest of the substances, both toluene and ethylbenzene, and xylenes, are markers for solvents and building products [52, 81, 82].

The studied receptor sites presented higher values of ratio of T/B than those found in the literature. Then values were 9.8, 15.3 and 10.7 in I, U, and R, respectively. Therefore, the emissions observed cannot be attributed to vehicular emissions, it suggests the industrial activities predominantly [32, 75]. Consequently, in all cases there was a significant association (p-value < 0.05) of toluene and benzene species, at no site was the spearman coefficient observed to exceed 0.7 (0.44, 0.67 and 0.70 for the U, R, and I sites, respectively).

The study of the spearman correlation was extended for the rest of the species by contemplating the observations of the campaign. As shown in Figure 2, all species with significant association (p-value > 0.05) have a positive relationship with each other, which is consistent with primary emissions. The correlations between the species suggested that there are two related groups (spearman's coefficient > 0.5). A group defined by aromatics, heptane and nonane, and a second group, defined by hexane, undecane and dodecane.

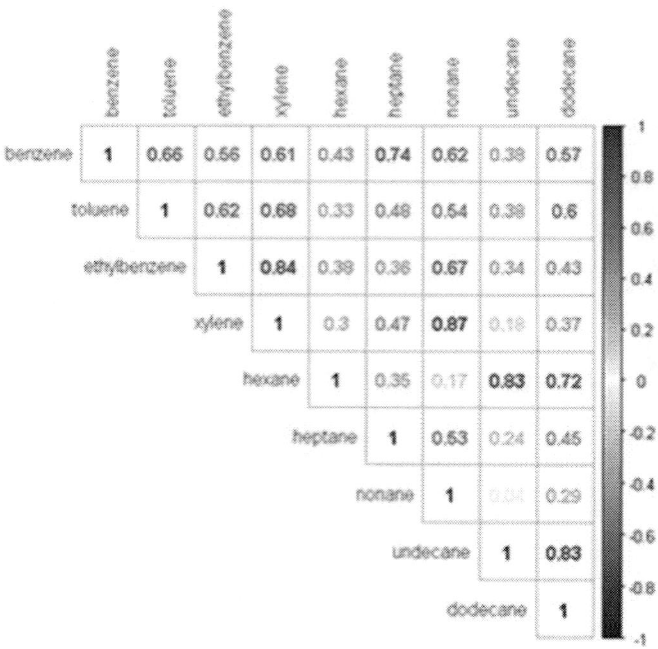

Figure 2. Association between observed species (spearman's coefficient, p-value < 0.05).

3.2. Source Apportionment

3.2.1. Identification of Emission Sources – Profiles and Contribution Sources

The Bartlett sphericity test (p-value < 0.05) and the Kaiser-Meyer-Olkin test (KMO > 0.7) were performed. PCA so it would be an efficient technique for reducing the dimensionality of the data set [58, 83].

Principal component analysis for the data set [69 x 9] is carried out using the correlation matrix. From the application, the eigenvalues and the variance explained by each of these are obtained (Table 4). Data reduction allows us to observe that there are two eigenvalues greater than unity, explaining 81.3% of the variance of the data. the first retains 58% of the variance of the data, while the second 23%, leaving 18.7% unexplained. In this context, the data are represented by two main identified components or factors, which are related to the possible emission sources of VOCs [42, 49, 64].

This result is consistent with the preliminary emissions inventory presented by the CIM (Environmental Research Center) and the works published by this group. Therefore, from previous studies, as mentioned in the characterization of the study site, the main sources of VOC emissions are those associated with the traffic and industrial activity of the petrochemical pole.

Table 4. PCA Summary: eigenvalues and explained variance calculated for principal components (PC) from data collected

	PC 1	PC 2	PC 3	PC 4	PC 5	PC 6	PC 7	PC 8	PC 9
Standard deviation	2.286	1.446	0.855	0.611	0.466	0.402	0.367	0.231	0.103
Proportion of Variance	0.581	0.232	0.081	0.041	0.024	0.018	0.015	0.006	0.001
Cumulative Proportion	0.581	0.813	0.894	0.936	0.960	0.978	0.993	0.999	1.000

Table 5. Factorial loadings obtained by the varimax rotation

	Factor 1	Factor 2
benzene	0.88	0.14
toluene	0.64	0.20
ethylbenzene	0.91	0.10
xylene	0.96	0.15
hexane	0.23	0.88
heptane	0.93	0.13
nonane	0.97	0.10
undecane	0.01	0.90
dodecane	0.16	0.90

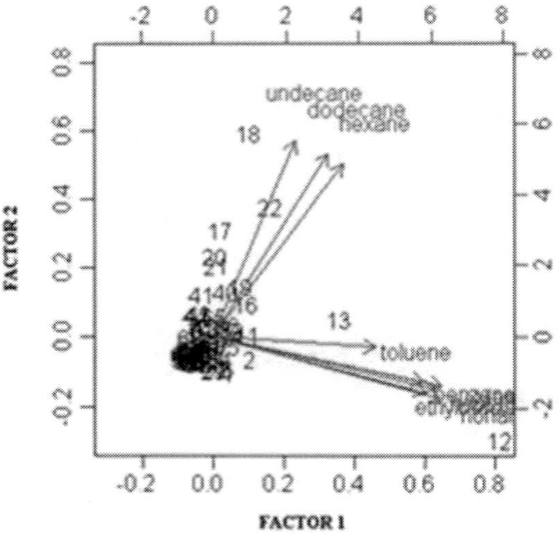

Figure 3. Biplot of multivariate data set resulting from the PCA.

Factorial loadings obtained in the PCA followed by the Varimax rotation account for the relationships between the principal components and the concentrations of the species. Rotated loads of the variables in each extracted factor give an idea of the identity of the sources (profile), so the higher the value of this load the greater the relationship with the possible source [53, 57, 83].

As shown in Table 5, the first factor is defined by benzene, toluene, ethylbenzene, xylene, heptane and nonane with values greater than 0.6. On the other hand, the second factor presents higher values for hexane, undecane and dodecane (greater than 0.8). The association pattern between the species agrees with what was observed when analyzing the spearman coefficients for the data set. Based on the literature, the first source is associated with emissions mainly developed by industrial activity and the second, with emissions related to vehicles [32, 66, 75].

This result can be observed graphically using a biplot that seeks to represent the set of observations and the variables of a matrix of multivariate data resulting from the PCA (Figure 3). In this graph, the dispersion that exists for the samples can be observed as the directions of the analyzed species. In the same direction and sense, the species present a higher correlation and can be described by the main components. Visually we can observe that there are two large patterns of species associations, the first quadrant is characterized by containing high concentrations of hexane, undecane and dodecane (observation n° 18) and the fourth quadrant is associated with high concentrations of the remaining species (observation n° 12).

Using the absolute factor scores obtained by PCA as independent variables, the dependent variable was the MLR run with the total elemental concentration (VOCs summary). The regression coefficients obtained by MLR were used to convert the absolute daily factors scores to estimate the contributions of the source mass [52, 53, 56, 59].

The application of MLR to the monitored sites allowed to observe high variability concerning the contribution of the sources. The first PC was dominant with 70.57% of the mass attributed, and the second one was 27.96%. Then, 1.46% of the mass contribution was represented for other sources that could not be identified with PCA.

3.2.2. Georeferencing of Emission Sources

Given the characteristics of the monitoring campaign, the value of the concentration obtained is a result of 8-hr integrated measurements. In order to make a correct application of hybrid models, it is necessary to have hourly measures to be analyzed together with local meteorological data, that is, the data must have the same temporal resolution. In this sense, we assumed that the integrated concentration obtained remained constant during the 8 hours of sampling. Similarly, the contributions of the factors (*G*) obtained by the PCA are considered constant and equal during the sampling period. Then the hourly contributions (G′) can be comparable and associated to the hourly meteorological data.

Regarding the meteorological data set, they were supplied by the Argentinean National Weather Service (Servicio Meteorológico Nacional, SMN). The wind speed is expressed in km hr^{-1} and the wind direction (°) in the azimuthal angle (this angle is measured clockwise from North). The measurements made in the station of the SMN are considered representative for the monitoring sites, according to the analysis carried out in the region studied [84]. The georeferenced representation that characterizes the NTA encourages one to use it as a tool for analysis of air quality observations, for its simple and visual output. In this study, an analysis of the back-trajectories of the PCA contributions for each factor was carried out.

To simplify the visualization, Figure 4 shows the representation of contributions (μg m^{-3}) of the first factor obtained from the previously applied tools, which are presumed corresponding to emissions related to industrial processes, while in Figure 5, the second factor corresponds to emissions associated with vehicular traffic.

Comparison of source contributions associated with emissions from industrial processes shows that the greatest contributions are presented near the Industrial receptor site, with respect to the remaining sites (U and R), given that they record higher levels in the chromatic scale. On the other hand, vehicular emissions have a greater spatial dispersion for the same contribution range (scale), as a consequence of the vehicular flow in the high-traffic routes included in the study region (Figure 1).

The contributions associated with emissions from industrial processes come from the petrochemical area (from 7.0 to 17.1 μg m^{-3}). This contribution could be observed in two polygons. One of them it′s from the Northwest to the North in a clockwise direction of the site with the higher contribution. The second one with lower contribution, it observed from the

North to the Northeast in a clockwise direction of the receptor industrial site (class: 7.0-10.4 μg m^{-3}).

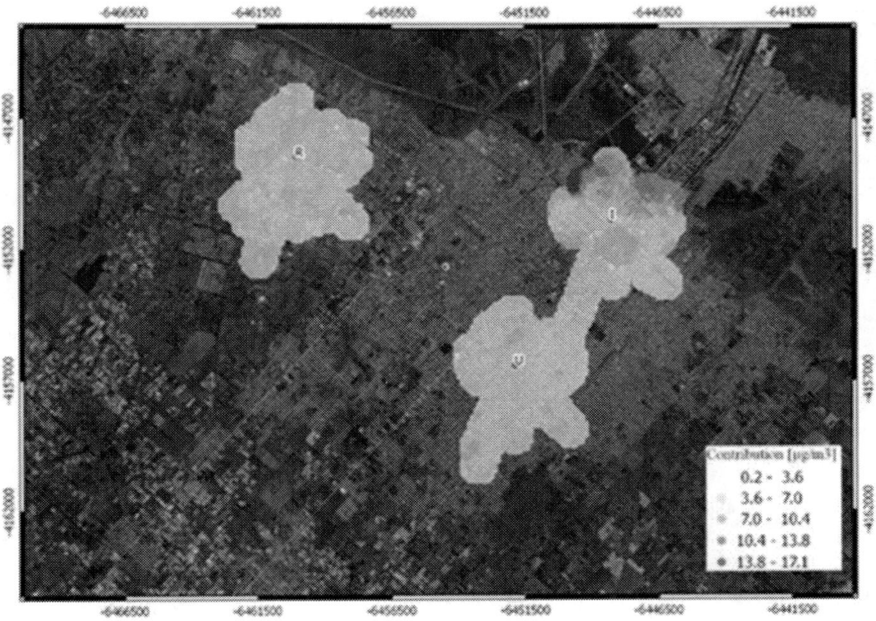

Figure 4. Spatial distribution of the contributions associated with emissions of industrial processes. I, R, and U correspond to the industrial, residential, and urban site, respectively.

Regarding the factor presumably associated with vehicle emissions, the contributions in proximity of the Industrial site (from 3.2 to 8.0 μg m^{-3}) are greater than those perceived in the Urban (class: 1.6-3.2) and Residential (class: 0.1-1.6) sites. This conclusion is visualized with the chromatic scales of Figure 5. In the Industrial site, the blue polygon (classes: from 3.2-4.8 to 6.4-8.0) projects larger contributions from the Northwest to the East of the site, in a clockwise direction. In this area are the following important roads: to the East, the Río de La Plata avenue that connects the city of La Plata with the Berisso District; to the Northeast, the provincial Route 215 (Vergara Road) that connects the city of la Plata with the Ensenada District; and to the Southeast, provincial Route 11 that connects the city of La Plata with the Atlantic Coast of Buenos Aires. All these roads are characterized by light and heavy vehicular traffic. In the Urban site, the highest contributions (blue polygon, class: 1.6-3.2) from the East and North are observed, associated with light and heavy vehicular traffic in the encirclement of La Plata, and from the commercial-administrative center of the city characterized mainly by light traffic, respectively. Analogously, contributions associated to the traffic cannot be appreciated in the vicinity of the residential site.

Figure 5. Spatial distribution of the contributions associated with traffic emissions. I, R, and U correspond to the industrial, residential, and urban site, respectively.

3.3. Risk Associated with Benzene as a Carcinogen

In order to quantitatively estimate the risk associated with chronic exposure to benzene, the Lifetime Cancer Risk "LCR" value was determined in each region for adult and children's exposure (residential, urban and industrial). Under conservative criteria, CDI (Table 6) is calculated with the median of benzene concentrations at each monitored site. Since both residential and urban sites present the same statistic, it is then expressed as "Residential/Urban" to simplify reading. The median concentrations used for the calculation are 0.12 μg m^{-3} for Residential/Urban sites and 0.28 μg m^{-3} for Industrial site.

Table 6. CDI and LCR values calculated for the monitored sites

CDI	Adults	Children
Industrial	$9.5\ 10^{-6}$	$7.7\ 10^{-7}$
Residential / Urban	$4.1\ 10^{-6}$	$3.3\ 10^{-7}$
LCR	Adults	Children
Industrial	$2.7\ 10^{-7}$	$2.2\ 10^{-8}$
Residential / Urban	$1.2\ 10^{-7}$	$9.6\ 10^{-9}$

The WHO considers LCR value within the range 10^{-5} and 10^{-6} as "acceptable," while the U.S. EPA recommends a CDI value lower than 10^{-6} [63, 64]. Taking into account these recommendations, the Gran La Plata would have acceptable values according to both U.S. EPA and WHO (Table 6).

The results (Table 6) show that the risk of exposure to benzene by children and adults meet the values recommended by U.S. EPA and WHO for the three regions studied, with CDI values in older adults in 1 order of magnitude than in children.

Although the values found are acceptable, it must be taken into account that the real risk is associated with the exposure of multiple pollutants (VOCs) that have not been taken into account for this study, such as styrene and tetrachlorethylene [42, 49, 54], which could raise the values of LCR in each study area.

Conclusion

In countries where the resources for air quality monitoring are limited and discrete monitoring systems are available as basic equipment, the experimental costs limit the number of measurements that can be obtained in a monitoring campaign. In this regard, this chapter shows the strengths of the integration of receptor models for the processing of small or medium databases. Simple, visual and low-cost computational outputs are useful for information processing and to stakeholders. We believe that when working with small database and a limited set of pollutants resultants of monitoring campaigns, integration of these tools is very useful because it allows to obtain information about the emission sources that describe the data in the receptor sites (quantity, characteristics and location).

In the first instance, a statistical analysis of the studied species not only allows comparative evaluation with historical data in the region, but also allows defining trends. Then, we were able to observe that the selected sites maintained the same behavior as in previous studies [42, 49, 54, 85-89]. In the international context, the level of benzene found was lower in comparison with other countries, while the rest of BTEX presented comparable values, with average levels observed near regions with industrial and petrochemical activity. In addition, these benzene levels are acceptable in terms of population health under the criteria of "lifetime cancer risk" according to the U.S. EPA and WHO. On the other hand, when working with VOCs, the relation of toluene and benzene was a useful tool to assess the influence of traffic in the region, that is, emissions related to traffic are predominant at a receptor site.

The PCA as receptor model, we were able to identify the potential emission sources from petrochemical area and vehicle fleet, being the first ones with the greatest contribution to the data set. Complementing with the NTA hybrid model, applied to the contributions of the associated samples affected by source, was able to represent the areas of origin of their contributions. Projecting the contributions of each of these sources allowed us to confirm the VOCs potential sources, which have been considered in the hypothesis of this case of study. The latter is useful for working spatially on a local scale, detecting which are the areas that contribute the most to the measurements observed at the receptor site.

Supported by the previous studies that our work group has been carrying out for 18 years [42, 49, 54, 85-89] given the lack of emission inventories in the region, we can affirm the existence of industry and vehicle fleet as relevant emission sources in the region. In our study, the integral use of receptor-oriented tools allowed the corroboration of the hypothesis that there are two types of main sources that contribute to our receptor sites, as well as the location of the critical emission areas.

If we had analyzed the database of this chapter with one of the receptor models applied, without considering the results obtained by the others, the description reached would have been incomplete, probably confusing and not useful to a stakeholder. The integration of the models has allowed to overcome the limitations of each of them with the strengths of the others [15, 23-24, 90]. Finally, let us remark that these conclusions could not have been reached through the individual and isolated application of the tools analyzed in this chapter, highlighting the combined use.

ACKNOWLEDGMENTS

The authors would like to thank the National Agency for Scientific and Technical Promotion (ANPCyT), the National Scientific and Technical Research Council (CONICET), the National University of La Plata and the Institute of Scientific and Technical Research for Defense (CITEDEF), which have financed this work. The National Meteorological Service (SMN) is also gratefully acknowledged for providing the meteorological information. We also want to extend special thanks to Dr. Pablo M. Jacovkis for accompanying this work.

REFERENCES

[1] Watson, John G., et al. (2008). "Source apportionment: findings from the US supersites program." *Journal of the Air & Waste Management Association* 58.2: 265-288.

[2] Behrentz, Eduardo, et al. (2009). "Elementos técnicos del plan decenal de descontaminación de Bogotá." *Secretaría Distrital de*

Ambiente, Universidad de los Andes, Alcaldía Mayor de Bogotá. [Technical elements of the ten-year plan for decontamination of Bogotá, *District Secretary for the Environment, University of the Andes, Bogotá Mayor's Office*].

[3] World Health Organization, WHO. (2002). "*Air quality guidelines for Europe*."

[4] Belis, Claudio A., et al. 2014. "European guide on air pollution source apportionment with receptor models." *European Commission, Joint Research Centre. Institute for Environment and Sustainability.*

[5] Hopke, Philip K. (2016). "Review of receptor modeling methods for source apportionment." *Journal of the Air & Waste Management Association* 66.3: 237-259.

[6] Taiwo, Adewale M., Roy M. Harrison, and Zongbo Shi. (2014). "A review of receptor modelling of industrially emitted particulate matter." *Atmospheric environment* 97: 109-120.

[7] United States Environmental Protection Agency, US EPA. "*Air Pollutant Receptor Modeling*." Accessed November 17. https://www.epa.gov/scram/air-pollutant-receptor-modeling.

[8] Lin, Ching-Ho, et al. (2004). "A method for locating influential pollution sources and estimating their contributions." *Environmental Modeling & Assessment* 9.2: 129-136.

[9] Tasic, Mirjana, et al. (2010). "Characteristics and application of receptor models to the atmospheric aerosols research." *Air Quality*. IntechOpen.

[10] Ethirajan, R., and S. Mohan. (2012). "Comparative evaluation of VOC source profiles developed by PMF and UNMIX models." *International Journal of Environmental Science and Development* 3.5: 450.

[11] Shi, Guo-Liang, et al. (2014). "A comparison of multiple combined models for source apportionment, including the PCA/MLR-CMB, Unmix-CMB and PMF-CMB models." *Aerosol and Air Quality Research* 14.7: 2040-2050.

[12] Urbančok, Dejan, Anthony JR Payne, and Richard D. Webster. (2017). "Regional transport, source apportionment and health impact

of PM10 bound polycyclic aromatic hydrocarbons in Singapore's atmosphere." *Environmental Pollution* 229: 984-993.

[13] Błaszczak, Barbara. (2018). "The Use of Principal Component Analysis for Source Identification of PM2.5 from Selected Urban and Regional Background Sites in Poland." *E3S Web of Conferences*. Vol. 28. EDP Sciences.

[14] Abbasi, Sajjad, and Behnam Keshavarzi. (2019). "Source identification of total petroleum hydrocarbons and polycyclic aromatic hydrocarbons in PM10 and street dust of a hot spot for petrochemical production: Asaluyeh County, Iran." *Sustainable cities and society* 45: 214-230.

[15] Rodelas, Roger Roig, et al. (2019). "Real-time assessment of wintertime organic aerosol characteristics and sources at a suburban site in northern France." *Atmospheric Environment* 203: 48-61.

[16] Kanabkaew, Thongchai, and Nguyen Thi Kim Oanh. (2011). "Development of spatial and temporal emission inventory for crop residue field burning." *Environmental Modeling & Assessment* 16.5: 453-464.

[17] Kim, Eugene, et al. (2004). "Analysis of ambient particle size distributions using unmix and positive matrix factorization." *Environmental science & technology* 38.1: 202-209.

[18] Hoque, Raza Rafiqul, et al. (2008). "Spatial and temporal variation of BTEX in the urban atmosphere of Delhi, India." *Science of the total environment* 392.1: 30-40.

[19] Chiou, Paul, et al. (2009). "Atmospheric aerosol over a southeastern region of Texas: chemical composition and possible sources." *Environmental Modeling & Assessment* 14.3: 333-350.

[20] Li, Mengmeng, et al. (2012). "Analysis of the transport pathways and potential sources of PM10 in Shanghai based on three methods." *Science of the Total Environment* 414: 525-534.

[21] Kumar, G. Praveen, and S. Palanivelraja. (2014). "Dispersion modelling of SO2 emission from a coal fired thermal power plant in Dadri, Uttar Pradesh." *International Journal of Engineering and Advanced Technology* 3.6: 237-241.

[22] Stojić, A., et al. (2016). "Comprehensive analysis of PM10 in Belgrade urban area on the basis of long-term measurements." *Environmental Science and Pollution Research* 23.11: 10722-10732.

[23] Squizzato, Stefania, et al. (2017). "Urban air quality in a mid-size city-PM2.5 composition, sources and identification of impact areas: From local to long range contributions." *Atmospheric Research* 186: 51-62.

[24] Argyropoulos, Georgios, et al. (2017). "Source apportionment of PM10 and PM2.5 in major urban Greek agglomerations using a hybrid source-receptor modeling process." *Science of the Total Environment* 601: 906-917.

[25] Han, Qinkai, et al. (2017). "Non-parametric hybrid models for wind speed forecasting." *Energy Conversion and Management* 148: 554-568.

[26] Jeong, Ukkyo, et al. (2017). "Assessing the effect of long-range pollutant transportation on air quality in Seoul using the conditional potential source contribution function method." *Atmospheric Environment* 150: 33-44.

[27] Henry, Ronald C., Subha Mohan, and Shahram Yazdani. (2019). "Estimating potential air quality impact of airports on children attending the surrounding schools." *Atmospheric Environment* 212: 128-135.

[28] Wei, Peng, et al. (2019). "Solar-powered air quality monitor applied under subtropical conditions in Hong Kong: Performance evaluation and application for pollution source tracking." *Atmospheric Environment* 214: 116825.

[29] Draxler, R. R., and G. D. Rolph. (2010). "*HYSPLIT* (*HYbrid Single-Particle Lagrangian Integrated Trajectory*) *model access via NOAA ARL READY website* (http://ready.arl.noaa.gov/HYSPLIT.php). *NOAA Air Resources Laboratory*." Silver Spring, MD 25.

[30] Uria-Tellaetxe, Iratxe, and David C. Carslaw. (2014). "Conditional bivariate probability function for source identification." *Environmental modelling & software* 59: 1-9.

[31] Cheng, Hanqi, Mitchell J. Small, and Natalie J. Pekney. (2015). "Application of nonparametric regression and statistical testing to identify the impact of oil and natural gas development on local air quality." *Atmospheric Environment* 119: 381-392.

[32] Zhang, Zhao Yang, Man Sing Wong, and Kwon Ho Lee. (2015). "Estimation of potential source regions of PM2.5 in Beijing using backward trajectories." *Atmospheric Pollution Research* 6.1: 173-177.

[33] Cahill, Thomas A., et al. (2016). "Transition metals in coarse, fine, very fine and ultra-fine particles from an interstate highway transect near Detroit." *Atmospheric Environment* 145: 158-175.

[34] Grange, Stuart K., Alastair C. Lewis, and David C. Carslaw. (2016). "Source apportionment advances using polar plots of bivariate correlation and regression statistics." *Atmospheric Environment* 145: 128-134.

[35] World Health Organization, WHO. "*Database on source apportionment studies for particulate matter in the air (PM10 and PM2.5)*." Accessed August 19. https://www.who.int/quantifying_ehimpacts/global/source_apport/en/.

[36] IQAir Air-Visual. (2018). "*World Air Quality Report: Region and City PM2.5 Ranking*." Accessed November 19. https://www.iqair.com/world-most-polluted-cities/world-air-quality-report-2018-en.pdf.

[37] World Health Organization. (2006). *Principles for evaluating health risks in children associated with exposure to chemicals*. Geneva: World Health Organization.

[38] Payne-Sturges, Devon, and Gilbert C. Gee. (2006). "National environmental health measures for minority and low-income populations: tracking social disparities in environmental health." *Environmental Research* 102.2: 154-171.

[39] Hashim, Dana, and Paolo Boffetta. (2014). "Occupational and environmental exposures and cancers in developing countries." *Annals of global health* 80.5: 393-411.

[40] INDEC. (2010). "Población total por sexo e índice de masculinidad, según edad en años simples y grupos quinquenales de edad." ["*Total population by sex and masculinity index, according to age in simple years and five-year age groups.*"]. Accessed August 17. http://www.indec.gov.ar/.

[41] *YPF. Downstream.*" Accessed January 18. https://www.ypf.com/LaCompania/Paginas/downstream.aspx.

[42] Colman Lerner, Jorge Esteban, et al. (2014). "Improvement of health risk factors after reduction of VOC concentrations in industrial and urban areas." *Environmental Science and Pollution Research* 21.16: 9676-9688.

[43] United States Environmental Protection Agency, US EPA. "*Manual AP-42: Compilation of Air Emissions Factors.*" Accessed February 18. https://www.epa.gov/air-emissions-factors-and-quantification/ap-42-compilation-air-emissions-factors.

[44] Franco, Juan Felipe, et al. (2015). "Characterization and source identification of VOC species in Bogotá, Colombia." *Atmósfera* 28.1: 1-11.

[45] Saeaw, Neungrothai, and Sarawut Thepanondh. (2015). "Source apportionment analysis of airborne VOCs using positive matrix factorization in industrial and urban areas in Thailand." *Atmospheric Pollution Research* 6.4: 644-650.

[46] Wang, Gang, et al. (2016). "Characteristics and source apportionment of VOCs in the suburban area of Beijing, China." *Atmospheric Pollution Research* 7.4: 711-724.

[47] Zhu, Tong, et al. (2012). "Impacts of megacities on air pollution and climate." *WMO, Geneva.*

[48] Wang, Shaojian, et al. (2020). "Strategizing the relation between urbanization and air pollution: Empirical evidence from global countries." *Journal of Cleaner Production* 243: 118615.

[49] Massolo, Laura, et al. (2010). "Indoor–outdoor distribution and risk assessment of volatile organic compounds in the atmosphere of industrial and urban areas." *Environmental toxicology* 25.4: 339-349.

[50] Mellado, Daniela. (2020). "*Estudio de zonas críticas de emisión con modelos de receptores en regiones urbanas con entornos industrializados*." PhD diss. Universidad Nacional de La Plata. ["*Study of critical emission zones using receptor models in urban regions with industrialized environments*." PhD diss., National University of La Plata].

[51] Kothai, P., et al. (2008). "Source apportionment of coarse and fine particulate matter at Navi Mumbai, India." *Aerosol and Air Quality Research* 8.4: 423-436.

[52] Geng, Fuhai, et al. (2009). "Analysis of VOC emissions using PCA/APCS receptor model at city of Shanghai, China." *Journal of atmospheric chemistry* 62.3: 229-247.

[53] de Paula, Pedro Henrique M., et al. (2015). "Biomonitoring of metals for air pollution assessment using a hemiepiphyte herb (Struthanthus flexicaulis)." *Chemosphere* 138: 429-437.

[54] Gioda, Adriana, et al. (2013). "Chemical constituents in clouds and rainwater in the Puerto Rican rainforest: potential sources and seasonal drivers." *Atmospheric Environment* 68: 208-220.

[55] Hopke, Philip K. (1985). *Receptor modeling in environmental chemistry*. John Wiley & Sons.

[56] Thurston, George D., and John D. Spengler. (1985). "A quantitative assessment of source contributions to inhalable particulate matter pollution in metropolitan Boston." *Atmospheric Environment* 19.1: 9-25.

[57] Mateus, Vinícius Lionel, et al. (2013). "Study of the chemical composition of particulate matter from the Rio de Janeiro metropolitan region, Brazil, by inductively coupled plasma-mass spectrometry and optical emission spectrometry." *Spectrochimica Acta Part B: Atomic Spectroscopy* 86: 131-136.

[58] Meyers, Lawrence S., Glenn Gamst, and Anthony J. Guarino. (2016). *Applied multivariate research: Design and interpretation*. Sage publications.

[59] Viana, Marcelo, et al. (2008). "Inter-comparison of receptor models for PM source apportionment: case study in an industrial area." *Atmospheric Environment* 42.16: 3820-3832.

[60] Henry, Ronald, et al. (2009). "Source region identification using kernel smoothing." *Environmental science & technology* 43.11: 4090-4097.

[61] Härdle, Wolfgang. (1990). *Applied nonparametric regression*. No. 19. Cambridge university press.

[62] IARC. (2012). "A review of human carcinogens: Chemical agents and related occupations." *IARC monographs on the evaluation of carcinogenic risks to humans* 100: 111-144.

[63] ASTM D3686-95. (1995). "Standard Practice for Sampling Atmospheres to Collect Organic Compound Vapors (Activated Charcoal Tube Adsorption Method)." *ASTM International*, West Conshohocken, PA. https://www.astm.org/DATABASE.CART/HISTORICAL/D3686-95.htm.

[64] Colman Lerner, Jorge Esteban, et al. (2012). "Characterization and health risk assessment of VOCs in occupational environments in Buenos Aires, Argentina." *Atmospheric environment* 55: 440-447.

[65] Xiong, Ying, et al. (2020). "Ambient volatile organic compounds (VOCs) in two coastal cities in western Canada: Spatiotemporal variation, source apportionment, and health risk assessment." *Science of the Total Environment* 706: 135970.

[66] Zheng, Huang, et al. (2020). "Compositions, sources and health risks of ambient volatile organic compounds (VOCs) at a petrochemical industrial park along the Yangtze River." *Science of The Total Environment* 703: 135505.

[67] Godoi, Ricardo H. M., et al. (2013). "Healthy environment-indoor air quality of Brazilian elementary schools nearby petrochemical industry." *Science of the total environment* 463: 639-646.

[68] López-Vargas, Rocío, et al. (2018). "Oxidative stress index is increased in children exposed to industrial discharges and is inversely correlated with metabolite excretion of voc." *Environmental and molecular mutagenesis* 59.7: 639-652.

[69] Bolden, Ashley L., Kwiatkowski, Carol F. and Colborn, Theo. (2015). "New look at BTEX: are ambient levels a problem?." *Environmental science & technology* 49.9: 5261-5276.

[70] Gallego-Díez, Mary Luz, Correa-Ochoa, Mauricio Andrés, and Saldarriaga-Molina, Julio César. (2016). "Validation of a methodology to determine Benzene, Toluene, Ethylbenzene, and Xylenes concentration present in the air and adsorbed in activated charcoal passive samplers by GC/FID chromatography." *Research journal of the Faculty of Engineering at the University of Antioquia* 79: 138-149.

[71] Serrano-Trespalacios, Paulina I., Louise Ryan, and Spengler, John D. (2004). "Ambient, indoor and personal exposure relationships of volatile organic compounds in Mexico City Metropolitan Area." *Journal of Exposure Science & Environmental Epidemiology* 14.1: S118-S132.

[72] An, Junlin, et al. (2014). "Characteristics and source apportionment of VOCs measured in an industrial area of Nanjing, Yangtze River Delta, China." *Atmospheric Environment* 97: 206-214.

[73] Song, Mengdi, et al. (2018). "Source apportionment and secondary transformation of atmospheric nonmethane hydrocarbons in Chengdu, Southwest China." *Journal of Geophysical Research: Atmospheres* 123.17: 9741-9763.

[74] Li, Bowei, et al. (2019). "Characterization of VOCs and their related atmospheric processes in a central Chinese city during severe ozone pollution periods." *Atmospheric Chemistry & Physics* 19.1.

[75] Barletta, Barbara, et al. (2008). "Ambient mixing ratios of nonmethane hydrocarbons (NMHCs) in two major urban centers of the Pearl River Delta (PRD) region: Guangzhou and Dongguan." *Atmospheric Environment* 42.18: 4393-4408.

[76] Khoder, Mamdouh Ibraheim. (2007). "Ambient levels of volatile organic compounds in the atmosphere of Greater Cairo." *Atmospheric Environment* 41.3: 554-566.

[77] Wang, Shaobin, H. M. Ang, and Moses O. Tade. (2007). "Volatile organic compounds in indoor environment and photocatalytic oxidation: state of the art." *Environment international* 33.5: 694-705.

[78] Miller, Lindsay, et al. (2011). "Spatial variability and application of ratios between BTEX in two Canadian cities." *The Scientific World Journal* 11.

[79] Hui, Lirong, et al. (2018). "Characteristics, source apportionment and contribution of VOCs to ozone formation in Wuhan, Central China." *Atmospheric Environment* 192: 55-71.

[80] Tsai, Stella Manchun, et al. (2003). "Characterization of non-methane hydrocarbons emitted from various cookstoves used in China." *Environmental science & technology* 37.13: 2869-2877.

[81] Liu, Ying, et al. (2008). "Source profiles of volatile organic compounds (VOCs) measured in China: Part I." *Atmospheric Environment* 42.25: 6247-6260.

[82] Yuan, Bin, et al. (2010). "Source profiles of volatile organic compounds associated with solvent use in Beijing, China." *Atmospheric Environment* 44.15: 1919-1926.

[83] Garson, D. G. (2008). "*Factor analysis: Statnotes. North Carolina State University Public Administration Program.*" www2.chass.ncsu.edu/garson/pa765/factor.htm.

[84] Ratto, Gustavo. (2016). "*Estudio de parámetros ambientales utilizando técnicas espectroscópicas, datos meteorológicos y métodos estadísticos.*" PhD diss., Universidad Nacional de La Plata. ["*Study of environmental parameters using spectroscopic techniques, meteorological data and statistical methods.*" PhD diss., National University of La Plata].

[85] Wichmann, Fernando A., et al. (2009). "Increased asthma and respiratory symptoms in children exposed to petrochemical pollution." *Journal of Allergy and Clinical Immunology* 123.3: 632-638.

[86] Colman Lerner, Jorge Esteban, et al. (2018). "Characterization and cancer risk assessment of VOCs in home and school environments in

gran La Plata, Argentina." *Environmental Science and Pollution Research* 25.10: 10039-10048.

[87] Represa, Soledad, et al. (2016). "Aplicación de tecnologías de sistemas de información geográfica (SIG) para la estimación del riesgo por exposición a compuestos orgánicos volátiles (COVs)." *Congreso de la Sociedad de Análisis de Riesgo Latinoamericana* (SRA-LA), Brasil, Mayo. ["Application of geographic information systems (GIS) technologies to estimate risk from exposure to volatile organic compounds (VOCs)." Paper presented at meeting for *Latin American Risk Analysis Society* (SRA-LA), Brazil, May].

[88] Gutiérrez, María de los Angeles, et al. (2020). "Monitoring human genotoxicity risk associated to urban and industrial Buenos Aires air pollution exposure." *Environmental Science and Pollution Research*: 1-12.

[89] Gutiérrez, María de los Angeles, et al. (2019). "Comparative study of tear lipid composition in two human populations with different exposure to particulate matter in La Plata, Argentina." *Environmental Science and Pollution Research* 6.7: 6948-6956.

[90] Landis, Matthew S., et al. (2019). "Source apportionment of an epiphytic lichen biomonitor to elucidate the sources and spatial distribution of polycyclic aromatic hydrocarbons in the Athabasca Oil Sands Region, Alberta, Canada." *Science of the Total Environment* 654: 1241-1257.

In: Air Pollution: Effects and Dangers ISBN: 978-1-53619-544-6
Editor: J. Esteban Colman Lerner

Chapter 8

Regional Reference Values for Spirometric Parameters in Young People of La Plata, Province of Buenos Aires, Argentina

Myrian E. Aguilar *** and Andrés Porta***
Centro de Investigaciones del Medioambiente (CONICET-UNLP),
La Plata, Argentina

Abstract

It is recommended that each population has its own reference values for studies of respiratory function, due to the ethnic, age, social, geographical and climatic characteristics of the different regions. Previous studies from the National University of La Plata show high levels of particulate matter and volatile organic compounds in La Plata and in areas near the Ensenada Petrochemical area, which correlate chronic exposure to pollutants linked to vehicular traffic and the chemical and petrochemical industry with adverse effects on development and lung function. Based on these recommendations, and the risk factors presented for respiratory function in our region, it is proposed to establish values of

* Corresponding Author's Email: myriamaguilar@quimica.unlp.edu.ar.

forced spirometry parameters in adolescents from 13 to 17 years of age in the city of La Plata, following the recommendations of the American Thoracic Society y la European Respiratory Society (ATS/ERS) based on anthropometric variables. A cross-sectional, observational, and descriptive study was carried out, in which 171 spirometries were analyzed in healthy young people between 13 and 17 years of age selected in the city of La Plata during 2018. The correlation between variables was analyzed to identify predictive variables of spirometry. To establish the prediction equations for the spirometric parameters, regression models were adjusted according to sex. The highest correlation was found with height and weight. Predictive equations for forced vital capacity (FVC) and forced expiratory volume in the first second (FEV1) were developed, separately for gender. In conclusion, it has been possible to establish regional reference equations, which show differences with respect to those proposed by the international bibliography, confirming the importance of having our own estimated.

1. INTRODUCTION

Non-communicable diseases (NCDs), also known as chronic diseases, carry a growing burden of mortality and morbidity throughout the world because of a combination of genetic, physiological, environmental, and behavioral factors. They are one of the major health and development challenges of the 21st century. From a total of 56 million deaths occurred worldwide during 2012, 38 million were due to NCDs, principally cardiovascular diseases, cancer, and chronic respiratory diseases. Nearly three quarters of these NCD deaths (28 million) occurred in low- and middle-income countries. The number of NCD deaths has increased worldwide and in every region since 2000, when there were 31 million NCD deaths[1, 2.]

In 2020, if trends continue, NCDs will account for 73% of deaths and 60% of the burden of disease. This rapid increase represents the main health challenge for world development. While the threat of NCDs in

---

[1]Noncommunicable-diseases, WHO. https://www.who.int/news-room/fact-sheets/detail/ non communicable-diseases

[2]World Health Organization. Global status report on noncommunicable diseases 2014. Geneva: WHO, 2014. https.who.int/iris/handle/10665/148114

developed countries has long been recognized, the prevalence of these diseases in developing countries is of growing concern, particularly due to conditions of vulnerability resulting from the lack of quality services, unsatisfied basic needs and water, soil, and air pollution [2,3].

On the other hand, asthma is the most common chronic disease among children. According to WHO data, it affects about 235 million people in the world, and about 4 million people die annually from chronic respiratory diseases, ranking third among the causes of death globally, after cardiovascular diseases and cancer. Among them, chronic obstructive pulmonary disease (COPD) is a complex disease, underdiagnosed, with high morbidity and mortality. It is estimated that it was responsible, in 2012, for about 3 million deaths worldwide [2, 4].

According to publications from the World Health Organization (WHO), NCDs represent 78% of total deaths in Argentina[4]. Also, in this country, chronic respiratory diseases constitute one of the main causes of mortality, accounting for 9% of deaths, many of which occur at productive ages. In turn, chronic respiratory diseases cause three percent (3%) of all hospitalizations, leading to a significant loss of days of useful life and generating increasing health costs. The highest burden of prevalence and morbidity and mortality is attributable to obstructive diseases[5].

The First National Survey of Asthma Prevalence in Adults carried out in our country between 2014 and 2015, reported that 6 out of every 100 people had an asthma attack in the last 12 months or was taking asthma medication at the time of being surveyed; while 9 out of 100 have or have ever had physician-confirmed asthma. In Argentina, asthma causes more than 400 annual deaths and more than 15,000 hospitalizations per year in

---

[3] CARMEN. A Multifactorial Noncommunicable Disease Reduction Action Kit Initiative, PAHO, October 2003. http://www.paho.org/hq/index.php?option=com_docman&task=doc_view&gid=16166&Itemid=

[4] Noncommunicable diseases country profiles 2018. Argentina. https://www.who.int/nmh/countries/en/

[5] National Administration of Institutes and Laboratories of Health (INER), Ministry of Health - Argentina. "Mortalidad por Asma de 5 a 39 Años en Argentina". Enero 2015. http://www.msal.gov.ar/ent/images/stories/vigilancia/pdf/2015-02_Boletin-Actualizacion-ASMA-2013.pdf.

public hospitals throughout the country. Ten percent (10%, N = 44) of deaths from asthma occur in the population aged 5 to 39 years.

A recent WHO report[6] states that in 2012 about 7 million people died because of exposure to air pollution, doubling previous estimates and confirming that air pollution constitutes the most important environmental health risk in the world. More than half of this disease burden falls on the populations of developing countries. This conclusion more than duplicates previous estimates and confirms that air pollution alone is the most significant environmental health risk in the world today.

There is abundant evidence, in the international bibliography, that correlates chronic exposure to pollutants related to vehicular traffic and the chemical and petrochemical industry, such as particulate matter (PM), metals, polycyclic aromatic hydrocarbons (PAHs) and volatile organic compounds (VOCs), with adverse effects on lung development and function, in addition to increased respiratory morbidity, frequently expressed in terms of asthma or COPD diagnosis, or in increased mortality [1-5].

Previous studies developed in La Plata (Buenos Aires Province, Argentina) and its surroundings, from the CIM (Center for Environmental Research) of the National University of La Plata and CONICET (National Council for Scientific and Technical Research of Argentina), allowed characterizing the quality of outdoor air, finding high levels of particulate material (PM_{10} and $PM_{2.5}$), PAHs and VOCs, in addition to adverse effects on the respiratory function of the population. The city of La Plata, characterized by an important level of urbanization, has a traffic made up of some 350,000 registered automobiles, in addition to those used for the transport of passengers or loads of the most varied nature with diesel engines. On the other hand, the Petrochemical area is located about 10 km away from the city of La Plata, which includes the most important oil refinery in the country, which processes about 38,000 m^3 day^{-1} of crude. Nearby, there are other subsidiary industries that produce aromatic

[6]WHO (2014). http://www.who.int/mediacentre/news/releases/2014/air-pollution/es/

compounds (benzene, toluene, and xylenes), aliphatics (pentane, hexane, and heptane), maleic anhydride and petroleum coke, among others [5-13].

Based on the data obtained regarding air pollution in the region and the effects determined on respiratory function, added the importance or the importance of spirometry as an indicator of effects on the health of the population exposed to air pollutants, and due to the lack of regional reference values, it is proposed to construct reference equations of the respiratory function parameters for young people of the region.

The importance of having regional reference values lies in developing a greater interpretive capacity, since usually the spirometric parameters that are reported result from the comparison of the determined data with reference values obtained through epidemiological studies approved for this purpose by specialized institutions such as the American Thoracic Society (ATS) and the European Respiratory Society (ERS). These studies are frequently carried out in central countries, with sociocultural and environmental characteristics different from those of our region [14, 15].

2. Methodology

2.1. Spirometric Parameters

Spirometry is the technique that measures respiratory flows and volumes useful for the diagnosis and monitoring of respiratory pathologies, allowing the diagnosis and classification of lung conditions, determining their severity and eventual evolution. The spirometry can be subdivided into:

- *Simple spirometry:* the patient is asked to expel all the air from his lungs after a maximum inspiration for as long as he needs to do so.
- *Forced spirometry:* maneuver that records the maximum volume of air that a subject can move from maximum inspiration to full exhalation. At the same time, the maximum expired volume is

recorded, which is related to the duration of the maneuver, and the different flows are measured.

Spirometry is indicated in the medical clinic for the diagnosis of the following cases: diagnosis of patients with respiratory symptoms, for preoperative risk assessment; assessment of the response to certain drugs or diseases with pulmonary involvement. The usually measured parameters in spirometry used to evaluate obstruction or restrictions in lung function are:

- *Forced Vital Capacity (FVC)*: amount of air expelled in one breath, after inhaling, filling your lungs to maximum capacity. A reduction in its value shows restrictive diseases. It is measured in liters.
- *Forced Expiratory Volume in the first second (FEV1)*: volume of air expelled during forced exhalation, after maximal inspiration, in the first second. A low value indicates an obstructive disease. It is measured in liters. In general, the FEV1 corresponds to the maximum volume of air exhaled at one time. The most used are FEV3 and FEV6, at 3 and 6 seconds, respectively.
- *FEV1/FVC (%):* is used in the diagnosis of restrictive and obstructive lung diseases. It represents the proportion of a person's vital capacity that can be overcome in the first second of exhalation.
- *Forced Expiratory Flow (FEF)*: expresses a percentage of vital capacity. The FEF25% −75% is the average flow from the moment 25% of the FVC is exhaled to 75% of the exhaled FVC. Instantaneous Expiratory flows (FEFx %) refer to the flow when the corresponding percentage of FVC (x %) has been exhaled, FEF25%, FEF50% and FEF75% ($l\ s^{-1}$).
- *Maximum Expiratory Flow PEF ($l\ s^{-1}$)* is the maximal speed of airflow as the patient exhales obtained from the peak in the flow-volume curve.

Before interpreting the results of spirometry, it is necessary to check if it has been performed correctly. In this context, the acceptability criteria proposed by the regulations for forced spirometry recommended by SEPAR (Spanish Society of Pulmonology and Thoracic Surgery), ERS or ATS are used, performing a minimum of three maneuvers and a maximum of eight when these are not deemed adequate [14-18]. A spirometry will be acceptable when it presents:

- *Good start:* requires a retrograde extrapolation volume less than 0.15 L or 5% of the FVC (0.08 L). The shape of the volume/time curve (V/T) should show a sharp rise, while in the Flow/Volume curve (F/V) the maximum forced expiratory flow (PEF or maximum FEF) should be achieved in less than 120 m s or before exhaling 15% of the Forced Vital Capacity (FVC).
- *Good finish:* asymptotic curves and not perpendicular or abrupt, presenting a final plateau in which there is hardly any variation in the volume of expired air (flow less than 30ml s^{-1}).
- *Artifact-free curves:* during the course, the curves should not present steps that show the presence of cough, Valsalva maneuver (glottis closure); or air leaks between the mouthpiece and the lips.

An additional acceptance criterion is the repeatability of the maneuver, for which the difference between the two best FVC and the acceptable forced expiratory volume in the first second (FEV1) must be less than 0.15 L or less than 15%. As an essential criterion, the two best expirations of the three best acceptable curves should not vary from each other by more than 200 ml of FVC or FEV1 [15, 17].

Generally, Spirometric values are expressed graphically (V/T and F/V curves), numerically and in percent with respect to theoretical reference values. The numerical results are of three types: reference value of the parameters (for age, sex, weight, and height), the result obtained for each parameter in the patient, and the percentage of the mean values obtained in the test in relation to the reference ones.

For people younger than 65 years old and whose height is not extreme, spirometry is considered "normal" when it presents values higher than the lower limit of the confidence interval (LIN), that is, around 80% of the theoretical value of FEV1, FVC and 0.7 for the FEV1/FVC ratio. When this is not fulfilled, we can be in the presence of alterations. An adequate interpretation of the spirometric values allows classifying the alterations in ventilatory function in:

- *Obstructive pattern* (Figure 1a): any condition that affects the lumen of the airways (either due to its excessive production of mucus, inflammation or bronchoconstriction, among others), increasing the resistance of the airways or decreasing the elastic recoil pressure (emphysema) produced, therefore, reduction of the maximum air flow that can be expelled by the lung in relation to the maximum volume that can expel. It is characterized by decreased FEV1, moderately reduced or preserved FVC, reduced FVC/FVC, (below 70% or below the 5th percentile of the predicted value) [19]. Regarding the morphology of the F/V curve, a normal or reduced peak and a sharp fall with an upward concavity curvature can be observed in its descending loop [16-20].
- *Restrictive pattern* (Figure 1b). Any condition that affects the capacity of the lungs to expand to receive a normal amount of air, thus limiting inspiration. It is characterized by presenting decreased FVC, normal, increased or slightly decreased FEV1, FVC/FVC greater than 75% and even around 90%. The morphology of the flow/volume curve presents a prominent peak, showing a convex shape [21-23].
- *Mixed pattern* (Figure 1c): the presence of obstructive and restrictive indicators was combined, presenting with a decreased FVC, decreased FEV1 and the FEV1/FVC ratio also decreased. The F/V curve looks like a "miniature," but not of the normal curve, but of the obstructive one.

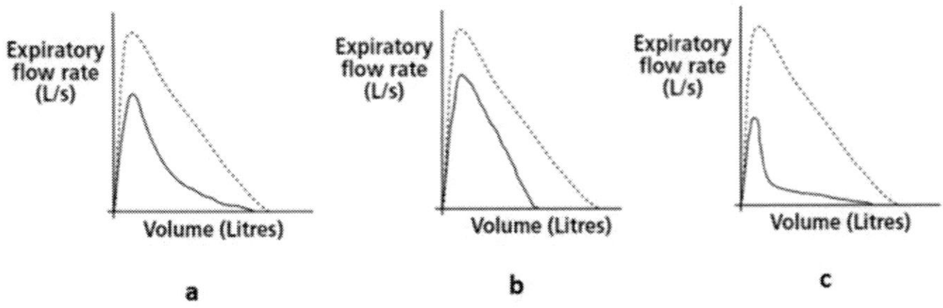

Figure 1. Obstructive pattern (a); Restrictive pattern (b), and Mixed pattern (c).

2.2. Methodological Development

2.2.1. Study Population

Based on the statistical data of the 2010 Argentinian census (INDEC), the potentially affected population amounts to 654,324 inhabitants of the La Plata district, and considering the age range under study between 13 and 17 years, it was constituted by 56,943 (28,033 women and 28,910 men).

In order to account for a representative youth population of the Municipality of La Plata, the National College of the city of La Plata (CNLP), dependent on the National University of La Plata, was selected, since it is attended by students from all over the region. Working in a school space allowed the articulation of the different methodological components of the research project: epidemiological surveys, spirometry, and space for exchange with young people, parents, and authorities [5, 24-26].

A prospective, cross-sectional, observational, and descriptive study was carried out, in which healthy young people were randomly selected from the different courses that attended the CNLP, aged between 13 and 17 years, for which they had the consent of the authorities of the CNLP. The research objectives and procedure of the pulmonary function test were explained to the participants, emphasizing that it was not an invasive study.

2.2.2. Epidemiological Surveys

To collect information on exposure and socio-economic factors, brief questionnaires were applied. They were developed based on international recommendations and criteria (ISSAC, GINA) and the team's own experience. These questionnaires include data on the participants, medical history (individual and family), characteristics of the area of residence, the dwelling, demographic data of the family group and the corresponding authorizations for conducting the survey and spirometry, with the approval of the Central Advisory Committee of Bioethics of the National University of La Plata. The forms for data loading were prepared using the Epi Info program (Epi7). All the information provided in the forms was strictly confidential.

The sample was selected from a random probability sampling, based on the population enrolled in the CNLP (approximately 1800 young people) and considering the following restrictions according to the bibliography [15-18]: type of error 1 (alpha) of the 5%, 95% confidence level (95% CI), standard deviation of 15% of the mean value of normal spirometry for the ages in the sample, precision 2.4 (confidence interval range closest to the deviation standard). The sample size calculates by Epidat 3.1 software, and estimating 20% of invalid spirometries [17], was 170 participating students.

2.2.3. Spirometries

Pulmonary function studies were carried out in CNLP students by measuring forced expiratory flows, total lung capacity and forced vital capacity, without a bronchodilator response. For this purpose, the forced spirometric tests were used, using a DATOSPIR-120, a portable spirometer appropriately calibrated, in accordance with international recommendations (ATS, ERS) [5, 13, 14, 19, 27-29].

The DATOSPIR-120 spirometer acquires and processes physical signals indicating a response proportional to lung functionality. The air incorporated during the maneuver causes an interruption of an infrared light beam whose sensor converts the received light into an electrical signal. The number of beam interruptions equals the volume of air entering the system. The spirometer has a control program responsible for ensuring that the treatment of spirometry signals complies with the applicable standards, according to the ATS and SEPAR criteria [15-17]. Furthermore, it has a microprocessor that calculates the values obtained by offering a diagnostic orientation using the Miller quadrant: a coordinate axis divided into quadrants according to the limit values of the variables studied (generally, FEV1/FVC and FVC). The analysis should be done in conjunction with the results [17]. The spirometric test was performed during school hours complying with the acceptability and reproducibility criteria of the maneuver according to the ATS and SEPAR protocols [5, 13, 14, 19, 26-29].

Initially, all the CNLP students were included, who voluntarily signed the informed consent/assent and whose parents expressed their willingness to accompany them with the assent/consent, through the epidemiological questionnaire, and healthy young people without respiratory, cardiovascular, and neuromuscular diseases were selected, or a history of lung surgeries. The volunteers had to meet the following characteristics: not require bronchodilator or beta-blockers medication, not be smokers, or ex-smokers, and not use illicit drugs.

Technical personnel, opportunely trained, carried out lung function studies by measuring forced expiratory flows, total lung capacity and forced vital capacity, without bronchodilator response, in accordance with international recommendations (ATS, ERS). In addition, each participant was measured (barefoot with the back resting on the measuring rod) and was weighed with light clothing.

Those students who, even having informed consent, could not perform spirometry due to lack of collaboration (poor effort, cough in three consecutive maneuvers, patient who did not understand the procedure) were excluded. Lack of repeatability of the maneuver: difference between the two best FVC and FEV1 greater than 0.15 liters; spirometer failure, either in the recording of any of the evaluated parameters or in the recording of the curves; missing in some of the anthropometric or spirometric parameters.

In each experimental unit (young participant) the anthropometric parameters were studied: age (years), weight (kilograms) and height (centimeters) and spirometric parameters (FVC (l), FEV1 (l), FEV1/FVC (%), FEF25-75% ($l\ s^{-1}$), PEF, FEV3, FEV6, FEF25, FEF50, FEF75), in addition personal information as address, school, date of birth. The data corresponding to each of these variables were used to create the database. All possible categories or modalities were defined for each qualitative variable, and for quantitative variables all possible values with their corresponding unit of measure.

2.2.4. Statistical Analysis

Statistical analysis was performed using the SPSS 20.0 program and the R programming language software version 3.4.1 for Windows. Table 1 shows the operational definitions corresponding to the selected variables:

- *Independent variables*: sex, weight, height, and age.
- *Dependent variables* (spirometric parameters): Forced Vital Capacity, FVC; Forced Expiratory Volume in 1 second, FEV1; FEV1/FVC ratio (%); Forced mid-expiratory flow between 25 and 75% of FVC, FEF25-75% ($l\ s^{-1}$).

Once it was decided which were the valid spirometries, preliminary analyzes were carried out to study the distributions and associations of the variables involved in the development of predictive models across the different sexes.

A descriptive analysis was performed by calculating the most appropriate statistics for the nature and scale of measurement considered. In this sense, the discrete variables were analyzed according to number of cases (N) and percentage (%), using absolute and relative frequencies, while the continuous variables were analyzed with measures of central tendency, such as the mean +/- standard deviation.

Table 1. Variables selected for statistical analysis

Variable	Definition	Dimension	Indicator	Scale
Participant	Youth (13 to 17 years old)	Health Status	Sick - Healthy	Nominal
Sex	Phenotypic characteristics that define the individual	Phenotype	Male Woman	Nominal
Weight	Measurement of the gravitational force exerted on the subject at study	Weight of the person	Weight in kilograms	Continuous numerical
Height	Quantification of the person's height	Height of person	Centimeters	Continuous numerical
Forced Vital Capacity (FVC)	Maximum volume expired after maximum inspiration	Liters measured with spirometer	Volume in liters	Numerical
Forced expiratory capacity in the first second (FEV1)	Peak volume expired during the first second from its onset, during a forced expiration started with full lung capacity	Liters measured in the first second of expiration by the spirometer	Volume in liters	Numerical
Forced expiratory flow after exhaling 25% and 75% of the FVC (FEF 25-75%)	Maximum flow reached when exhaled between 25% and 75% of the FVC	Measured by spirometer	Liters per second	Numerical
Study zone	Geographic space defined according to the predominant emission source	Air quality	Urban city Industrial	Nominal

The consistency of the data was evaluated to verify that they correspond only to the possible values of the corresponding variables, which implied identifying values outside the scale of the variable and those that, despite being within the measurement scale, they were conceptually impossible for the property measured in the study population within established knowledge. Several regression models were run, including logarithmic transformations of the spirometric parameters and quadratic terms of the predictor variables (age, weight, and height). The best-fitting model was analyzed for each sex.

To decide the final model, statistical criteria of regression analysis were used, such as the coefficient of determination (r^2), standard deviation of the residuals, and homogeneity of the variances.

3. Results and Discussion

A total of 294 young people who responded to the epidemiological survey were included and 240 of them gave their consent/assent to carry out the pulmonary function test. Spirometries were performed on 208 students, who had a minimum of 3 respiratory function tests. 37 spirometries were discarded because they presented some exclusion factor. Statistical analysis was performed with 171 spirometries that met the selection and acceptability criteria from ATS and SEPAR. Of the 171 spirometries, 45.03% (N = 77) corresponded to women and 54.97% (N = 94) to men (Table 2).

Table 2. Sex distribution by age, La Plata, 2018

Age (years)	Women (N=77)	Male (N=94)
13	12 (46.15%)	14 (53.85%)
14	21 (50%)	21 (50%)
15	30 (41.10%)	43 (58.90%)
16	11(47.85%)	12 (52.17%)
17	3 (42.86%)	4 (57.14%)
Total	77 (45.03%)	94 (54.97%)

Table 3. Anthropometric and respirometric descriptions by sex, 13 to 17 years old, La Plata, 2018

Women (N = 77)	Min	1st.Qu	Median	Average	3rd Qu.	Max	SD
Age (years)	13.35	14.50	15.16	15.18	15.73	17.24	1.03
Weight (kg)	39.00	49.00	54.00	55.70	58.00	105.00	10.74
Height (cm)	148.00	157.00	160.00	160.90	165.00	183.00	0.06
FVC (l)	2.17	2.78	3.17	3.19	3.53	5.36	0.58
FEV1 (l)	1.84	2.64	2.90	2.90	3.17	4.17	0.44
FEV1/FVC	63.96	86.63	93.25	91.38	97.69	100.00	7.85
$FEF_{25\%-75\%}$ (l/s)	1.44	3.06	3.50	3.44	3.92	6.10	3.98
FEF25%	2.64	4.44	5.23	5.08	5.68	8.88	1.06
FEF50%	1.87	3.13	3.70	3.65	4.20	6.39	0.77
FEF75%	0.88	1.86	2.22	2.22	2.56	4.15	0.59
Age (years)	13.00	14.00	15.00	14.69	15.00	17.00	0.95
Weight (kg)	44.00	56.25	61.50	64.06	69.75	103.00	11.84
Height (cm)	153.00	167.00	171.00	171.30	176.00	189.00	0.07
FVC (l)	2.41	3.54	4.02	4.06	4.57	5.81	0.68
FEV1 (l)	2.21	3.26	3.61	3.64	4.05	5.17	0.61
FEV1/FVC	65.46	86.86	89.73	89.92	94.94	100.00	6.73
$FEF_{25\%-75\%}$ (l/s)	1.79	3.57	4.14	4.85	4.74	7.18	0.95
FEF25%	2.64	5.46	6.12	6.25	7.08	9.13	1.89
FEF50%	1.88	3.75	4.43	4.44	5.11	7.59	1.00
FEF75%	1.20	2.06	2.46	2.67	3.21	5.65	0.88
Totals (N = 171)	Min	1st.Qu	Median	Average	3rd Qu.	Max	SD
Age (years)	13.35	14.50	15.16	15.18	15.73	17.24	0.99
Weight (kg)	44.00	56.25	61.50	64.06	69.75	103.00	12.07
Height (cm)	148.00	160.00	166.00	166.60	173.00	189.00	0.08
FVC (l)	2.17	3.13	3.58	3.67	4.16	5.81	0.77
FEV1 (l)	1.84	2.83	3.30	3.07	3.67	5.17	0.65
FEV1/FVC	63.96	86.74	91.42	90.58	96.31	100	7.27
$FEF_{25\%-75\%}$ (l/s)	1.44	3.21	3.78	3.85	4.37	7.81	0.94
FEF25%	2.64	4.97	5.67	5.73	6.44	9.13	1.27
FEF50%	1.87	3.45	4.06	4.09	4.65	7.59	0.99
FEF75%	0.88	1.94	2.34	2.47	2.90	5.65	0.79

FVC (l): Forced Vital Capacity, FEV1 (l): Forced Expiratory Volume in the first second, FEF25% - 75% (l/s): Forced meso-expiratory flow between 25 and 75% of the FVC, FEF (X%): Forced expiratory flow when X% of the FVC has been exhaled, PEF (l/s): Peak Expiratory Flow, FEV3 (l): Forced Expiratory Volume in the first three seconds, FEV6 (l): Expiratory Volume Forced at six seconds. SD: Standard deviation.

The descriptive analysis allowed characterizing the distribution of the population (Table 3) in study separated by gender, based on the anthropometric parameters (age, height, and weight) and spirometric parameters.

From the analysis of the variable age according to sex, it was observed that men presented a lower median value (15.16 > 15.00), not showing symmetry of the data in relation to it in both cases. The Weight variable showed a higher median for Men (61.50 > 54.00), besides the Women presented less dispersion in 25% respect of the lower data, while Men presented greater dispersion in 25% of the upper data. Regarding Height, it was observed that Women had a lower median (161.00 < 170.00). In both cases, they presented greater dispersion of the data in the range that represents 25% of the upper data.

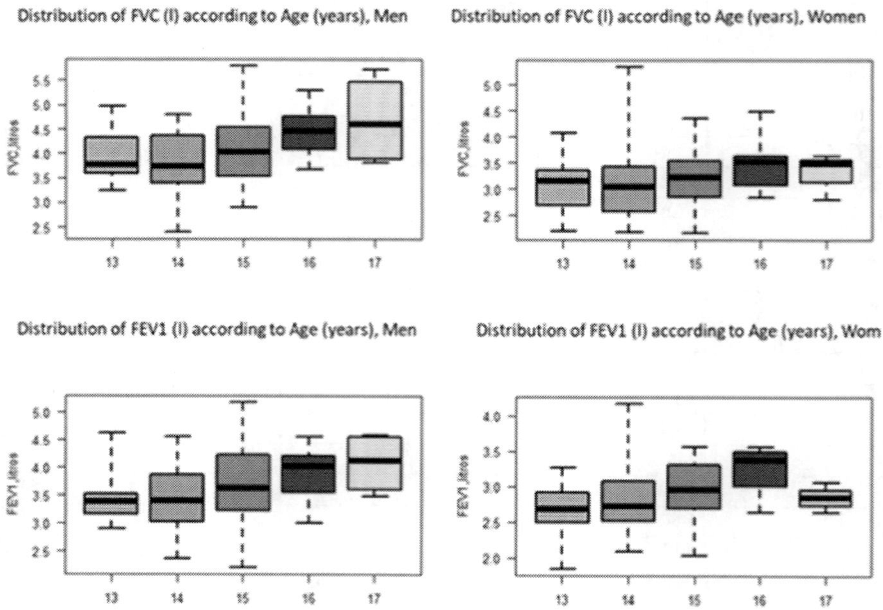

Figure 2. FVC (l) and FEV1 (l) by age (years), according to sex (La Plata, 2018).

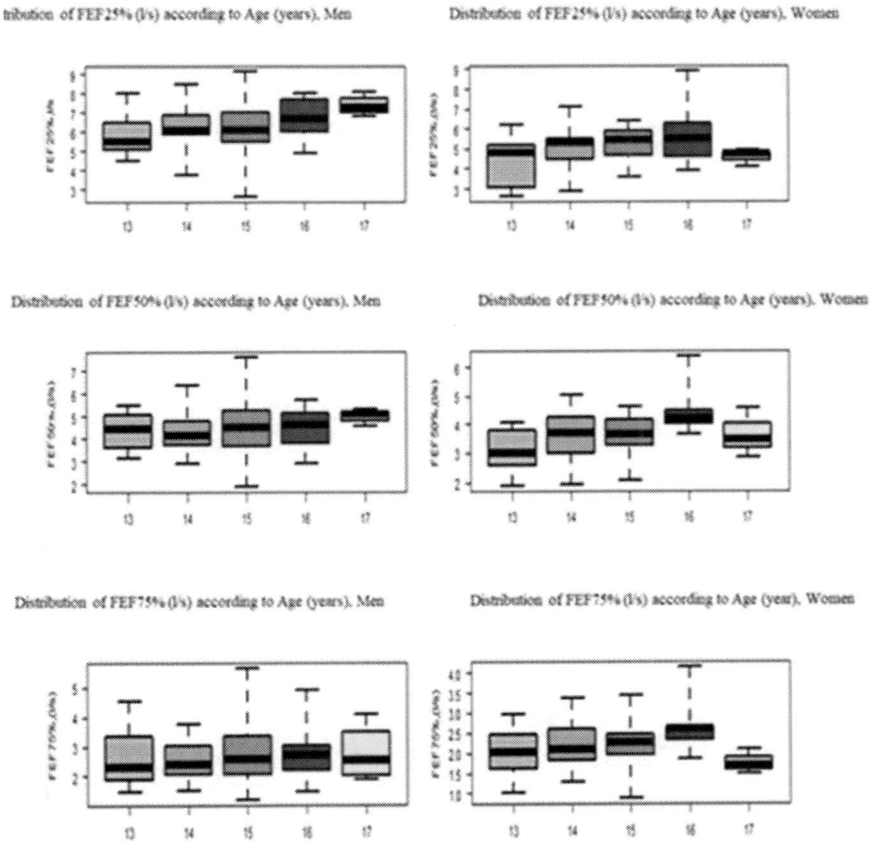

Figure 3. Other Spirometric Flows by age (years), according to sex. (La Plata, 2008).

From the analysis of FVC according to Sex (Table 3), a higher value was observed in the median for Men (4.06 > 3.19), as well as for the FEV1 parameter (3.64 > 2.90). Both groups presented similar variation of the data around the median (Figure 2).

The remaining spirometric parameters according to age are shown in in Figure 3.

To identify the predictor variables, the Pearson or Spearman correlation analysis was performed, as appropriate. The normality of the distribution was analyzed using the Kolmogorov-Smirnov test. The means were compared by means of a parametric (Student's t) or non-parametric (Wilcox) test according to each variable under study.

Table 4. Correlation analysis (La Plata, 2018)

Women (N = 77)		FVC	FEV_1	PEF	FEF $_{25-75}$	FEF/FVC
AGE	r=	0.1727	0.3084	0.3452	0.3734	0.1445
	p-value	0.133	< 0.05	< 0.05	< 0.05	0.2098
WEIGHT	r=	0.5532	0.4750	0.2619	0.2486	-0.1816
	p-value	< 0.05	< 0.05	< 0.05	< 0.05	0.1138
HEIGHT	r=	0.5189	0.5781	0.4281	0.2539	-0.0749
	p-value	< 0.05	< 0.05	< 0.05	< 0.05	0.5173
Men (N = 4)		FVC	FEV_1	PEF	FEF $_{25-75}$	FEF/FVC
AGE	r=	0.3077	0.2523	0.3162	0.0912	-0.1123
	p-value	< 0.05	< 0.05	< 0.05	0.3816	0.2811
WEIGHT	r=	0.4857	0.4623	0.3532	0.2682	-0.1257
	p-value	< 0.05	< 0.05	< 0.05	< 0.05	0.2272
HEIGHT	r=	0.5166	0.5530	0.3590	0.4000	0.0186
	p-value	< 0.05	< 0.05	< 0.05	< 0.05	0.8582

The normal distribution of the anthropometric and spirometric parameters was corroborated using the Kolmogorov-Smirnov test. The correlation coefficients (Pearson or Spearman) between spirometric (FVC, FEV1, FEF 25% -75%) and anthropometric parameters are shown in Table 4.

The regression model accepted for each spirometric parameter was the one that for each step, the change in the coefficient of determination (r^2) was significant, explaining the greater variation of the dependent variable. Subsequently, the standard error of the estimate (SEE) was analyzed, and estimates of the regression coefficients, the coefficient of determination (r^2), were obtained, while the goodness of fit was examined. The comparison with the international spirometric values was made using the equations proposed by Casan et al. [30], using the t-test for comparison of samples and the Levene test to analyze the equality of the variances.

Table 5 shows the results of the multiple regression analysis of the spirometric parameters, for adolescents aged 13 to 17 years from the CNLP Plata, after analyzing several regression models, including logarithmic transformations of the spirometric parameters and quadratic terms of the predictor variables using the stepwise procedure.

Table 5. Prediction equations of spirometric parameters according to Sex, 13 to 17 years old, La Plata, 2018

Men, 13 to 17 years old (N = 94)								
	α	β_1	β_2	β_3	β_4	r^2	*RSD*	*p-valor*
FVC(l)	-5.531455	-	0.099782	-0.0006152	0.033916	0.3359	0.5629	4.52E-05
FEV1(l)	-5.297178	-	0.0763619	-0.0004754	0.0354224	0.3534	0.4969	1.39E-05
Women, 13 to 17 years old (N = 77))								
	α	β_1	β_3	β_4	β_5	r^2	*RSD*	*p-valor*
FVC(l)	-5.010469	-	0.0761919	-0.0004526	0.0336479	0.3836	0.4649	9.29e-08
FEV1(l)	-3.86719	0.080334	0.0090687	-	313.5059	0.3926	0.3519	5.51E-05
* *Pulmonary Function Parameter* = $\alpha + \beta 1$ *Age* + $\beta 2$ *Weight* + $\beta 3$ *Weight*^{2} + $\beta 4$ *Height*								

FVC (l): Forced Vital Capacity, FEV1 (l): Forced Expiratory Volume in the first second, RSD: Residual standard error, r^2: coefficient of determination, p-value < 0.05 significantly different from zero.

The analysis of the proposed equation for the FVC parameter for the case of Men indicates that in our model the ordinate to the origin is worth -5.531455 (p-value < 0.05) with a p-value = 0.003221. The coefficient for the variable weight, that of the variable weight squared and height are also significantly (p-value < 0.05). From the ANOVA analysis, it is observed that the Weight variable largely explains the variability of the FVC values, while the Height variable is less explanatory.

The proposed equation for the FEV1 parameter for the case of Men: indicates that in our model the ordinate to the origin is -5.297178 and was significantly different from zero (p-value = 0.00146); as well as the coefficient for the variable Weight, Weight squared and Height, with a p-value = 0.03772, p-value = 0.06309, p-value = 2.81E-05 respectively. By ANOVA results was observed the Weight variable explains to a great extent (58.17%) the variability of the FEV1 values, followed of the variable Height (48.11%). both with a level of significance < 0.001. For its part, the Weight squared explains 15.19% with a significance level < 0.05.

In the case of women from 13 to 17 years old from the city of La Plata, the analysis of the equations indicates that: for the proposed model, the ordinate to the origin was -5.010469 and was significantly different from zero (p-value = 0.001090). The coefficient for the variable Weight, Weight squared, and Height are also significantly different from zero, with a p-value = 0.025200, p-value = 0.075811, p-value = 6.19E-04, respectively. From the ANOVA results, it is evident that the Weight variable explains to a great extent (50.41%) the variability of the FVC values. While that variability explains of the variables Height and squared Weight were 28.23% and 21.56%, respectively.

When analyzing the model proposed for the FEV1 variable for Women, we observe that the ordinate to the origin takes a value of -3.86719 and is significantly different from zero (p-value = 7.65E-04). The coefficients for the variable Age, Weight, Height are also significantly different from zero, with a p-value = 0.048137, p-value = 0.024722, p-value = 1.59E-05, respectively. The variable Height explains 26.48% the variability of the FEV1 values. While the Weight variable explains a little less (17.78%) and Age explains 14.16%.

Table 6 shows the mean values with their corresponding standard deviation calculated from the equations obtained in this study. In addition, they were compared with the equations by Casan et al. [30] (resulting from a multicenter study carried out in Barcelona with young people from 6 to 20 years of age), and the prediction equations obtained for Caucasian subjects under 20 years of age in the third National Health and Nutrition Examination Survey [31].

Table 6. Comparison of spirometric parameters from the equations developed and those proposed by Casan et al. [30] and NHANES III, according to sex

	Current study	NHANES III [31] caucasian	Casan et al. [30]
Male (N = 94)			
FVC(l)	4.059 ± 0.394	3.702 ± 0.906	4.184 ± 0.705
FEV1(l)	3.644 ± 0.361	2.788 ± 0.398	3.523 ± 0.521
Women (N = 77)			
FVC(l)	3.192 ± 0.363	3.689 ± 0.0.899	4.172 ± 0.694
FEV1(l)	2.897 ± 0.277	2.787± 0.395	3.516 ± 0.513

CONCLUSION

Spirometry is the technique that measures the respiratory flows and volumes useful for the diagnosis and monitoring of respiratory diseases, allowing the diagnosis and classification of pulmonary conditions, determining their severity and eventual evolution.

Given that spirometric values normally refer to a comparison with reference values obtained through epidemiological studies of countries with sociocultural and environmental characteristics different from that of our region, having equations with regional reference values will allow a greater interpretive capacity of spirometric surveys. As air quality changes in the study area, tests of greater sensitivity will be required to determine health effects. In this sense, obtaining regional reference values regarding spirometric parameters will allow an improved, more specific, and sensitive evaluation of the effects associated with exposure to air pollutants on lung function.

Multiple linear regression models were studied, according to Sex, with the independent variables Age, Height and Weight, for each lung function parameter: FVC, FEV1. Multiple regression models were fitted including quadratic terms in age, weight, and height. However, in some cases, these models did not show a better goodness of fit than the models that only incorporated non-quadratic terms, so the best fit was achieved using the quadratic term of the Weight variable for some of the respiratory parameters.

Although in some cases it has only been possible to correlate with an anthropometric parameter, this situation is common in publications on the specialty. And in this study, correlations have been found with up to the three parameters, in some cases.

It has been possible to calculate regional reference equations for the FVC and FEV1 parameters, with values lower than those proposed by Casan et al. [30], both for men and women between 13 and 17 years old in the city of La Plata. However, when comparing with the equations obtained in the Third National Health and Nutrition Examination Survey (NHANES III) for Caucasian subjects under 20 years of age, the values obtained show differences with respect to the equations proposed in the present study, which confirms the importance of having specific reference values for each region [31].

The results show that there are variations that justify the need to develop specific regional formulas. We have found a similar result for children aged 6 to 12 years, in 2016.

It would be useful in a future study to validate the prediction equations found, establish their reliability, relevance, and performance to detect respiratory functional alteration. In addition, a comparison with other reference equations such as those published by Hankinson et al. [31] and Knudson et al. [32] would be interesting.

Acknowledgments

The authors would like to thank the Ministry of Health of Argentina, the Scientific research Commission of the Buenos Aires Province, and the National University of La Plata. We also want to extend special thanks to Dr. Juan Ditondo for accompanying this work.

References

[1] WHO. (2013). *Review of evidence on health aspects of air pollution – REVIHAAP Project*. WHO European Centre for Environment and Health, Bonn, WHO 2013. 309 pp. www.euro.who.int/__data/assets/pdf_file/0004/193108/ REVIHAAP-Final-technical-report.

[2] Guarnieri M., JR. Balmes. (2014). Outdoor air pollution and asthma. *Lancet*, 383(9928): 1581–92.

[3] Couto Gomes E., G. Florida-James (2014). Lung Inflammation, Oxidative Stress and Air Pollution. In: *Lung Inflammation*, Kian Chung Ong, Ed. Chap. 1. http://www.Intechopen.com/books/

statistics/lung-inflammation/lung-inflammation-oxidative-stress-and-air-pollution.

[4] Romieu I., et al. 2012. Multicity study of air pollution and mortality in Latin America (the ESCALA study). *Research report Health Effects Institute*, 171: 5-86.

[5] Wichmann FA., et al. 2009. Increased asthma and respiratory symptoms in children exposed to petrochemical pollution. *J. Allergy and Clinical Immunology,* 12 (3): 632-638.

[6] Gutiérrez MA. 2016. Doctoral thesis. *Faculty of Exact Sciences*, UNLP. http://sedici.unlp.edu.ar/bitstream/handle/10915/52404.

[7] Massolo LA. 2004. Doctoral thesis. *Faculty of Exact Sciences*, UNLP. http://sedici.unlp.edu.ar/handle/10915/2267.

[8] Giuliani DS., 2020. Doctoral thesis. *Faculty of Exact Sciences*, UNLP. http://sedici.unlp.edu.ar/handle/10915/110949

[9] Mueller A, et al. 2009. Risk assessment of airborne particles and volatile organic compounds from industrial areas. *Industrial Pollution including Oil Spills*. H. Newbury & W. De Lorne, Ed., 101-41.

[10] Colman Lerner JE., et al. 2014. Improvement of health risk factors after reduction of VOC concentrations in industrial and urban areas, *Environmental Science and Pollution Research,* 21, 9676-9688.

[11] Colman Lerner JE, et al., 2012. Characterization and health risk assessment of VOCs in occupational environments in Buenos Aires, Argentina. *Atmospheric Environment*, 55: 440-447.

[12] Gutierrez MA., Colman Lerner JE., Giuliani DS., Porta A., Andrinolo D (2019). Comparative study of tear lipid composition in two human populations with different exposure to particulate matter in La Plata, Argentina. *Environmental Science and Pollution Research,* 26: 6948–6956.

[13] Colman Lerner JE, Morales A, Aguilar M, Orte M, Giulani D, Ditondo J, Dodero V, Massolo L, Sánchez EY, Matamoros N, Porta A, 2014. *Enviromental Impact II*, 181: 659-670.

[14] Sood A., et al. 2007. Effect of change of reference standard to NHANES III on interpretation of spirometric 'abnormality'.

International Journal of Chronic Obstructive Pulmonary Disease, 2(3): 361-67.

[15] Quanjer PH., et al., 2012. Multi-ethnic reference values for spirometry for the 3-95 year age range: The Global Lung Function 2012 Equations. *European Respiratory Journal*, 40(6): 1324-1343.

[16] Spirometry regulations (2013 revision). SEPAR. *Módulo 63* [*Module 63*]. ISBN: 978-84-940708-0-8.

[17] Crapo R., et al. 1990. Normal spirometric values in healthy Hispanic Americans. *Chest,* 98(6): 1435-39.

[18] Pereira CA, Sato T, Rodrigues SC. New reference values for forced spirometry in white adults in Brazil. *Brazilian Journal of Pulmonology,* 2007, 33(4): 397-406.

[19] Pakhale S., et al. 2009. Comparison of percent predicted and percentile values for pulmonary function test interpretation. *Can. Respir. J.,* 16(6): 189–193.

[20] González Villaescusa MC. 2009. Evaluación de las pruebas funcionales respiratorias más frecuentes. Espirometría forzada [Evaluation of the most frequent respiratory function tests. Forced spirometry]. *Avances en Diabetología*, 25: 455-58.

[21] Gutiérrez M., et al. 2014. Proposición de nuevas ecuaciones para calcular valores espirométricos de referencia en población chilena adulta [Proposal of new equations to calculate reference spirometric values in Chilean adult population]. SER. *Revista Médica de Chile*, 2014; 142: 143-152.

[22] Perez-Yarza EG., et al. 2009. Espirometría forzada en preescolares sanos bajo las recomendaciones de la ATS/ERS: estudio CANDELA [Forced spirometry in healthy preschoolers under the ATS / ERS recommendations: the CANDELA study]. *Anales de Pediatría,* 70(1): 3–11.

[23] Alvarez G., et al. 2004. Comparison of spirometric reference values in Chilean children., *Revista Médica de Chile*, 132: 1205-1210.

[24] Miller M., et al. 2005. General considerations for lung function testing. *European Respiratory Journal*, 26: 153–161.

[25] Pértegas Díaz S., Pita Fernández S. 2002. Determination of the master size for calculating the significance of the linear correlation coefficient. Clinical Epidemiology and Biostatistics Unit. Complexo Hospitalario Juan Canalejo. A Coruña. *Cadernos de Atención Primaria*; 9: 209-211.

[26] Multiple regression analysis. Chapter 4. *Computer Statistics: cases and examples with SPSS.* http://faculty.washington.edu/htamura/qm500/king/Ch04.pdf.

[27] Chapman RS. et al. 2003. Influences of asthma and household environment on lung function in children and adolescents: the third national health and nutrition examination survey. *American Journal of Epidemiol*, 158(2): 175-89.

[28] Miller MR et al. 2005. Standardization of spirometry. *European Respiratory Journal*, 26: 319-338.

[29] *Manual ESPIROMETRO DATOSPIR-110/120 511-800-MU1*. https://www.sibelmed.com/wp-content/uploads/2015/07/Manual_uso_espirometro_Datospir_110_120_es.pdf.

[30] Casan P., Roca J., Sanchis J. Spirometric response to a bronchodilator. Reference values for healthy children and adolescents. *Bulletin europeen de physiopathologie respiratoire,* 1983, 19(6): 567-9.

[31] Hankinson JL., et al. 1999. Spirometric reference values from a sample of the general U.S. population. *American Journal of Respiratory and Critical Care Medicine*, 159(1): 179-87.

[32] Knudson RJ. et al. 1983. Changes in the normal maximal expiratory flow-volume curve with growth and aging. *The American Review of Respiratory Disease*, 127(6): 725-34.

[33] Pellegrino R, Viegi G, Brusasco V, et al. Interpretative strategies for lung function test. *European Respiratory Journal*, 26(5): 948-968.

[34] Pérez-Padilla R, Valdivia G, López MV, et al. (2006). Spirometric reference values in 5 large Latin American cities for subjects 40 years of age or older. *Archivos de Bronconeumología* 42(7): 317-25.

[35] Rosental M, Bain S, Helms P, Deninson D, Bush A, Warner J. (1993). Lung function in white children aged 4 to 19 years: I-Spirometry. *Thorax* 48:794-802.

[36] F. González Barcalaa C, Cadarso Suárezb L, Valdés Cuadrado R, Leisc R, Cabanasc R, Tojo (2008). Reference values of respiratory function in children and adolescents (6-18 years) in Galicia. *Archivos de Bronconeumología*; 44(6):295-302.

[37] Morato MD, Rodríguez E, González Pérez-Yarza J. Emparanza Knörr A, Pérez Legorburu, Aguirre Conde A, Delgado Rubio (1999). Spirometric values in healthy children in an urban area of the Basque Autonomous Community. *Anales Españoles de Pediatría*. Vol. 51 Nº 1.

[38] Rodriguez Martines C, Sossa M, Falla S. (2005). Spirometry reference values in healthy children and adolescents in the city of Bogota. *Revista Colombiana de neumología,* 17, 3:153-163.

[39] Facchini F, Fiori G, Bedogni G, GallettiL, Isnagulov O, Ismagulova A,Sharmanov T, Tsoy I, Belcastro M, Rizzoli S, Goldoni M (2007). Spirometric reference values for children and adolescents from Kazakhstan. *Annals of Human Biology*, 34(5): 519–534.

[40] Kaditis A, Gourgoulianis K, Tsoutsou P, Papaioannou A, Fotiadou A, Messini C, Samaras K, Piperi M, Gissaki D, Zintzaras E, Paschalis Molyvdas A (2008). Spirometric values in Gypsy (Roma) children. *Respiratory Medicine* 102, 1321e1328.

[41] Meng-Chiao Tsai, Mei-Jy Jeng, Hua-Lun Chang, Pei-Chen Tsao, Chia-Feng Yang,Yu-Yun Peng, Yu-Sheng Lee, Wen-Jue Soong Ren-Bin Tang (2010). Spirometric Reference Equations for Healthy Children Aged 6 to 11 Years in Taiwan. *Journal of the Chinese Medical Association*, Vol 73 No 1:21-28.

[42] García-Ríoa F, Calle M, Burgos F, PCasand P, del Campoe F, Galdizf J, Giner g J, González N, Mangado H, Ortegai F, Puente Maestuj L (2013). Espirometría [Spirometry]. *Archivos de Bronconeumología*, 49(9):388–401.

[43] Hansen J; Xing-Guo Sun, and Karlman Wasserman, (2007). Spirometri criteria for airway obstruction. *CHEST*, 131 – 2.

[44] Pérez Padilla JR (2014). The need for reference values for respiratory function testing. *RAMR*, 1:4-6 - ISSN 1852 - 236X.

[45] Miller M, Crapo R, Hankinson J, Brusasco V, Burgos F, Casaburi R, Coates A, EnrightP, van der Grinten C, Gustafsson P, Jensen R, Johnson D, MacIntyre N, McKay R, Navajas D, Pedersen O, Pellegrino P, Viegi G and Wanger J (2005). General considerations for lung function testing. *European Respiratory Journal*, 26: 153–161.

[46] Muiño A, Akiki A, Marquez N, Moreira I, Pinchak M, Fernández E, González K, Silvera I (2013). Aproximación a los valores de referencia de la capacidad residual funcional en una población de niños y adolescentes [Approximation to the reference values of functional residual capacity in a population of children and adolescents]. *Archivos de Pediatría del Uruguay*, 84(S1): S18-S27.

[47] Álvarez Cáceres R (2007). *Statistics applied to health sciences*. ISBN: 978-84-7978-823-0.

[48] Years Purushottam Pramanik, Indra Narayan Ganguli, Manas Ghosh (2015). *Predicted Equations of Pulmonary Function Indices for East Indian Adolescent Boys Aged 10-18*. IOSR-JDMS e-ISSN: 2279-0853, p-ISSN: 2279-0861. Volume 14, Issue 2 Ver. IV, PP 65-69.

[49] Quanjer, Gerard J. Borsboom J, Jana Kivastik, Peter J, Merkus J, Hankinson L, Danny Houthuijs, Bert Brunekreef, Gabriele Ihorst, and Joachim Ku¨hr (20089. Cross-sectional and Longitudinal Spirometry in Children and Adolescents Interpretative Strategies Philip H. *American Journal of Respiratory and Critical Care Medicine,* Vol 178. pp 1262–1270.

[50] *Manual de SEPAR de Procedimientos*. Módulo 3 [*SEPAR Procedures Manual*. Module 3]. ISBN Módulo 3: 84-7989-155-6.

[51] Moore, V. C (2012). Spirometry: step by step. *European Respiratory Society*, .8(3), 233–240.

[52] Alexandraki S, Koutrilieris M, Siafaka N, Katsardis C (2010). *Spirometric Reference Values in Greek Children and Adolescents*. 24: 195-200.

[53] Gutierrez M, Valdivia G,Villarroel L, Contreras T, Cartagenes C, Lisboa C (2010). Nomograms of spirometric reference equations SER 2009. *Revista Chilena de Enfermedades Respiratorias*, 26: 9-15.
[54] Chhabra SK, Vijayan V, Rahman M, Mittal V and Singh P (2012). Regression Equations for Spirometry in Children Aged 6 to 17 Years in Delhi Region. *The Indian Journal of Chest Diseases & Allied Sciences*, Vol. 54:59-63.
[55] Pereira CA, Sato T, Rodrigues SC (2007). New reference values for forced spirometry in white adults in Brazil. *Jornal Brasileiro de Pneumologia*, 33(4): 397-406.

In: Air Pollution: Effects and Dangers ISBN: 978-1-53619-544-6
Editor: J. Esteban Colman Lerner

Chapter 9

CONTRIBUTIONS OF DATA SCIENCE TO AIR QUALITY MANAGEMENT: AN INTEGRATED AIR POLLUTION DECISION SUPPORT SYSTEM

Natacha S. Represa*[1,2,*]*, Jesús Palomar-Vázquez*[1]*,
***Alfonso Fernández-Sarría*[1] *and Andrés Porta*[2]**
[1]Geo-Environmental Cartography and Remote Sensing Group,
Polytechnic University of Valencia, Valencia, Spain
[2]Centro de Investigaciones del Medioambiente (CONICET-UNLP),
La Plata, Argentina

ABSTRACT

Air pollution in urban environments is a serious problem affecting human health and the environment. A wide range of tools are needed to identify air pollution problems and to test the effectiveness of any air quality action plan. This document describes a new decision support

* Corresponding Author's Email: solrepresa@quimica.unlp.edu.ar.

system for linking existing air quality tools through a modular methodology which allows the air quality and health effects health impact assessment. The system is designed to incorporate multi-source, multitype and multi-scale databases, queries and visualization, and thus take advantage of ground-based stations, remote sensing technology, and atmospheric chemical transport models. Also, allows linking the air quality with health impact assessment through tools proposed by the WHO as environmental burden disease and lifetime cancer risk.

1. INTRODUCTION

Air pollution in urban environments represents a significant threat to human health, property and the environment [1, 2]. The majority of the world's cities have pollution problems, both in developed and developing countries [3]. The scenario of concentrated urban growth places transport as the primary source of anthropogenic pollution [4]. However, in cities with mixed land use, industrial activity and power generation systems have a very important impact on air quality.

Air pollution has been linked to a wide range of health effects, including lung disease, heart problems, asthma, and lung cancer [5, 6]. Epidemiological studies in this field provide direct evidence on the magnitude of chronic effects, estimated to cause 1.4% of the global burden of deaths, which occur mostly in subjects under 60 years [7]. Urban and industrial areas are also major sources of potent greenhouse gases, contributing to climate change [8].

Air quality is a wicked problem that requires immediate intervention [9]. Even in countries with monitoring networks, air quality measurement capabilities are weak and poorly coordinated with management. There is a notable absence of international policy goals and policy stagnation, and the air pollution crisis in a growing number of cities is alarming [8].

Air quality management (AQM) is the activities aimed at reducing air pollution to protect human health and the environment in a specific geographical area. Effective AQM involves the availability of systematic assessment tools to address urban air pollution problems [10]. There is a requirement to have the skills, expertise, and capacity to provide the scientific support to policy, its implementation and delivery that builds on the requisite underlying research, and development [9].

Decision-making is an essential part of AQM [11]. While there are different approaches, evidence-based decision making has better results [12]. In an increasingly complex world, data science makes it easier to extract meaning from complex data [9]. Data science can also help by providing a structured work methodology that simplifies data analysis [13].

This paper describes a conceptual framework for developing a new air pollution decision support system (APDSS) for local AQM. This work's main contribution is to link existing air quality instruments through a modular methodology that allows the assessment of air quality effects and health effects. The proposed heuristics uses a design based on data science, facilitating knowledge from the data. In the next section, first, a general overview of the proposed framework will be provided. Then, we present a review of the difficulties in assessing air quality and the most critical antecedents for decision support systems in AQM. After that, we will focus on the contribution of data science and big data to air quality study. Finally, each building block will be described in detail, focusing on input, functionality, and output.

1.1. Air Quality Management and Decision Making

Air pollution presents a great spatial and temporal variation [14]. Levels are strongly dependent on emission sources and meteorological and geographic conditions [15]. Ground monitors are sparsely located, even in countries with extensive monitoring networks, and remain nonexistent throughout much of the world [16].

Remote sensing is the only technology capable of providing global coverage of atmospheric pollutants with sufficient spatial resolution. However, sensors measure pollutants in the atmospheric column, requiring additional information to infer concentrations at ground level [17]. The advent of low-cost sensors provides interesting prospects for increasing surveillance networks, but questions remain about their reliability, durability, and accuracy [18]. Also, without sufficient information, the modeling of contaminants is inaccurate [19, 20].

Accordingly, [21] identified three main challenges for AQM: i) it is burdensome and complex to develop appropriate datasets, models, and interpretative outputs needed for dispersion modeling, health impact assessments, and other purposes; ii) the difficult to identifying emissions reduction strategies that avoid litigation and satisfy the public and other stakeholders; iii) the health impacts and benefits of emission reduction strategies are rarely assessed as part of the proper AQM.

For this complex problem, decision science proposes the use of systems that facilitate the management, reducing time in decision making, and facilitating the identification of the most optimal strategy. A decision support system (DSS) is an information system that supports business or organizational decision-making activities.

Many approaches have been used to address the challenges mentioned above [21]. Presented a modeling framework that incorporates dispersion modeling, health impact assessments and optimization analyses into AQM [22]. Developed an air quality monitoring and analysis system for the urban environment using remote sensing measures [23]. Proposed a system based on CALPUFF dispersion model, digital maps, and related databases to estimate air pollutants' emissions and spatial distribution [24]. Brought together air quality modeling and mathematical programming techniques to determine optimal bundles of air pollution control options. All these systems depend on partial information sources for air pollution monitoring and analysis. Also, the use of models in DSS can be computationally intensive, requiring estimates at high resolution over long periods.

Data-driven decision support models are also challenged by the difficulties in handling dynamic and uncertain features of real-world environmental systems. As conditions for environmental management change with time, demanding a periodic update. Data science brings about new environmental study challenges, particularly around complexity, spatial and temporal reasoning, and managing uncertainty [25].

1.2. Data Science Contributions to Air Quality Study

With the many models available to assess concentrations in time and space, automatic measurements of atmospheric pollutants by ground-based monitoring or remote sensing provide an enormous amount of data. The massiveness of data transforms our understanding of environmental systems and generates new paradigms [26].

Big Data (BD) is the name coined to speak of enormous volumes of data that are created in real-time, diverse in variety, structured and unstructured in nature, exhaustive in scope, developed to capture populations or entire systems, with an acceptable mesh spatial resolution, relational in nature, flexible in structure, extensible to new variables and rapidly scalable in size [26, 27]. The challenge is to deal with abundance, variety, disorder, and dynamism, and the fact that much of this data answers anachronistic questions or is a product of other activities or experiences [28, 29]. In turn, the BD presents the challenge of adopting an epistemological framework in which the phenomenon's complexity fits into the study and the tools for addressing it [26].

In contrast to the current tendencies towards empiricism, data science, or data-driven science, poses a different approach to traditional deductive design. Its approach seeks to generate new hypotheses from the data [30]. This viewpoint is possible by incorporating an inductive mode in the research design that allows the identification of new relevant questions to be examined and tested.

The data science approach contributes to the study of air quality by providing a structured working methodology that organizes and simplifies the steps from data acquisition to information production for end users. Data science's epistemological strategy poses a process of knowledge discovery guided by the existing theory that guides the process [28, 31, 32].

Data Mining is the set of tools used to extract (or produce) information from large volumes of data through automation [13]. Different models of mining processes have been developed for data to facilitate the work with BD. A process model can be defined as the set of framework activities and tasks to do the work, including inputs and outputs for each task [33]. The ultimate goal of a process model is to make the process manageable, repeatable, and quantifiable. Examples of these are the pioneering KDD (Knowledge Discovery in Databases) model developed by Fayyad et al. [34-36] and the widely used CRISP-DM model [37] for the development of data mining models.

While data mining process models define what to do, the methodologies define how to do it. A methodology is a constitutive part of a process model, in that, along with the tasks, inputs, and outputs, how they are carried out [38,39]. Mariscal et al. [39] present a series of phases that are repeated in the different data mining methodologies; these are the problem specification, data preparation, modeling, model evaluation, and production of information (or new hypotheses) as an advance of knowledge in the field Sivarajah et al. [27] and Shah et al. [40].

1.3. A General Overview of the APDSS Framework

The framework presented here is an easy-to-apply guide to obtain a complete air quality analysis based on the information available. For ease-of-reading, a workflow is presented that organizes and synthesizes the necessary steps in air quality assessment.

The workflow was generated from a series of design rules. As general characteristics, it was thought that the methodology could be easily integrated with working models, not excluding previous technologies and that it could be adapted to new advances in the field. We sought to integrate algorithms that were robust, innovative, and easily applicable, without requiring long processing times for maintenance, and that made intensive use of available data by taking advantage of global sources of information.

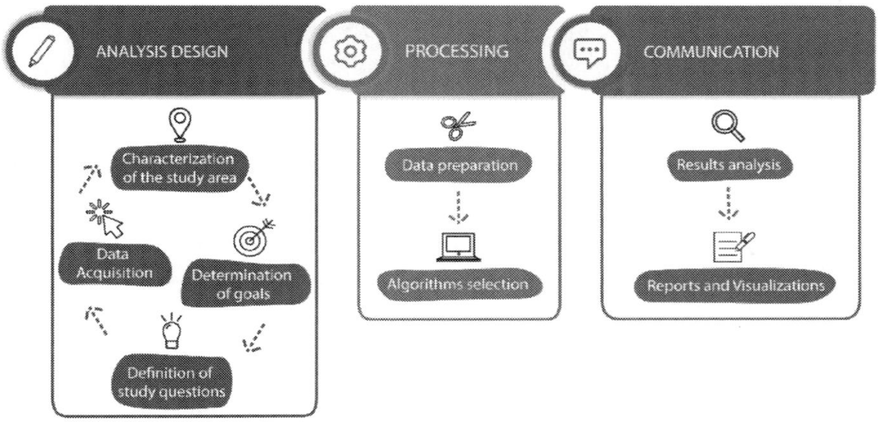

Figure 1. Diagram of the modules and tasks of the DSSAQ.

A workflow poses a sequence of operations related to each other and is performed recursively to obtain a product or result. Workflow occupies a central place in automation processes since tasks are structured according to data and control structures' dependence. In our case, workflow describes the aspects of the relevant process and can be managed independently and require specific skills to be carried out. We will call these sections, which can be thought of as autonomous in the process, functional modules [41, 42].

The ability to standardize the process of analyzing air quality data in modules offers several advantages. As a first aspect, it gives us the opportunity to integrate new processes and capabilities with those already in place. It allows us to recognize the previous organization giving value to other experiences and knowledge. Secondly, it is possible to increase the response's specificity by adding new resources or data. As a third aspect, it offers the possibility of adapting the methodology to the user by developing slight changes according to the final recipient [43]. A final aspect is that the modular process allows for updating versions without losing functionality. Furthermore, modify or replace the modules' components when the technology on which it is based evolves, without sacrificing the entire process [44].

2. A Detailed Description of the APDSS Framework Blocks

In this section, the three building blocks of the APDSS framework will be discussed in detail, considering their input, functionality, tasks, and output. The functionality is the central part of the description and defines the cause-effect relationship between input and output. Briefly, the main characteristics of each of the blocks are.

The first module focuses on the study area's knowledge, the collection of previous studies and available data, the analysis of the regulations, the setting of an objective, and the selection of the study questions. This stage limits the scope of analysis to be used in the next module. For its execution, expert knowledge in management and air quality is required.

Module two covers the pre-treatment and processing of the data. This stage begins once the questions to be answered have been selected. While there are general issues in data preparation, specific issues need to be considered depending on the analysis algorithms that are selected. This section requires mainly skills in data analysis and engineering.

Finally, module three focuses on the analysis of the results and their communication. This step leads to the visualization of results and the generation of reports. Communication of results as a fundamental part of AQM incorporates the population for better risk management. This section requires skills in scientific communication.

2.1. Analysis Design

The essential function of the ANALYSIS DESIGN block is to establish the main goals to be achieved in air quality assessment. This stage provides as outputs the working questions and the data sources needed for their resolution.

This module is an iterative sequence of steps that orders the necessary actions to define the working questions that guide the analysis. This block is a recursive stage, therefore, dynamic and of continuous improvement. It is crucial to carry out periodic reviews of the goals and sources of information available to keep our system updated with scientific and technological advances.

2.2. Definition and Characterization of the Study Area

As the first input, we need to define the extension of the analysis domain. It is defined according to the project's magnitude and the presence of people susceptible. The domain may need to be adjusted (expanded) based on preliminary analysis results, for example, to find the point of maximum impact.

Each analysis scenario has particular characteristics that define its dynamics. However, because of the physical and chemical laws that govern the phenomena studied, there are common issues that recur in each study site. The identification of them in this instance provides us with crucial information to contextualize the results.

The study area's characterization requires knowledge of demographics, economic activities, land uses, meteorology of the region, local geography, degree of urbanism or land cover, climatic events, and legislation. These elements may be obtained from scientific literature, documentation from environmental agencies, and the media, also through visits and interviews.

2.3. Determination of Goals

The target setting represents a starting point for air quality assessment. Different purposes are depending on the specific needs and circumstances. The goals established define the study questions that guide the work.

The goals of the AQM can be limited to a characterization of air quality in a single time at a specific site or cover a more extended period and area (the study of a pollutant's temporal evolution over the years or a region). The goals that were identified as the most frequently raised in the study of air quality are:

- Know issues of land surveillance
- Determine air quality at a measurement site
- Evaluate the temporal evolution of the pollutants
- Assess the spatial distribution of contaminants.

2.4. Definition of Study Questions

Data science is considered an analysis process based on answering questions through data. Based on this, we made a series of guiding questions for the analysis. These questions represent a specific set of queries in terms of entities and relationships.

Table 1 serves as the basis for the air quality study approach. The questions allow us to discern what data we should collect and what types of analyses should or can be conducted. Different questions require diverse types of data, which makes a difference in data collection.

In practice, however, not all the required information is always available. This situation leads to the need to review the questions to be addressed to obtain a final set of questions for each case study.

2.5. Data Acquisition

The data acquisition consists of identifying and obtaining the data required for analysis. One difficulty encountered was the lack of availability of information. In turn, the quality of the data used is of great importance in our evaluation's conclusions. Data quality marks the scope of our work, and the uncertainties of the methods or the lack of data limit the conclusions we could obtain.

Table 1. Goals and questions for air quality analysis

Goals	Questions
Know issues of land surveillance	- What design does the monitoring network follow? - What are the techniques used for monitoring? - What is the quality of the information collected by the monitoring network? - Is there redundancy among the data collected by the stations? - What are the characteristics of the area where the monitoring stations are located? - How do local winds influence the levels recorded at the stations? - From which sector of the wind do the highest concentrations arrive?
Determine air quality at a measurement site	- What are the most frequent levels of contaminants? - Did the concentrations of the contaminants exceed the legal limits? - Do the measured values of the contaminants put the population at risk?
Evaluate the temporal evolution of the pollutants	- At what time of day is the concentration of the contaminants higher? - Do the maximum concentrations vary throughout the week? - At what time of year do peak concentrations occur? - How have contaminant levels changed over the years?
Assess the spatial distribution of contaminants	- Where do maximum concentrations occur? - Is there a spatial pattern for the maximum concentrations? - What is air quality like where it is not measured? - Does the air quality in the region represent a risk to the population?

Data may come from different sources, such as official monitoring stations, satellite images, or low-cost sensors [45, 46]. However, each data source has its limitations. This step aims to identify the available data sources and recognize their possible applications [47].

As a starting point, previous work at the study site should be reviewed to analyze previously consulted data sources. This task includes reviewing the academic literature, government reports, social institutions' productions, and reviewing the region's epidemiological literature. This step is relevant to evaluate quantitative data availability [47].

2.6. Data Preparation

The purpose of this step is to guarantee the quality of the data obtained. A prerequisite for successful analysis is data without anomalies and free of inconsistencies. In turn, knowing the limitations of the data makes it easier to interpret the results.

Data preparation consists of different techniques such as 1) noise cleaning, 2) imputation of lost or missing values, 3) aggregation and 4) data transformation and normalization [34-36]. Some analyses require the mandatory application of all these techniques since, without them, the statistical algorithms' results would offer spurious results [48]. However, it is advisable to be conservative about the modifications made to the operational data, and it is convenient to change the techniques before applying transformations or making imputations.

2.7. Algorithms Selection

Once the data has been acquired and pre-processed, the analysis and processing phase begins. The questions that arise from the defined objectives guide this stage. This strategy converts data extraction exercise into relevant conclusions for reports and action.

Table 2. Dimensions of air quality analysis: study questions, specific goals, and techniques

<table>
<tr><td>Questions</td><td>Specific objectives</td><td>References</td></tr>
<tr><td>What design does the monitoring network follow?</td><td>Analyze the spatial distribution pattern of the monitoring stations through point pattern analysis.</td><td>49</td></tr>
<tr><td>What are the techniques used for monitoring?</td><td rowspan="2">Describe the equipment used by each parameter and analyze the quality of the data obtained by the monitoring stations.</td><td rowspan="2">50, 51</td></tr>
<tr><td>What is the quality of the information collected by the monitoring network?</td></tr>
<tr><td>Is there redundancy among the data collected by the stations?</td><td>Evaluate the redundancy among the monitoring stations by means of a correlation analysis and analysis of main components.
Identify the stations that have the same behavior through hierarchical cluster analysis.</td><td>43, 52-57</td></tr>
<tr><td>What are the characteristics of the monitoring stations area?</td><td>Analyze the area of influence of the monitoring stations. Analyze the winds in the seasons.</td><td>54, 58-60</td></tr>
</table>

Questions	Specific objectives	References
How do local winds influence the levels recorded at the stations?	Analysis of contributions by means of pollution rose and map of emission sources.	54, 58-60
From which sector of the wind do the highest concentrations arrive?		
What are the most frequent levels of contaminants?	Perform an exploratory analysis. Descriptive analysis with respect to guide levels. Study of the frequency and analysis of the accumulated frequency of concentrations. Analysis of daily, monthly, and annual behavior.	61-63
Did the pollutant concentrations exceed the legal limits?		
Do pollutant concentrations put the population at risk?		
At what time of day are the pollutant concentrations higher?		
Do the maximum concentrations vary throughout the week?	Perform an exploratory analysis. Descriptive analysis with respect to guide levels. Study of the frequency and analysis of the accumulated frequency of concentrations. Analysis of daily, monthly, and annual behavior.	61-63
At what time of year do peak concentrations occur?		
How have pollutant concentrations changed over the years?	Evaluation of the long-term tendency.	64, 65
Where do maximum pollutant concentrations occur?	Build a spatial model with the ground measurements. Make maps of air quality and health risk areas. Analyze the spatial and temporal behavior of the variables from the satellite.	66-72
Is there a spatial pattern for the maximum pollutant concentrations?		
What is air quality like where it is not measured?		
Does the air quality in the region represent a risk to the population?		

Table 2 shows the specific objectives that emerge from the questions. Objectives have been addressed in the literature by different techniques, which constitute entire fields of study. Here we mention the most accepted algorithms employed.

2.8. Communication

The COMMUNICATION block represents the results publication framework, that is, the set of techniques/approaches used to present the data and generate knowledge to make decisions on emission reduction measures, on activity changes, or on direct concentration reductions. The input for this block is the results obtained from the application of the algorithms selected in the "ANALYSIS DESIGN" block. The functionality of this block should suggest answers to the decision maker, providing outputs of easy interpretation.

2.8.1. Results analysis

The analysis of the results obtained often caused a return to the pre-processing phase to obtain more data or more attributes. It is essential to have the pre-processed data set, the corresponding metadata elaborated, and the algorithms were chosen for success. This final stage of the process is crucial in our work.

2.8.2. Reports and Visualizations

As a final step, reports and visualizations need to be prepared to highlight the results. The communication is the biggest challenge of the process since it requires showing information obtained from large data sets of different types and multiple sources to technical or non-expert personnel who do not always know the complexity of the data or the analyses performed [72]. Therefore, the reports must be understandable to a broad audience and self-descriptive. This step is the only part of the analytical process that reaches the audience. Even when the conclusions may be evident, the public also plays a central role in interpreting the results.

The message should not sound apocalyptic, conveying a threat to which no can cope. On the contrary, the message is more effective when the perception of danger does not exceed the perception of resources or capacities to deal with it [73]. Poorly presented conclusions can obscure key issues and skew public perceptions [74]. Effective communication requires the identification of specific points of the analysis we want to show. In this sense, it is necessary to leave aside the elaborate exploratory graphs to understand the data set's deeper relationships. Likewise, visualizations designed exclusively for exploratory analysis are discouraged [75].

The visual representation selected should provide a clear picture of the patterns and relationships discovered, facilitating the audience's interpretation and understanding of the information. It is advisable that the data presentation can be presented in narrative form and with an appropriate visual presentation [74].

The following are some examples of the most commonly used visualizations already introduced in the text. The most used graphic to visualize the relationships between different variables is the dendrogram. This type of representation is particularly useful for presenting the arrangement of clusters produced by hierarchical grouping techniques. It may be convenient to show the monitoring stations' grouping on a map for representation in space [76, 77].

The time series plots serve to visualize the time pattern. It is common to aggregate the data to show behavior during the week and months. It is also useful to show the series trend of the period analyzed [78]. To allow daily, monthly, and annual comparisons of the air's variability, it is convenient to use a calendar table for easy viewing [22].

The pollution roses represent the pollution levels observed by each monitoring station according to wind direction (Figure 2). The model of the contamination's rose is based on the simple idea of classifying the concentration measurements obtained at a given site for a given period according to the direction in which the wind is blowing at the time of measurement [79]. This graph provides a relative notion of the location of pollution sources.

Maps are an effective way to visualize and communicate the spatial variation of concentrations of contaminants. Heat maps identify areas of higher concentration [80]. However, if a color scale is not selected appropriately, the interpretation becomes difficult. Therefore, the monochromatic palette is preferable to show the areas of overcoming, and the multicolored one is used to show the differences in the territory.

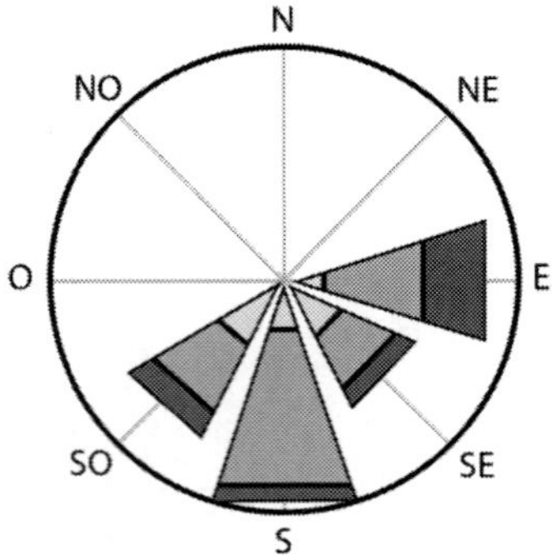

Figure 2. Example of pollution rose.

Finally, to reinforce the presentation's narrative, the measured or predicted concentrations are compared with air quality standards. There are differences between air quality standards and the World Health Organization (WHO) recommendations. The standards of air quality are often conditioned by economic policy issues and take longer to be updated, while the WHO recommendations are based on scientific research that supports and are reviewed when evidence is gathered sufficiently [81].

2.8.2.1. Indexes Development

The communication of air quality information to the public and policymakers is mostly done by using indicators and indexes because it facilitates the public interpretation [82]. While an indicator is a statistical parameter to obtain and share information for decision or policy making [83], an index is an aggregation of indicators, which is simplified to make it easier to communicate to the public than multiple values. Indexes are often non-dimensional and can provide consistency to a reported value for multiple air pollutants with different units (e.g., 0-100).

For public reporting of forecasts or real-time, an air quality index (AQI) is appropriate, where multiple severity ranges, depending on the level of response required by different groups affected. The AQI usually converts a contaminant concentration into a number on a known scale, and contaminant levels are classified as "very high," "high," "moderate and "low" [84, 85]. The breakpoints in the categorization vary from one pollutant to another [86]. A variant can be seen in Olvera-García et al., [87], where a diffuse analysis was applied to generate AQI weights of contaminants with more significant health effects.

Reporting through an AQI has three major advantages. Firstly, this ensures people's care through advice that enables them to adapt their behavior or take mitigation measures to avoid health effects. Secondly, they serve to condense complex data by facilitating decision making. A third reason is to draw the public's attention to the problem of air pollution, generating a higher level of awareness by reporting on air quality daily [88].

Despite the advantages, AQIs have certain limitations in their use because the factors that intervene between air quality and health effects are multiple and diverse, interact in ways that are still poorly understood, and their exact nature is difficult to quantify. Besides, the index's value refers to a point in time when the health effects occur at different exposure times. For example, an AQI based on short term effects time for health can indicate "good levels" of air quality, implicitly pointing that air quality is not a problem, and at the same time, exceed the annual limit values.

Simultaneously, not all pollutants harmful to health are regulated and controlled and therefore are commonly excluded from the indexes. Another drawback is that few indexes are based on local epidemiological data and lack specificity [89]. These issues make the use of an AQI difficult. An interesting example is the one presented by the World Air Quality Index project [90], where the AQI is based on the WHO recommendations for their preparation [91], establishing a chromatic scale for each of the levels.

2.8.2.2. Air Quality and Risk Maps

As a last resort, air quality and risk maps are widely used to design and apply territorial policies (Figure 3). The visualization through maps facilitates spatial pattern identification, allowing to delimit the areas that require further intervention. Maps can be made daily, monthly, or annually, depending on their purpose [69]. Although it is convenient to work with daily maps for forecasting purposes because of the predictions, for epidemiological purposes, it is convenient to have monthly maps or annually, according to the minimum level of aggregation of health data.

While the concentration maps show the contaminants' levels in the territory, air quality maps present a categorization of concentrations at which the population is exposed and relativizes it according to the levels recommended by the WHO [92].

Similarly, the risk from exposure to contaminants is calculated on the map's concentration values [93]. In the literature, we find many works where relative risk (RR) functions for air pollution have been obtained [7, 94]. The RR is the probability of health effects occurring in a population exposed to a higher level of air pollution than that considered to be background [7]. The RR has an intuitive interpretation, making it easier to identify the areas at greatest risk of disease.

Risk maps are cartographic products that may be visualized within a Geographic Information System (GIS), allowing its integration with other layers of information depending on the application you want to give, such as sources of contamination or the available emergency services [68, 95].

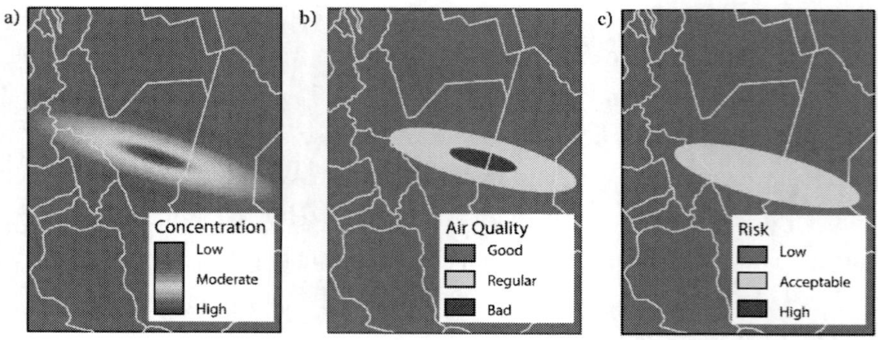

Figure 3. Example of a) concentration map, b) air quality map and c) risk map.

Conclusion

This document has presented a framework for knowledge discovery to support the decision-making process using the most relevant literature on the subject. The approach is based on the iterative combination of blocks; it leads the analysis from identifying the domain to the stakeholders' publication of reports. This proposal highlights the potential value of data-based approaches to air quality assessment and decision making. It can help decision-makers to identify desired pollution-abatement strategies with minimized costs and maximized environmental efficiencies.

REFERENCES

[1] Gurjar, B. R., Butler, T. M., Lawrence, M. G. and Lelieveld, J., 2008. Evaluation of emissions and air quality in megacities. *Atmospheric Environment*, 42(7), pp. 1593-1606.

[2] Krzyzanowski, M., & Cohen, A. (2008). Update of WHO air quality guidelines. *Air Quality, Atmosphere & Health*, 1(1), 7- 13.

[3] Osseiran, N., & Chriscaden, K. (2016). Air pollution levels rising in many of the world's poorest cities. *World Health Organization.*

[4] Nieuwenhuijsen MJ (2016). Urban and transport planning, environmental exposures and health-new concepts, methods and tools to improve health in cities. *Environmental Health* 15(Suppl 1): 38, 161-171.

[5] Brunekreef B, Janssen NA, de Hartog J, Harssema H, Knape M, van Vliet P. (1997). Air pollution from truck traffic and lung function in children living near motorways. *Epidemiology* 8:298–303.

[6] Manisalidis I, Stavropoulou E, Stavropoulos A and Bezirtzoglou E (2020) Environmental and Health Impacts of Air Pollution: A Review. *Frontiers in Public Health* 8:14. doi: 10.3389/fpubh.2020.00014.

[7] Ostro, B., & World Health Organization (2004). *Outdoor air pollution: assessing the environmental burden of disease at national and local levels.*

[8] Baklanov, A., Molina, L. T. and Gauss, M., (2016). Megacities, air quality and climate. *Atmospheric Environment*, 126, pp. 235-249.

[9] Monks, P. S. and Williams, M. L., (2020). What does success look like for air quality policy? A perspective. *Philosophical Transactions of the Royal Society A*, 378(2183), p. 20190326.

[10] Franco, J. F., Gidhagen, L., Morales, R. and Behrentz, E., 2019. Towards a better understanding of urban air quality management capabilities in Latin America. *Environmental Science & Policy*, 102, pp. 43-53.

[11] Fisher, B., (2003). Fuzzy environmental decision-making: applications to air pollution. *Atmospheric Environment*, 37(14), pp. 1865-1877.

[12] Smith, K. and Haux, T., (2017). 8 Evidence-Based Policy-Making (EBPM). *Handbook of Social Policy Evaluation*, 141.

[13] Represa, N. S., Fernández-Sarría, A., Porta, A. and Palomar-Vázquez, J., (2020). Data mining paradigm in the study of air quality. *Environmental Processes*, 7(1), pp. 1-21.

[14] Fu, Weicong, Chen, Ziru, Zhu, Zhipeng; Liu, Qunyue, Qi, Jinda; Dang, Emily; Wang, Minhua; Dong, Jianwen. (2018). "Correction: Fu et al. Long-Term Atmospheric Visibility Trends and Characteristics of 31 Provincial Capital Cities in China during 1957–2016. Atmosphere, 9, 318" *Atmosphere* 9, no. 10: 389. https://doi.org/10.3390/atmos9100389.

[15] Borge R., WJ. Requia, C. Yagüe, l. Jhund, P. Koutrakis. (2019). Impact of weather changes on air quality and related mortality in Spain overa 25 year period [1993–2017]. *Environmental international*, 133: 105272. https://doi.org/10.1016/j.envint.2019.105272.

[16] Martin, R. V., Brauer, M., van Donkelaar, A., Shaddick, G., Narain, U. and Dey, S., (2019). No one knows which city has the highest concentration of fine particulate matter. *Atmospheric Environment*: X, 3, p. 100040.

[17] Wei, X., Chang, N. B., Bai, K. and Gao, W., (2020). Satellite remote sensing of aerosol optical depth: advances, challenges, and perspectives. *Critical Reviews in Environmental Science and Technology*, 50(16), pp. 1640-1725.

[18] Snyder, E. G., Watkins, T. H., Solomon, P. A., Thoma, E. D., Williams, R. W., Hagler, G. S., Shelow, D., Hindin, D. A., Kilaru, V. J. and Preuss, P. W., (2013). The changing paradigm of air pollution monitoring. *Environmental Science & Technology*, 47(20):11369-77. doi: 10.1021/es4022602.

[19] Bai L., Wang J., Ma X., Lu H. (2018). Air Pollution Forecasts: An Overview. *Int. J. Environmental Research and Public Health*, 15, 780; doi: 10.3390/ijerph15040780.

[20] Derwent D., A. Fraser, J. Abbott, M. Jenkin, P. Willis, T. Murrells (2010). *Evaluating the Performance of Air Quality Models.* Department for Environment, Foods, and Rural Affairs (DEFRA). United Kingdom. https://uk-air.defra.gov.uk/assets/documents/reports/cat05/1006241607_100608_MIP_Final_Version.pdf.

[21] Milando, C. W., Martenies, S. E. and Batterman, S. A., (2016). Assessing concentrations and health impacts of air quality management strategies: Framework for Rapid Emissions Scenario and Health impact ESTimation (FRESH-EST). *Environment international*, 94, pp. 473-481.

[22] Leung, Y., Leung, K. S., Wong, M. H., Mak, T., Cheung, K. Y., Lo, L. Y., Yi, W. Y. and Dong, Y. L., (2018). An integrated web-based air pollution decision support system–a prototype. *International Journal of Geographical Information Science*, 32(9), pp. 1787-1814.

[23] Elbir, T., (2004). A GIS based decision support system for estimation, visualization and analysis of air pollution for large Turkish cities. *Atmospheric Environment*, 38(27), pp. 4509-4517.

[24] Vlachokostas, C., Achillas, C., Moussiopoulos, N., Hourdakis, E., Tsilingiridis, G., Ntziachristos, L., Banias, G., Stavrakakis, N. and Sidiropoulos, C., (2009). Decision support system for the evaluation of urban air pollution control options: application for particulate pollution in Thessaloniki, Greece. *Science of the Total Environment*, 407(23), pp. 5937-5948.

[25] Blair, G., Henrys, P., Leeson, A., Watkins, J., Eastoe, E., Jarvis, S. and Young, P., (2019). Data science of the natural environment: a research roadmap. *Frontiers in Environmental Science*, 7, p. 121.

[26] Kitchin, R. (2014). Big Data, new epistemologies and paradigm shifts. *Big data & society*, 1(1), 2053951714528481.
[27] Sivarajah, U., Kamal, M. M., Irani, Z., & Weerakkody, V. (2017). Critical analysis of Big Data challenges and analytical methods. *Journal of Business Research*, 70, 263-286.
[28] Han, J., Pei, J., & Kamber, M. (2011). *Data mining: concepts and techniques*. Elsevier. ISBN: 9780123814807.
[29] Hastie, T, Tibshirani, R, Friedman, J (2009) *The Elements of Statistical Learning: Data Mining, Inference, and Prediction*, 2nd ed. New York: Springer.
[30] Kelling, S, Hochachka, W, Fink, D (2009) Data-intensive Science: A new paradigm for biodiversity studies. *BioScience* 59(7): 613–620.
[31] Bellinger, C., Jabbar, M. S. M., Zaïane, O., & Osornio-Vargas, A. (2017). A systematic review of data mining and machine learning for air pollution epidemiology. *BMC public health*, 17(1), 907.
[32] Chen, Y., Wang, L., Li, F., Du, B., Choo, K. K. R., Hassan, H., & Qin, W. (2017). Air quality data clustering using EPLS method. *Information Fusion*, 36, 225-232.
[33] Pressman, R. S. 2005. *Software Engineering: A Practitioner's Approach, 6th edition*. McGraw-Hill Science.
[34] Fayyad, U. M., Piatetsky-Shapiro, G. & Smyth, P. (1996). From data mining to knowledge discovery: an overview, Advances in Knowledge Discovery and Data Mining, 1–34. *American Association for Artificial Intelligence.*
[35] Fayyad, U., Piatetsky-Shapiro, G. & Smyth, P. (1996). The KDD PROCESS for extracting useful knowledge from volumes of data. *Communication of the ACM* 39, 27–34.
[36] Fayyad, U., Piatetsky-Shapiro, G., Smith, P. & Uthurusamy, R. (1996). *Advances in Knowledge Discovery and Data Mining.* AAAI/MIT Press.
[37] Chapman, P., Clinton, J., Kerber, R., Khabaza, T., Reinartz, T., Shearer, C. & Wirth, R. (2000). CRISP-DM 1.0 Step-by-Step Data Mining Guide. *Technical report*, CRISP-DM.

[38] Yang, Q. & Wu, X. (2006). 10 challenging problems in data mining research. *International Journal of Information Technology and Decision Making* 5(4), 597–604.

[39] Mariscal, G., Marban, O. and Fernandez, C., (2010). A survey of data mining and knowledge discovery process models and methodologies. *The Knowledge Engineering Review*, 25(2), p. 137.

[40] Shah, T., Rabhi, F., & Ray, P. (2015). Investigating an ontology-based approach for Big Data analysis of inter-dependent medical and oral health conditions. *Cluster Computing*, 18(1), 351–367.

[41] Georgakopoulos, D., Hornick, M., & Sheth, A. (1995). An overview of workflow management: From process modeling to workflow automation infrastructure. *Distributed and Parallel Databases*, 3(2), 119–153.doi: 10.1007/bf01277643.

[42] Russell, N., Ter Hofstede, A. H., Edmond, D., & van der Aalst, W. M. (2005). Workflow data patterns: Identification, representation and tool support. In *International Conference on Conceptual Modeling* (pp. 353-368). Springer, Berlin, Heidelberg.

[43] Wang, Z., King, E., Smith, G., Bellgard, M., Broomhall, M., Chedzey, H., ... & Schibeci, D. (2011). *RS-YABI: A workflow system for Remote Sensing Processing in AusCover*.

[44] Tansley, S., & Tolle, K. M. (2009). *The fourth paradigm: data-intensive scientific discovery (Vol. 1)*. A. J. Hey (Ed.). Redmond, WA: Microsoft research.

[45] Candia, A., Represa, S. N., Giuliani, D., Luengo, M. Á., Porta, A. A. and Marrone, L. A., (2018), November. Solutions for SmartCities: proposal of a monitoring system of air quality based on a LoRaWAN network with low-cost sensors. In *2018 Congreso Argentino de Ciencias de la Informática y Desarrollos de Investigación* (CACIDI) (pp. 1-6). IEEE.

[46] Morawska, L., Thai, P. K., Liu, X., Asumadu-Sakyi, A., Ayoko, G., Bartonova, A., Bedini, A., Chai, F., Christensen, B., Dunbabin, M. and Gao, J., 2018. Applications of low-cost sensing technologies for air quality monitoring and exposure assessment: How far have they gone?. *Environment international*, 116, pp. 286-299.

[47] Limaye, V. S., Knowlton, K., Sarkar, S., Ganguly, P. S., Pingle, S., Dutta, P., Tiwari, A., Solanki, B., Shah, C., Raval, G. and Kakkad, K., (2018). Development of Ahmedabad's Air Information and Response (AIR) Plan to Protect Public Health. *International Journal of Environmental Research and Public Health*, 15(7), p. 1460.

[48] García Ferreyra, M. F., Curci, G., & Lanfri, M. (2016). First Implementation of the WRF-CHIMERE-EDGAR Modeling System Over Argentina. *IEEE Journal of Selected Topics in Applied Earth Observations and Remote Sensing*, 9(12), 5304-5314.

[49] Monogan III, J. E., Konisky, D. M. and Woods, N. D., (2013), May. Strategic placement of air polluters: an application of point pattern models. In *State Politics and Policy Conference*.

[50] Mura, I., Franco, J. F., Bernal, L., Melo, N., Díaz, J. J. and Akhavan-Tabatabaei, R., (2020). A Decade of Air Quality in Bogotá: A Descriptive Analysis. *Frontiers in Environmental Science*, 8, p. 65.

[51] Năstase, G., Şerban, A., Năstase, A.F., Dragomir, G. and Brezeanu, A.I., (2018). Air quality, primary air pollutants and ambient concentrations inventory for Romania. *Atmospheric Environment*, 184, pp.292-303.

[52] Dominick, D., Juahir, H., Latif, M. T., Zain, S. M. and Aris, A. Z., (2012). Spatial assessment of air quality patterns in Malaysia using multivariate analysis. *Atmospheric environment*, 60, pp. 172-181.

[53] Lu, W. Z., He, H. D. and Dong, L. Y., (2011). Performance assessment of air quality monitoring networks using principal component analysis and cluster analysis. *Building and Environment*, 46(3), pp. 577-583.

[54] Pearce, J. L., Beringer, J., Nicholls, N., Hyndman, R. J. and Tapper, N. J., (2011). Quantifying the influence of local meteorology on air quality using generalized additive models. *Atmospheric Environment*, 45(6), pp. 1328-1336.

[55] Pires, J. C. M., Sousa, S. I. V., Pereira, M. C., Alvim-Ferraz, M. C. M., Martins, F. G., (2008). Management of air quality monitoring using principal component and cluster analysis – part I: SO2 and PM10. *Atmospheric Environment* 42 (6), 1249–1260.

[56] Pires, J. C. M., Sousa, S. I. V., Pereira, M. C., Alvim-Ferraz, M. C. M., Martins, F. G., (2008). Management of air quality monitoring using principal component and cluster analysis – part II: CO, NO2 and O3. *Atmospheric Environment* 42 (6), 1261–1274.

[57] Yang, F., Tan, J., Zhao, Q., Du, Z., He, K., Ma, Y., Duan, F. and Chen, G. J. A. C., (2011). Characteristics of PM 2.5 speciation in representative megacities and across China. *Atmospheric Chemistry and Physics*, 11(11), pp. 5207-5219.

[58] Lin, C. Y., Liu, S. C., Chou, C. C. K., Huang, S. J., Liu, C. M., Kuo, C. H. and Young, C. Y., (2005). Long-range transport of aerosols and their impact on the air quality of Taiwan. *Atmospheric Environment*, 39(33), pp. 6066-6076.

[59] Rosario, L., Pietro, M. and Francesco, S. P., 2016. Comparative analyses of urban air quality monitoring systems: passive sampling and continuous monitoring stations. *Energ Procedia*, 101(2), pp. 321-8.

[60] Vardoulakis, S., Valiantis, M., Milner, J., & ApSimon, H. (2007). Operational air pollution modelling in the UK—Street canyon applications and challenges. *Atmospheric Environment*, 41(22), 4622-4637.

[61] Gokhale, S. and Khare, M., (2007). Statistical behavior of carbon monoxide from vehicular exhausts in urban environments. *Environmental Modelling & Software*, 22(4), pp. 526-535.

[62] Ratto, G., Videla, F., Almandos, J. R., Maronna, R., & Schinca, D. (2006). Study of meteorological aspects and urban concentration of SO2 in atmospheric environment of La Plata, Argentina. *Environmental monitoring and assessment*, 121(1-3), 327-342.

[63] Reddy, B. S. K., Kumar, K. R., Balakrishnaiah, G., Gopal, K. R., Reddy, R. R., Sivakumar, V., Lingaswamy, A. P., Arafath, S. M., Umadevi, K., Kumari, S. P. and Ahammed, Y. N., (2012). Analysis of diurnal and seasonal behavior of surface ozone and its precursors (NOx) at a semi-arid rural site in southern India. *Aerosol and Air Quality Research*, 12(6), pp. 1081-1094.

[64] Guerreiro, C. B., Foltescu, V. and De Leeuw, F., (2014). Air quality status and trends in Europe. *Atmospheric environment*, 98, pp. 376-384.

[65] Guo, J. P., Zhang, X. Y., Wu, Y. R., Zhaxi, Y., Che, H. Z., La, B., Wang, W. and Li, X. W., (2011). Spatio-temporal variation trends of satellite-based aerosol optical depth in China during 1980–2008. *Atmospheric Environment*, 45(37), pp. 6802-6811.

[66] Beauchamp, M., de Fouquet, C. and Malherbe, L., (2017). Dealing with non-stationarity through explanatory variables in kriging-based air quality maps. *Spatial Statistics*, 22, pp. 18-46.

[67] Gurjar, B. R., Jain, A., Sharma, A., Agarwal, A., Gupta, P., Nagpure, A. S. and Lelieveld, J., (2010). Human health risks in megacities due to air pollution. *Atmospheric Environment*, 44(36), pp. 4606-4613.

[68] Matejicek, L., (2005). *Spatial modelling of air pollution in urban areas with GIS: a case study on integrated database development.*

[69] Saisana, M., Dubois, G., Chaloulakou, A., & Spyrellis, N. (2004). *Classification criteria and probability risk maps: Limitations and perspectives.*

[70] Wong, D. W., Yuan, L. and Perlin, S. A., (2004). Comparison of spatial interpolation methods for the estimation of air quality data. *Journal of Exposure Science & Environmental Epidemiology*, 14(5), pp. 404-415.

[71] Yang, X., Zheng, Y., Geng, G., Liu, H., Man, H., Lv, Z., He, K. and de Hoogh, K., 2017. Development of PM2. 5 and NO2 models in a LUR framework incorporating satellite remote sensing and air quality model data in Pearl River Delta region, China. *Environmental pollution*, 226, pp. 143-153.

[72] Yeganeh, B., Hewson, M. G., Clifford, S., Knibbs, L. D., & Morawska, L. (2017). A satellite-based model for estimating PM2. 5 concentration in a sparsely populated environment using soft computing techniques. *Environmental Modelling & Software*, 88, 84-92.

[73] Brulle, R. J. (2010). From environmental campaigns to advancing the public dialog: Environmental communication for civic engagement. *Environmental Communication*, 4(1), 82-98.

[74] Ma, K. L., Liao, I., Frazier, J., Hauser, H., & Kostis, H. N. (2012). Scientific storytelling using visualization. *IEEE Computer Graphics and Applications*, 32(1), 12-19.

[75] Knaflic, C. N. (2015). *Storytelling with data: A data visualization guide for business professionals.* John Wiley & Sons.

[76] Xie, Y., Zhao, L., Xue, J., Gao, H. O., Li, H., Jiang, R., Qiu, X. and Zhang, S., (2018). Methods for defining the scopes and priorities for joint prevention and control of air pollution regions based on data-mining technologies. *Journal of cleaner production*, 185, pp. 912-921.

[77] Wang, C., Zhao, L., Sun, W., Xue, J. and Xie, Y., (2018). Identifying redundant monitoring stations in an air quality monitoring network. *Atmospheric Environment*, 190, pp. 256-268.

[78] Elangasinghe, M. A., Singhal, N., Dirks, K. N., Salmond, J. A., & Samarasinghe, S. (2014). Complex time series analysis of PM10 and PM2. 5 for a coastal site using artificial neural network modelling and k-means clustering. *Atmospheric Environment*, 94, 106-116.

[79] Duboue, M. (1978). Pollution roses: a simple way of interpreting the data obtained by air pollution measurement systems in the proximity of refineries. In *Studies in Environmental Science* (Vol. 1, pp. 133-136). Elsevier.

[80] Austin, E., Coull, B. A., Zanobetti, A., & Koutrakis, P. (2013). A framework to spatially cluster air pollution monitoring sites in US based on the PM2.5 composition. *Environment international*, 59, 244-254.

[81] Zhang, N. N., Ma, F., Qin, C. B., & Li, Y. F. (2018). Spatiotemporal trends in PM2. 5 levels from 2013 to 2017 and regional demarcations for joint prevention and control of atmospheric pollution in China. *Chemosphere*, 210, 1176-1184.

[82] Zhang, Y., Bocquet, M., Mallet, V., Seigneur, C., & Baklanov, A. (2012). Real-time air quality forecasting, part I: History, techniques, and current status. *Atmospheric Environment*, 60, 632-655.

[83] Hsu, A., Reuben, A., Shindell, D., de Sherbinin, A., & Levy, M. (2013). Toward the next generation of air quality monitoring indicators. *Atmospheric environment*, 80, 561-570.

[84] Sadat, Y. K., Nikaein, T., & Karimipour, F. (2015). Fuzzy spatial association rule mining to analyze the effect of environmental variables on the risk of allergic asthma prevalence. *Geodesy and Cartography*, 41(2), 101-112.

[85] Engel-Cox, J., Oanh, N. T. K., van Donkelaar, A., Martin, R. V., & Zell, E. (2013). Toward the next generation of air quality monitoring: Particulate Matter. *Atmospheric Environment*, 80, 584-590.

[86] Mintz, D. (2012). Technical assistance document for the reporting of daily air quality-the air quality index (aqi): Us environmental protection agency. *Office of Air Quality Planning and Standards*, 1-26.

[87] Olvera-García, M. Á., Carbajal-Hernández, J. J., Sánchez-Fernández, L. P., & Hernández-Bautista, I. (2016). Air quality assessment using a weighted Fuzzy Inference System. *Ecological informatics*, 33, 57-74.

[88] Van den Elshout, S., Léger, K., & Heich, H. (2014). CAQI common air quality index—Update with PM2. 5 and sensitivity analysis. *Science of the Total Environment*, 488, 461-468.

[89] Stieb, D. M., Burnett, R. T., Smith-Doiron, M., Brion, O., Shin, H. H., & Economou, V. (2008). A new multipollutant, no threshold air quality health index based on short-term associations observed in daily time-series analyses. *Journal of the Air & Waste Management Association*, 58(3), 435-450.

[90] World Air Quality Index project. (2015). *World Health Organization: 2014 Air Pollution Ranking*. 22 noviembre, 2019, de https://aqicn.org/faq/2015-05-16/world-health-organization-2014-air-pollution-ranking/.

[91] Krzyzanowski, M., Apte, J. S., Bonjour, S. P., Brauer, M., Cohen, A. J. and Prüss-Ustun, A. M., (2014). Air pollution in the mega-cities. *Current Environmental Health Reports*, 1(3), pp. 185-191.

[92] Hasenfratz, D., Saukh, O., Walser, C., Hueglin, C., Fierz, M., Arn, T., Beutel, J. and Thiele, L., (2015). Deriving high resolution urban air pollution maps using mobile sensor nodes. *Pervasive and Mobile Computing*, 16, pp. 268-285.

[93] Soares, A., & Pereira, M. J. (2007). Space–time modelling of air quality for environmental-risk maps: A case study in South Portugal. *Computers & Geosciences*, 33(10), 1327-1336.

[94] Pope III, C. A., Burnett, R. T., Thun, M. J., Calle, E. E., Krewski, D., Ito, K., & Thurston, G. D. (2002). Lung cancer, cardiopulmonary mortality, and long-term exposure to fine particulate air pollution. *Jama*, 287(9), 1132-1141.

[95] Chattopadhyay, S., Gupta, S. and Saha, R. N., (2010). Spatial and temporal variation of urban air quality: a GIS approach. *Journal of Environmental Protection*, 1(03), p. 264.

In: Air Pollution: Effects and Dangers ISBN: 978-1-53619-544-6
Editor: J. Esteban Colman Lerner

Chapter 10

DATA SCIENCE APPROACH FOR URBAN MONITORING NETWORK ASSESSMENT WITH APPLICATION IN BUENOS AIRES

Natacha S. Represa*[1,2,*]*, María F. García Ferreyra*[3]*,
***Gabriela Abril*[4] *and Andrés Porta*[1]**

[1]Centro de Investigaciones del Medioambiente (CONICET-UNLP), La Plata, Argentina.
[2]Geo-Environmental Cartography and Remote Sensing Group, Polytechnic University of Valencia, Valencia, Spain.
[3]Comisión Nacional de Actividades Espaciales (CONAE), Falda del Cañete, Argentina
[4]Autoridad de Cuenca Matanza Riachuelo (ACUMAR), Ciudad Autónoma de Buenos Aires, Argentina

ABSTRACT

Low air quality is a public health problem with severe global repercussions. This work presents a data science approach for obtaining

[*] Corresponding Author's Email: solrepresa@quimica.unlp.edu.ar.

relevant information from an urban monitoring network assessment. The long-term concentration of carbon monoxide (CO), nitrogen dioxide (NO_2), and particulate matter (PM_{10}) were examined at three stations (Córdoba, Centenario, and La Boca) in Buenos Aires for the period 2009-2017. The methodology allowed the evaluation of the monitoring network, the analysis of extreme events, and local meteorology's influence on the recorded levels. The description of the hourly, daily, and seasonal behavior of pollutants and the determination of concentration trends using the Mann-Kendall trend test were also included. The analyzed pollutants have reached high maximum values, which poses a risk to the exposed population's health. All stations showed a significant reduction in NO_2 concentrations, but only two stations had a reduction in PM_{10}. We observed significant increases in CO levels. These results provide evidence of air quality degradation and provide critical information for regional government policy development.

1. INTRODUCTION

Urban air quality has become an issue of global concern in the last decades [1]. Growth in populations, vehicular traffic, industrialization, and energy usage have led to increased air pollution levels and subsequent public health effects at the urban, regional and global scales [2-4].

Numerous studies have reported that human exposure to air pollutants may cause serious health concerns such as eye irritation, breathing difficulty and cardiovascular problems [5, 6]. Besides, long-term exposure may damage the immune, neurological, reproductive, and respiratory systems, cause lung cancer and even premature death [7-11]. Furthermore, poor air quality contributes to climate change and negatively impacts ecosystems [12-14].

Carbon monoxide (CO), nitrogen dioxide (NO_2) and particulate matter (PM) are pollutants with particular relevance in urban areas. The rapid photochemical reaction between tropospheric ozone (O_3) and nitrogen monoxide (NO) produces NO_2 during fossil fuels combustion. While CO is a product of incomplete combustion as well as photochemical breakdowns of hydrocarbons [5]. Finally, the PM is a complex and varying mixture of particles suspended in the breathing air, which are produced by an assortment of natural and anthropogenic activities, such as factories, power plants, refuse incinerators, motor vehicles, construction activity, biomass burning, and natural windblown dust [15, 16].

Air quality management programs provide essential tools to prevent health impacts. With an important relevance, the air monitoring in large cities provides the levels of air pollutants for analysis [17, 18]. The large amount of data generated presents a challenge in its processing and interpretation. Moreover, decision-makers are interested in knowing the risk of predictable diseases to manage the resources available better [19].

Data science is the science of extracting meaning from complex data, hence supporting decision-making in an increasingly complex world [20]. Exploratory data analysis is one of the first steps when the objective is the characterization of a database. The exploratory analysis consists of step series, such as identifying the variables and the type of data, transforming the variables, and studying missing values, outliers, and trends. This approach provides a greater degree of knowledge, favoring the elaboration of research questions. Data science is emerging as a major new area of study, with important implications for diverse areas as trade, logistics, and health [21-23]. However, incorporating data science into environmental management is late compared to the study of other sciences [24, 25].

This paper proposes a set of data science techniques to achieve the integration of meteorological and air pollutant data to assess long-term air quality in the context of a risk analysis. The assessment aims to examine and describe the variations in concentrations of the above pollutants, their temporal trends, and seasonality of air quality and determine the risk. Also, this work presents and discusses for the first time an exhaustive analysis of hourly concentrations of NO_2, CO, and PM_{10} over eight consecutive years (2009-2017) in Buenos Aires, Argentina.

2. Materials and Methods

2.1. Study Area Description

Argentina has yet few studies on air quality [26-28]. Most research studies have been focused on topics including impact assessment of specific sources [29-34], characterization of multi-elemental aerosol composition [35-38], estimation of emission inventories [27, 3 9-41], implementation of pollutants transport models within the atmosphere [41-44], development of bio-monitoring studies [45-47] and source identification through receptor models [48]. However, these studies do not provide specific methodology to assess the trend of air quality nor the seasonal behavior of atmospheric pollutants.

Systematic monitoring of air pollutants is also sparse in space and time in Argentina. Most of the time, measurements belong to scientific campaigns to study a specific event [49]. These measurements are made during a limited period and respond to specific objectives, making it difficult to use them for other purposes. Even so, the most prolonged air quality continuous monitoring in Argentina has been carried out in the Buenos Aires city by the Environmental Protection Agency (APrA) of the Government of the Autonomous City of Buenos Aires (CABA).

Buenos Aires monitoring network has three sites that measure hourly concentrations of NO_2, CO, and PM_{10} since 2009. Measurements analysis and meteorological data provide relevant information for air quality management.

Buenos Aires, the capital of Argentina, is located at 34.61°S, 58.37°W, on the western riverside of the Río de la Plata (Figure 1). It is a shallow, large-scale, coastal plain estuary that covers an approximate area of 35 000 km^2 [50]. The whole terrain is flat with only small height differences of less than 30 m [51].

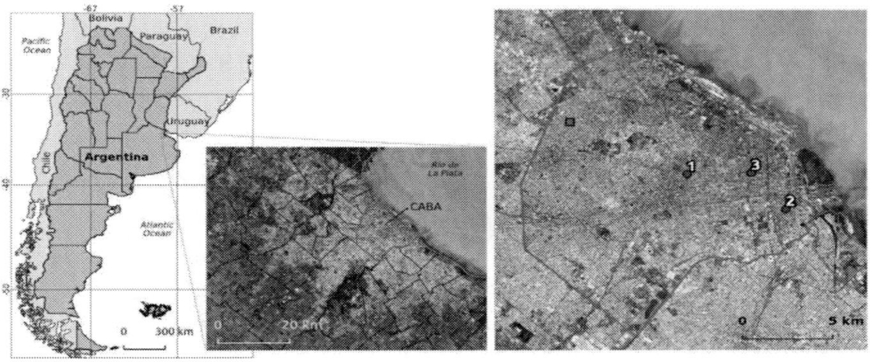

Figure 1. Map of the Metropolitan Area of Buenos Aires showing the location of the three monitoring sites within CABA (red circles: 1- Centenario station 2- La Boca station 3- Córdoba station) and the weather station (blue square).

Buenos Aires is immersed in a continuous urban area, known as the Metropolitan Area of Buenos Aires. With a population density of 3 600 inhabitants per km^2, it is the most populated area of the country [52], the tenth megalopolis in the world and the third in Latin America. Arkouli et al. [51] have pointed out that the primary sources of air pollution in Buenos Aires are traffic and the industrial center that includes power plants to the southeast of the city.

2.2. Air Quality Data

Air quality data was provided by the government of the city of Buenos Aires through the website[1]. Nowadays, the air quality monitoring network has three EPA type stations, which provide hourly data for NO_2, CO and PM_{10} following Title 40, Part 53 of the Code of Federal Regulations (40 CFR Part 53, USA). The analysis techniques used are Gas Phase Chemiluminescence for NO_2, Non-Dispersive Infrared Photometry for CO and Beta Attenuation Monitoring for PM_{10}. The period analyzed in this paper extends from 2009-10-01 to 2017-10-01.

---

[1] https://data.buenosaires.gob.ar/, last visited March 2018.

The stations are in diverse areas of the city of Buenos Aires and monitor different scenarios: Centenario (34.60° S, 58.43° W), Córdoba (34.60° S, 58.39° W) and La Boca (34.63° S, 58.37° W), as shown in Figure 1. Córdoba station is placed on a high traffic avenue, in a commercial area of high-rise buildings. Moreover, La Boca station is in the Matanza-Riachuelo basin area, close to a freeway, where the Dock Sud port facility operates, this is a mixed zone of high vehicular flow and stationary sources (such as oil refineries, two thermoelectric plants, waste management plants, deposits of hydrocarbons and non-flammable liquid products, chemical industries, among others). While La Boca and Córdoba stations have been in operation since 2009, Centenario station began operating in 2005. It is on the outskirts of Centenario park, an area of dense vegetation, in a commercial area with a medium vehicular flow and low influence of stationary sources. A fourth monitoring station, Palermo, which has operated from 2002 to 2010, has been shut down due to technical problems. Its data was not included in this study since the records do not wholly coincide in time with the other stations.

The calibration program consists of two-point calibrations every 7/15 days, and multipoint calibrations every three months, or when the equipment requires maintenance. All calibrations are carried out against gaseous standards according to EPA protocol, making dilutions. Regarding particulate material, flow calibrations and head cleaning are monthly operations, and every 2 months, a two-point calibration is performed, according to the operation of the equipment. All the equipment is subject to an annual preventive and corrective maintenance program according to the specifications of each manufacturer.

The concentration units reported by APrA were maintained since the pollutants were not compared with each other. On the other hand, the hourly data communicated by APrA for the CO concentration corresponds to the moving average of the previous 8 hours; for NO_2, the previous 60 minutes, and for PM_{10}, the previous 24 hours.

2.3. Meteorological Data

Meteorological data contributed to the interpretation of the pollutant concentrations observed at the urban-traffic station (Córdoba), at the urban-background station (Centenario), and the urban-industrial station (La Boca). Given the unavailability of measurements at continuous monitoring stations, meteorological data provided by the National Weather Service (SMN) were used. We worked with hourly surface data for wind speed and direction. The SMN dataset comes from the Buenos Aires Observatory station (34.58° S 58.50° W), the nearest meteorological station located at approximately 7, 10 and 13 km away from the air quality monitoring stations (Figure 1).

2.4. Data Analysis and Mining

As a first step, we performed a descriptive statistical analysis of the air quality measurements using hourly data. The time series analysis allowed the detection of events where the pollutants concentrations reached extremely high values. The wind frequencies for the study area were also analyzed, and the wind rose for the different seasons was plotted. Likewise, the hourly, monthly, and annual variation of the pollutant concentrations were studied. We performed the seasonal analysis following the parameters used by the SMN, which defines the seasons according to climatic conditions. Therefore, summer was considered from December to February, autumn from March to May, winter from June to August, and spring from September to November.

The monitoring network was evaluated through an analysis of the redundancy of the information collected. Pearson's correlation coefficient provided an assessment of the relationship between the data recorded in each monitoring station. Then, we evaluated the redundancy between stations by hierarchical cluster analysis, using the Euclidean distance as the agglomeration method.

We calculated significant upward or downward trends with a non-parametric method based on Mann-Kendall with Theil-Sen slope [53]. Theil-Sen is a non-parametric regression technique that is insensitive to outliers and provides the average slope of all slopes calculated from each possible pair of values [54]. The Theil–Sen estimator of a set of points (x_i, y_i) is the median of the slopes $(y_j - y_i)/(x_j - x_i)$ for all pairs of sample points. The confidence interval for the slope estimation is given by the range containing the average 95% slope of the lines given by the pairs of points.

Finally, we calculated the annual exceedances of WHO recommendations and the relative risk (RR) associated with PM_{10} levels. The RR of premature deaths from short-term PM_{10} exposure was calculated according to Ostro & World Health Organization [55] using the formula (equation 1):

$$RR = e^{\beta (x - x_0)} \tag{1}$$

where $\beta = 0.0008$ and $x_0 = 10$ μg.m^{-3}. Once the RR has been determined, the attributable percentage of deaths (APD) can be known, which is the percentage of deaths that could be avoided if the levels of the contaminant were reduced (equation 2):

$$APD = \frac{RR - 1}{RR} \times 100 \tag{2}$$

We conducted all the statistical elaborations of this study using the programming language R, working mainly with the packages “open-air” [56] and “ggplot2” [32].

3. Results and Discussion

3.1. Descriptive Statistics of the Pollutant's Measurements

This study analyzed a dataset of three airborne pollutants (CO, NO_2, and PM_{10}), recorded at urban background sites in Buenos Aires. The period explored in this work ranges from October 1st, 2009 to October 1st, 2017, with 70 115 records for all pollutants and stations.

While Córdoba station has the lowest number of records (only 65% for the entire period), La Boca station is the most complete (76%). The most considerable lack of measurements at Córdoba station occurred during 2013, with less than 50% of the hourly data kept. At the same time, all stations were missing entries for the last months of 2015. For PM_{10} measurements, records are only available from 2011 onwards.

Table 1 reports the results of the descriptive statistics on hourly measurements obtained for the different atmospheric pollutants for the whole measurement period. Analyzing CO for the entire period, we observed that while maximum values (50.0, 47.3 and 44.7 ppm for Centenario, Córdoba and La Boca stations, respectively) differ between the three stations by more than twice their standard deviations (0.9, 1.1, and 1.0 ppm), medians and means (0.5 and 0.6 ppm) are equal for Centenario and Córdoba stations, while are lower for La Boca station (0.2 and 0.3 ppm). As for NO_2, the median concentrations are within those expected for urban areas [57-59], in the range of 17-23 ppb. The wide dispersion of NO_2 data from Centenario station before the interruption of measurements during 2013 is remarkable. Regarding the median for PM_{10} concentrations, the three monitoring stations presented close results (25-26 μg m^{-3}). For all chemical species, the minimum values recorded were below the detection limit of the instrument.

Table 1. Descriptive statistics of concentration values obtained for the different atmospheric pollutants in each of the stations, for the analyzed period (DL = detection limit)

Pollutant	Station	Min	Median	Mean	Max	SD
CO (ppm)	Centenario	<DL	0.5	0.6	50.0	0.9
	Córdoba	<DL	0.5	0.6	47.3	1.1
	La Boca	<DL	0.2	0.3	44.7	1.0
NO_2 (ppb)	Centenario	<DL	18.0	19.6	155.0	9.7
	Córdoba	<DL	23.0	26.7	283.0	16.9
	La Boca	<DL	17.0	19.6	129.0	11.8
PM_{10} ($\mu g\ m^{-3}$)	Centenario	<DL	26.0	30.0	427.0	17.9
	Córdoba	<DL	25.0	28.5	253.0	16.2
	La Boca	<DL	26.0	29.3	387.0	17.2

Table 2. Seasonal meteorological parameters: wind speed ($m\ s^{-1}$), ambient temperature (°C) and relative humidity (%) for the period analyzed

	Wind speed ($m\ s^{-1}$)			**Temperature (°C)**			**Relative humidity (%)**		
Season	***Min***	***Mean***	***Max***	***Min***	***Mean***	***Max***	***Min***	***Mean***	***Max***
autumn	0.0	2.0	10.3	3.2	18.2	34.8	24	75	100
spring	0.0	2.6	11.4	3.4	18.0	35.2	20	69	100
summer	0.0	2.1	13.3	9.7	24.6	38.6	18	66	100
winter	0.0	2.6	12.8	-0.6	12.2	30.5	21	76	100

3.2. Meteorological Analysis

As mentioned in the literature review, the region has a humid subtropical climate [60]. Table 2 shows the seasonal statistical summary of meteorological parameters, including wind speed, ambient temperature, and relative humidity. Winters and summers are humid, with an average annual humidity of 71%.

Winds had an annual calm rate of 3.8% and an annual average wind speed of 2.3 m s^{-1}. Throughout the year, the prevailing winds came from the NE, except in winter when the wind from the NW sector became more frequent (Figure 2). The most considerable mean seasonal value of wind speed was during the summer (13.3 m s^{-1}). In previous works, during this season, the largest mean daily seasonal values of the recirculation factor have been encountered [61, 62].

It is important to note that the low seasonal variability can affect the wet deposition process, especially for PM_{10} [63]. Variations in wind speed also influence the airborne pollutant levels. Prolonged calm periods provide atmospheric stability that prevents the dispersion of pollutants. In this context, persistent concentrations of pollutants are expected.

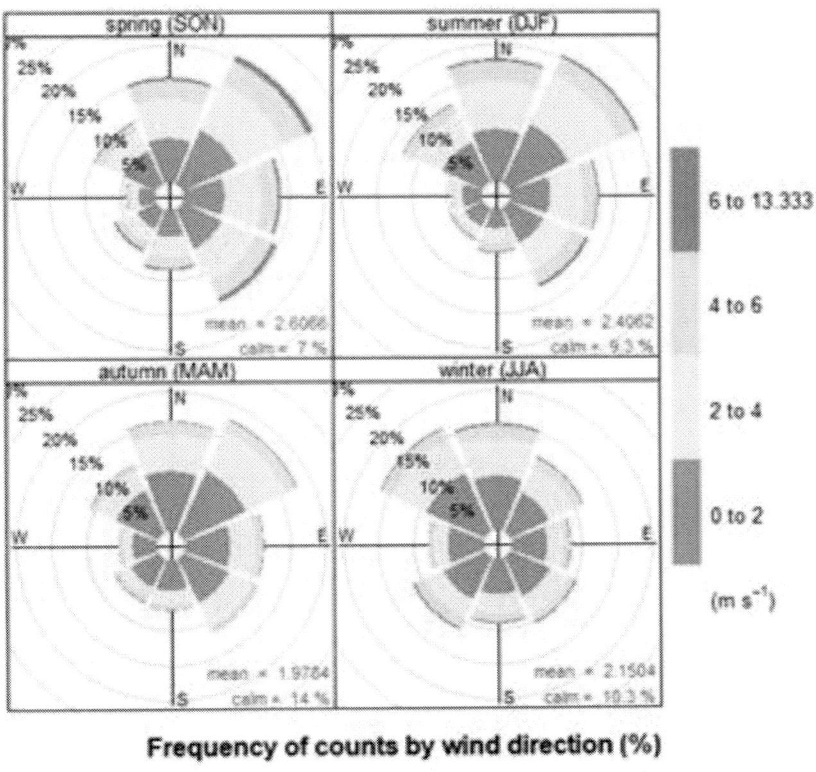

Figure 2. Seasonal wind roses in the period 2009 -2017. Wind speeds into the intervals shown by the scale. The grey circles show the frequency of counts by wind direction.

3.3. Impact of Local Sources Assessment

Pollution roses using the concentration for the different pollutants at each station are shown in Figure 3. In this way, the wind directions that contributed most to the total average concentrations allow the identification of the provenance sectors of the most relevant sources.

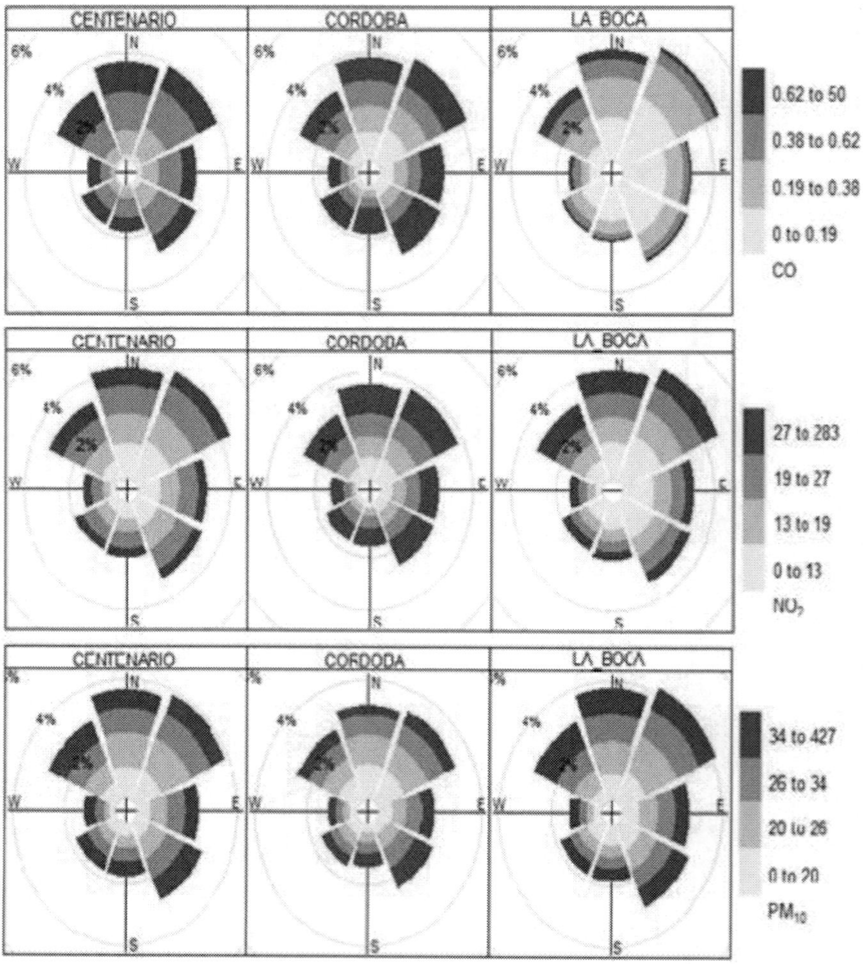

Figure 3. Pollution roses from monitoring stations for CO (ppm), NO_2 (ppb) and PM_{10} ($\mu g\ m^{-3}$). The frequency of counts by wind direction (%).

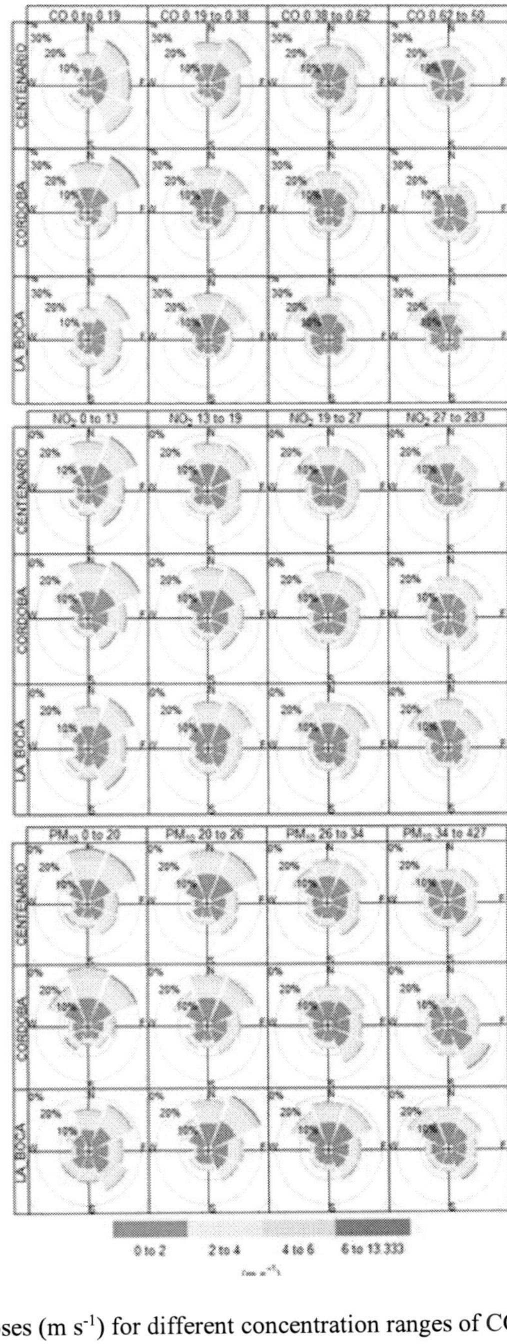

Figure 4. Wind roses (m s^{-1}) for different concentration ranges of CO (ppm), NO_2 (ppb) and PM_{10} ($\mu g\ m^{-3}$). The frequency of counts by wind direction (%).

At Centenario station, the highest concentrations of CO and NO_2 arrived more frequently from the NW sector, even though N and NE sectors made a considerable contribution, while the main PM_{10} contributions came from the NW and SE sectors. As for Córdoba station, the highest concentration of CO contributions came from SE, while the highest NO_2 concentrations came from N, NE, E and SE, and as for PM_{10}, the highest concentrations came from SE. Finally, at La Boca station the highest concentrations of CO and NO_2 came from NW > N > NE, while the contributions of PM_{10} were also higher from NW, however, the contributions of N, NE and SE were comparable.

There were no sectors that contributed substantially to the average concentrations of the three stations. As described above, these monitoring stations are in the urban centre, where airflow is low-moderate, and there are different sources with relatively similar intensity. Because of these characteristics, the proposed methodology has some limitations [64] that were overcome by the analysis of wind roses for different concentration ranges (Figure 4). Considering the influence of wind speed in the concentration of atmospheric pollutants, the highest concentrations of CO and NO_2 occur when winds are light or calm, while for PM_{10}, more intense winds contribute to dust resuspension events.

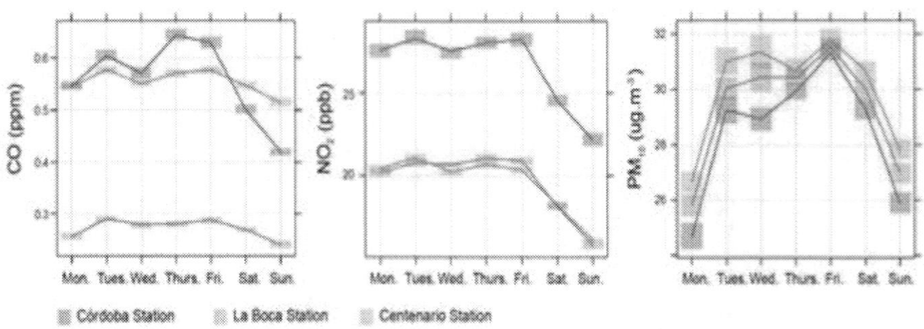

Figure 5. Daily averages of the hourly concentrations and 95% confidence interval calculated for the study period at each station.

3.4. Weekly, Daily and Hourly Analysis

The weekly and daily behavior of the pollutants are the results of the interaction of different factors, including the interaction between the emission intensity and photochemistry processes [14]. In this sense, the levels of pollutants analyzed decreased during the weekend and non-working hours when productive activities and vehicular traffic decline. Figure 5 shows the weekly behavior of pollutants by the monitoring station. In the case of CO, Centenario and Córdoba stations showed similar concentrations throughout the week, while in La Boca the measurements were lower. Córdoba station had a sharp decrease in CO levels towards the weekend when for the other stations the levels slightly changed. That indicates a strong relationship between the levels of CO and the productive activities that take place surrounding this station. Photolytic degradation of fuel aldehydes may cause high concentrations of CO [14].

NO_2 presents a different pattern, where all stations reported a sharp reduction in levels by the end of the week. Córdoba station showed higher concentrations of NO_2 than La Boca and Centenario stations. This result fulfils the prediction since Cordoba station is in a high transit point of the city of Buenos Aires, with majority contributions of pollutants from transport.

The weekly behavior of NO_2 and CO shows that while transit is the only source of CO at Córdoba station, a secondary source with a high CO contribution appears throughout the week at Centenario station. Due to its proximity, a possible source is the Luis Pasteur Zoonosis Institute, where animal cremation takes place. Contrary to the other pollutants, PM_{10} concentrations were similar during the week at the three monitoring stations, with lower average levels on Mondays and maximum levels on Fridays.

The hourly behavior of CO and NO_2 presented typical patterns for urban areas such as peaks related to high vehicular flow during the midday and afternoon. In Figure 6, it is possible to see an increase in concentrations from 6 am until noon when concentrations decrease slightly. Afterwards, concentrations increase again, reaching a maximum daily concentration around 6 pm, and then the levels fall again. The decrease between the two peaks could be related to different factors, among which we can highlight: lower emissions, enhanced atmospheric dispersion potential, and photochemical processes, particularly for NO_2. This type of behavior in urban areas has been described in other publications on the subject [65, 66]. A brief previous study in Buenos Aires pointed to a peak in NO_2 concentrations at 2 am [27] that we did not detect in this historical review report.

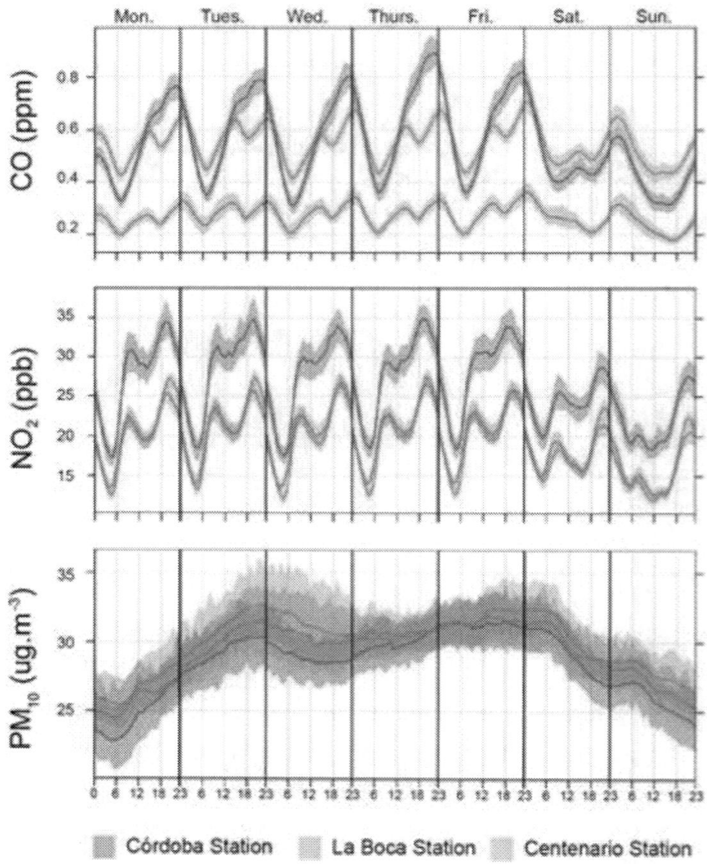

Figure 6. Hourly variation (local time) of the pollutant's levels for each day of the week and the 95% confidence interval in the mean.

The days of the weekend presented a characteristic behavior. The maximum values moved towards the first hours of the day. Particularly on Sundays, CO levels reached minimum values between 16-18 h while for NO_2 it occurs at 12-14 h.

The PM_{10} showed with higher intensity an accumulation capacity: concentrations increased at the beginning of the week, then remained at an average of 30 $\mu g\ m^{-3}$ on weekdays and decreased in the early hours of Saturday. The unusual behavior observed for PM_{10} may be due to pre-treatment data that includes the 24-hour moving average of the concentrations, which makes the variations that may be present within this time range more subtle. That is an important aspect to highlight, as it implies continuous exposure to a higher level of particulate matter than the average annual level recommended by the WHO [67].

3.5. Seasonal Variations

The seasonal averages of CO and NO_2 in all seasons (Figure 7) followed the pattern of decline winter > autumn > spring > summer, with significant differences between them ($p < 0.001$). Comparison of the seasonal mean data for CO showed the maximum peak in winter (0.8 ppm) at Córdoba station and a minimum value in summer (0.2 ppm) at La Boca station. In turn, NO_2 presented a maximum of seasonal averages during the winter (33.7 ppb) at Córdoba station, while the minimum seasonal value (15.7 ppb) was during the summer at La Boca station. This type of behavior is consistent with those reported by other authors in different cities [65, 66]; in winter, weather conditions and the high emission contribute to increased high levels of pollutants [68].

Nonetheless, PM_{10} concentrations had a different pattern for seasonal averages, with the lowest concentrations in autumn. PM_{10} reached the maximum seasonal mean value (33.7 $\mu g\ m^{-3}$) during winter in Centenario station and a minimum (26.2 $\mu g\ m^{-3}$) in autumn in La Boca station.

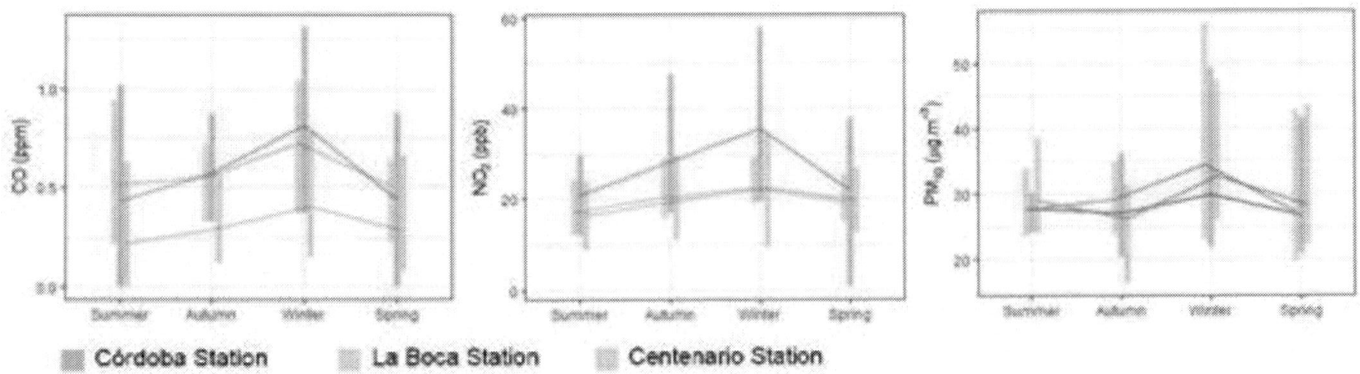

Figure 7. The average concentration of pollutants by season at each monitoring station.

While CO and NO_2 showed a characteristic behavior associated with increased winter emissions, reduced solar radiation, and lowered mixing layer height, PM_{10} concentrations did not show a marked seasonal decline (Figure 7). That may be due to the constant relative humidity and low surface wind speed throughout the year that keeps aerosol particles within the region of origin.

3.6. Monitoring Network Redundancy

The current monitoring network for NO_2, CO and PM_{10}, was designed according to the methodology proposed by Venegas & Mazzeo [69]. They obtained the multi-pollutant urban air-quality monitoring network through air quality simulation and objective analysis based on monitoring tasks to detect concentrations greater than reference concentration levels (RCL). Spatial representativeness of detected exceedances was evaluated by analyzing air quality simulations of cases with concentration > RCL at any monitoring site and obtaining the areas in the city where concentrations exceeded RCL with 70%, 80%, and 90% of probability.

If the city of Buenos Aires were divided into quadrants, the three monitoring stations are in quadrant I, with a distance between each other of 4 km (Centenario and Córdoba), 3.5 km (Córdoba and La Boca) and 6.6 km (Centenario and La Boca) and located from the Rio de La Plata at 2 km (La Boca), 4 km (Córdoba) and 6 km (Centenario) away. Since stations are relatively close to each other and within urban areas, one of the main interests of this study has been to evaluate the representativeness of the monitoring network. The first step was to evaluate the strength of the relationship between the stations for each of the pollutants. In all cases, the correlations were significant and high. For CO, the three stations showed an $r > 0.9$ ($p < 0.05$). The same happened for PM_{10}, with $r > 0.85$ ($p < 0.05$) in all cases. A lower correlation presented NO_2, where the highest relationship was between Centenario and La Boca ($r = 0.69$, $p < 0.05$) and the lowest for Centenario and Córdoba ($r = 0.56$, $p < 0.05$). As well, we applied hierarchical cluster analysis to group monitoring sites based on the similarity of the pollutant concentration values (Figure 8). CO and PM_{10} showed similar results; Centenario and Córdoba stations constituted a single cluster. However, Centenario and La Boca stations made a cluster for NO_2. Results indicate that the stations have similar measurements for the different pollutants, although none can be discarded as they record different atmospheric conditions and degrees of exposure of these pollutants.

3.7. Long-Term Trends (2010-2017)

For the quantification of trends over time we calculate the non-parametric Theil-Sen estimator of the slope at the monthly mean, and to examine changes in slope, we analyze the trends of the pollutants for each wind direction (Table 3).

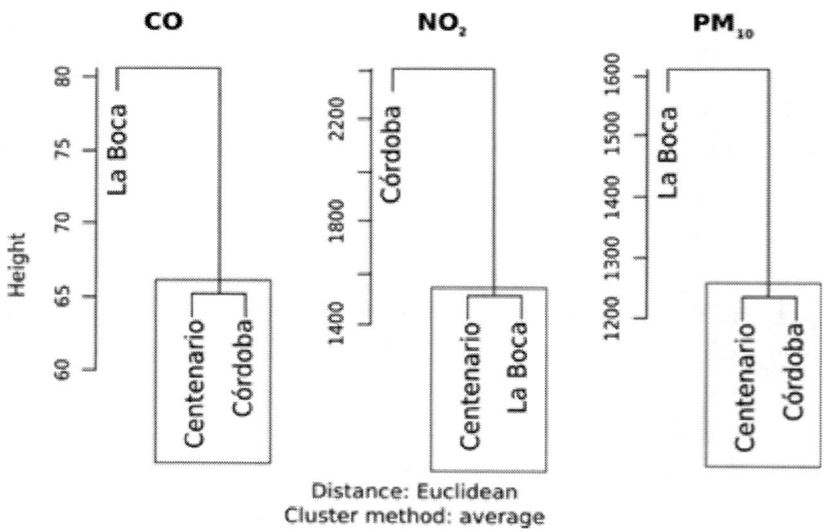

Figure 8. Cluster dendrograms for each pollutant.

Table 3. Significative long-term trends for different wind directions with 95% confidence interval in brackets

NW	N	NE
W		E
SW	S	SE

	Centenario			**Córdoba**			**La Boca**		
CO (ppm year^{-1})	-0.03 [-0.06, -0.01]**	-0.02 [-0.05, 0]**	-0.02 [-0.04, 0]**	0.04 [0.01, 0.06]*	*ns*	0.05 [0.02, 0.07]*	*ns*	*ns*	0.03 [0.02, 0.04]*
	ns		-0.03 [-0.05, -0.01]**	*ns*		*ns*	0.03 [0.01, 0.05]*		0.03 [0.02, 0.04]*
	ns	*ns*	*ns*	*ns*	*ns*	*ns*	0.04 [0.02, 0.05]*	0.03 [0.01, 0.04]*	0.03 [0.01, 0.04]*
NO_2 (ppb year^{-1})	-1.78 [-2.25, -1.27]*	-1.44 [-1.85, -0.97]*	-1.32 [-1.73, -0.97]*	-1.69 [-2.35, -1.06]*	-1.48 [-2.34, -0.74]*	-1.68 [-2.61, -1.04]*	-2.51 [-3.21, -1.84]*	-2.15 [-2.63, -1.60]*	-1.87 [-2.35, -1.29]*
	-1.23 [-1.66, -0.77]*		-1.06 [-1.48,-0.61]*	-1.18 [-1.91, -0.62]*		-1.12 [-2.31, -0.32]**	-1.67 [-2.32, -0.91]*		-1.47 [-1.88, -0.87]*
	-1.10 [-1.48, -0.70]*	-1.36 [-1.82, -0.87]*	-1.01 [-1.45, -0.57]*	-1.26 [-1.97, -0.66]*	-1.59 [-2.32, -0.83]*	-1.25 [-2.23, -0.25]**	-1.17 [-1.71, -0.56]*	-1.25 [-1.70, -0.69]*	-1.29 [-1.82, -0.66]*
PM_{10} (µg.m^{-3} year^{-1})	-2.68 [-3.74, -1.82]*	-2.00 [-2.77, -1.31]*	-1.63 [-2.22, -1.09]*	-1.79 [-2.55, -0.97]*	-0.91 [-1.80, -0.19]**	*ns*	*ns*	*ns*	*ns*
	-1.72 [-2.83, -0.82]*		*ns*	-1.74 [-2.93, -0.70]*		*ns*	*ns*		*ns*
	-2.04 [-3.16, -1.11]*	-2.15 [-3.24, -1.32]*	-1.92 [-2.79, -1.09]*	-1.79 [-2.98, -0.84]*	-1.78 [-2.76, -0.71]*	-1.13 [-1.93, -0.13]***	*ns*	*ns*	*ns*

ns non-significant

* $p < 0.001$; ** $p < 0.05$

The CO trend had a positive slope in Córdoba and La Boca stations. For the Córdoba station, contributions were significant from the NW and NE. Regarding La Boca station, the trend was increasing with significant contributions from the NE, E, SE, S, SW, and W. Although the trends were positive, the projection of the CO concentration does not come close in the short term to the limit value recommended by the WHO for the protection of the health of the population [70]. On the other hand, the decreases in the monthly average were consistent from the NW, N, NE, and E sectors in Centenario station.

NO_2 concentrations had a clear downward trend with high statistical significance over the period studied at all stations, and all wind directions. PM_{10} monthly mean concentrations also presented an overall decreasing trend at Córdoba (though not for NE and E directions) and Centenario (not for E direction) stations. However, the trend concentration for La Boca station was not significant in any wind direction. Car-fleet technology improvements, such as the use of catalytic converters in exhaust pipes and the less frequent use of diesel-fueled vehicles that emit a higher fraction of NO_2 compared to gasoline vehicles may lead to decrease NO_2 and PM_{10} concentration [66].

3.8. Particular Events

Extreme air pollution events are of concern as they represent a risk to human health and can produce changes in ecosystems [71]. Therefore, the extreme values of the period analyzed were studied in their context, analyzing the particularities that happened within the previous days. The extreme events in the eight years of measurements were identified and characterized as natural or anthropic events.

Regarding CO, there are three different occasions when concentrations exceeded the limit values (25 ppm h^{-1}) indicated by WHO, endangering the health of exposed people [70]. These events occurred on October 25-26, 2010, June 23-24, and December 23-24, 2014.

Of these, the most important event took place from June 23rd to 24th, 2014, at noon, and the three monitoring stations detected it. Figure 9 illustrates the levels of CO measured at the different sites at the same times. It is possible to see in the time series that the levels reached by CO are extremely high concerning the concentrations recorded during the previous day. In turn, we observed that elevated concentrations extend over several hours. The synchronicity of the registers and the continuity in the measurement of the high values indicate that the equipment recorded these values without any operating errors.

The highest record of CO was at the Centenario station at 3 am on June 24th (50 ppm), and noon on the same day at the Córdoba station (47 ppm). At La Boca station, an average concentration of 32 ppm was maintained throughout the event, contrasting with the regular measurements of 0.3 ppm. The climatic conditions associated with this event were an average wind speed of 0.9 m s^{-1}, an average temperature of 14 ºC and relative humidity of 94%. Afternoon on June 24th, CO levels returned to their previous levels.

Furthermore, high CO concentrations were measured by a single station at two other times during the period analyzed. On October 25th and 26th, 2010, La Boca station recorded an event that reached a concentration of 45 ppm. For its part, Córdoba station measured high values from December 23rd to 24th, 2014, registering 33 ppm of CO as the highest value.

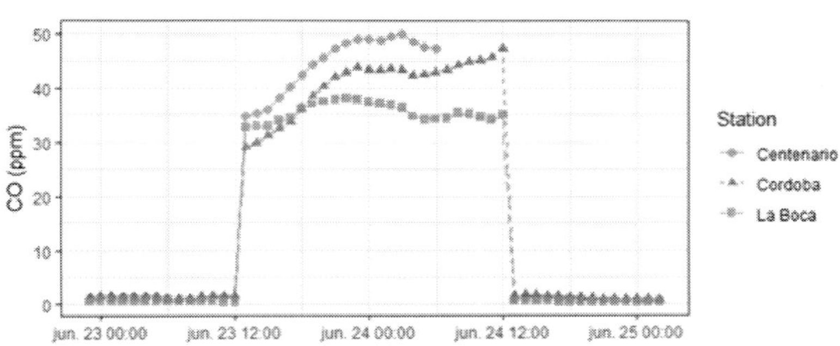

Figure 9. Time series of CO concentrations recorded during a particular event on June 23 - 24, 2014.

There was no mention of these events in previous scientific works or public/institutional reports. The non-existence of public records of accidental releases hinders an explanation for these measurements. At the same time, given the insufficient systematic records of the health system, it is not possible to analyze the implications of these occurrences for the exposed population.

Extreme values of NO_2 are frequent in the period analyzed, reaching extreme high concentrations that exceed WHO recommended levels [67]. These extreme levels (> 104 ppm for short-term) occurred 232 times at Córdoba station, seven times at La Boca station and two times at Centenario station. The fact that Cordoba station is on a high-traffic avenue circumscribed by high buildings may explain the number of excesses. Road traffic and heating emit nitrogen oxides which accumulate in the air due to the lack of ventilation in the city. Within the atmosphere, and in the presence of organic compounds and solar radiation, NO reacts with ozone to form secondary NO_2. Therefore, NO_2 concentrations can fluctuate considerably over time due to the strong influence of anthropogenic and natural factors.

Additionally, periods of increased PM_{10} concentrations were consistent with events of volcanic activity. The maximum PM_{10} values recorded were in July 2011, at Centenario and La Boca stations, and in October of the same year at the three stations (Figure 10). These dates are consistent with the months of the eruption of the Puyehue-Cordón Caulle Chilean volcanic complex in the Andes Mountains, located at 1 400 km from Buenos Aires. The eruption started on June 4th, 2011 and lasted several months emitting ash. The ash had circumpolar transportation, travelling east, and passing over South Africa, South of Australia, and New Zealand to reach Chile again after ten days [72, 73].

Another period of high PM_{10} concentrations coinciding with volcanic activity occurred in April 2015, when a swarm of volcanic-tectonic earthquakes, including the Calbuco volcano, erupted after decades of inactivity. Several days later, the winds carried away the ashes of the volcano's eruption and on Saturday, April 25th, the ash cloud reached the city of Buenos Aires and southern Uruguay [74].

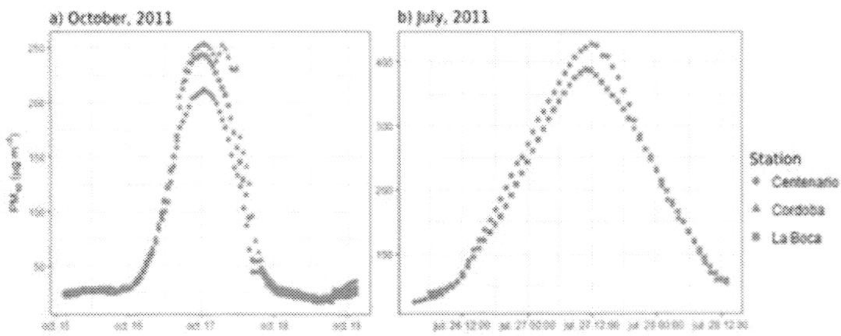

Figure 10. Time series of the PM_{10} concentrations measured at particular events.

It is important to note that the patterns and mechanisms of dust entrainment, transport, and deposition remain under-explored subjects for this region [75-77]. Previous research has shown that Patagonian dust can travel thousands of kilometers [59, 78, 79], and in certain weather conditions, gets transported in an easterly/north-easterly direction [80, 81]. In this sense, it would be interesting to promote studies in the region that can relate measurements of air quality with the transport of Patagonian dust.

3.9. Air Quality Risk Assessment

Numerous epidemiological studies indicate that CO, NO_2, and PM_{10} are good indicators of the relationship between air pollution and health. NO_2 is associated with adverse health effects, including mortality and morbidity [11]. Particles with a diameter of fewer than 10 μm constitute the so-called inhalable fraction, with severe effects on human health [10, 11]. PM_{10} exceeded the WHO recommended annual mean value of 20 $\mu g.m^{-3}$ for every year at the three monitoring stations. According to current inventories, the main responsible for the total emissions of inhalable particles is the traffic (72%), followed by power plants (18%) [68]. Annual average NO_2 concentrations were higher than maximum value compatible with adequate health protection suggested by WHO (21 ppb) for at least one of the monitoring sites in 2009, 2010, 2011, 2012, 2013 and 2017 (Figure 11).

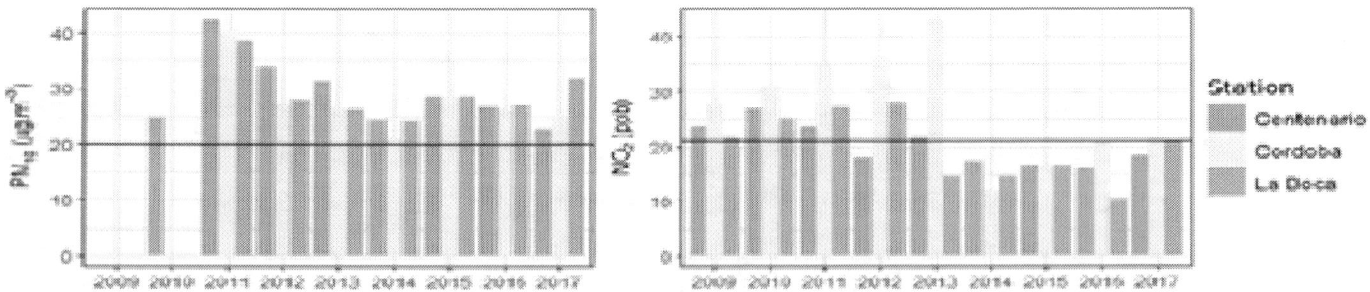

Figure 11. Annual averages of (a) PM_{10} and (b) NO_2 concentrations and WHO guideline values.

Table 4. Attributable percentage of deaths from short-term PM_{10} exposure (%)

	Centenario	*Cordoba*	*La Boca*
2011	2.6	2.4	2.3
2012	1.9	1.3	1.4
2013	1.7	1.3	1.3
2014	1.2	0.9	1.1
2015	1.5	1.5	1.5
2016	1.3	1.3	1.4
2017	1.0	1.2	1.7

The relative risks (RR) obtained from the PM_{10} levels registered in the stations were in the range 1.01 - 1.03. These results represent between 1 to 3% of increased risk of premature death. In terms of the annual attributable percentage of deaths (APD), the levels were higher in 2011 where they reached 2.6% in Centenario station (Table 4).

4. Conclusion

The incorporation of the data science approach in the study of air quality has allowed an exhaustive analysis of the hourly measurements collected in the city of Buenos Aires. The city has an extension of 203 km^2 and three million inhabitants approximately. Its complexity is given by the confluence of multiple emission sources, mainly from vehicle traffic. In this work, we made the first extensive temporary study of CO, NO_2, and PM_{10} concentrations by the period 2009 - 2017. We conducted a complete statistical analysis of the levels measured in 3 sites (Centenario, Córdoba and La Boca stations) to evaluate the local air quality. That included a description of the hourly, daily and seasonal behavior of the pollutant levels, an analysis of the influence of local meteorology on the levels recorded and an assessment of extreme events. Finally, trends in concentrations over time were studied, and possible causes discussed.

Concerning hourly, daily, and seasonal patterns, CO and NO_2 showed typical behavior in urban areas where maximum peaks coincide with periods of increased production activity and increased combustion processes (mobile and fixed sources). The PM_{10} data used in this study were the moving averages within 24 hours, which affected the analysis of hourly behavior. Despite this, PM_{10} concentrations remained without many changes throughout the workweek between values of 30-35 μg m^{-3} and decreased during the weekend. On the other hand, it did not show a pronounced seasonality. For all pollutants, La Boca station has a lower average concentration of pollutants, possibly related to its location on the periphery of the urban area.

The stations showed a high degree of redundancy in the information collected. The exploratory statistical analysis allowed identifying similar CO concentrations throughout the study period, with averages of 0.5 ppm for the Centenario and Córdoba stations and 0.2 ppm for La Boca. Average NO_2 concentrations were within the expected range for urban areas, between 17 and 23 ppb. For PM_{10}, the average concentration value was 25-26 μg m^{-3} for the three monitoring stations. Even so, having a monitoring network of 3 stations is not enough to represent the different atmospheric conditions and degrees of exposure of most of its inhabitants. Updated emission inventories and atmospheric simulations are needed to assess the feasibility of expanding the monitoring network, identifying new representative sites. Furthermore, considering that the advances in epidemiological studies reveal that fine particles have more adverse health effects since they can easily be deposited deeply in the human lungs, thus increasing the risk for cardiorespiratory diseases, the expansion of the air quality monitoring network for the city of Buenos Aires should include monitoring of $PM_{2.5}$.

When studying the pollutant trends in the period 2009-2017, CO concentrations increased by 9 and 17% at Córdoba and La Boca stations, respectively. In turn, NO_2 concentrations decreased significantly at all stations by 5-7%. For PM_{10}, the trend showed a decrease between 3 and 5% in the stations of Córdoba and Centenario. It is possible to attribute this negative trend to the replacement of the vehicle park, incorporating personal vehicles with catalyst technologies.

The different pollutants have reached extremely high maximum values during the period assessed, and it was possible to identify the events involved with these high concentration values detected, which were associated with particular situations such as the eruption of a volcanic complex in the Andes Mountains.

This long-term assessment reveals annual concentrations of PM_{10} and NO_2 critical to the health of the population. On the other hand, the trends of these concentrations have presented a promising future, which requires the development of public policies that contribute to this sense. Finally, this study shows the need to expand the monitoring network, incorporate wind measurements at the monitoring sites, conduct a local emissions inventory to advance understanding of peak episodes, and include the determination of $PM_{2.5}$ to improve risk estimation.

5. ACKNOWLEDGMENTS

The authors gratefully acknowledge the support of the Argentinean Scientific and Technical Research Council, the Polytechnic University of Valencia, Spain, and the National University of La Plata, Argentina. They also acknowledge the National Weather Service belonging to the Argentine Ministry of Defense, for supplying the meteorological information and the Environmental Protection Agency of the Government of the Autonomous City of Buenos Aires for making air quality data available. Finally, this work was supported by the National Agency for Scientific and Technological Promotion (PICT 2015-0618).

6. REFERENCES

[1] World Health Organization. (2013). *Review of evidence on health aspects of air pollution – REVIHAAP Project*. WHO European Centre for Environment and Health, Bonn, WHO 2013., 309 pp.

[2] Akimoto, H. (2003). Global air quality and pollution. *Science*, 302(5651), 1716-1719.

[3] Molina, M. J. & Molina, L. T. (2004). Megacities and atmospheric pollution. *Journal of the Air & Waste Management Association*, 54(6), 644-680.

[4] Gurjar, B. R., Jain, A., Sharma, A., Agarwal, A., Gupta, P., Nagpure, A. S. & Lelieveld, J. (2010). Human health risks in megacities due to air pollution. *Atmospheric Environment*, 44(36), 4606-4613.

[5] Kampa, M. & Castanas, E. (2008). Human health effects of air pollution. *Environmental Pollution*, 151, 362-367.

[6] Brook, R. D., Rajagopalan, S., Pope, C. A., Brook, J. R., Bhatnagar, A., Diez-Roux, A. V. & Peters, A. (2010). Particulate matter air pollution and cardiovascular disease. *Circulation*, 121(21), 2331-2378.

[7] Zanobetti, A. & Schwartz, J. (2009). The effect of fine and coarse particulate air pollution on mortality: a national analysis. *Environmental health perspectives*, 117(6), 898.

[8] Rückerl R., Schneider, A., Breitner, S., Cyrys, J. & Peters, A. (2011). Health effects of particulate air pollution: a review of epidemiological evidence. *Inhalation Toxicology*, 23(10), 555–592.

[9] Lippmann Morton. (2014). Toxicological and epidemiological studies of cardiovascular effects of ambient air fine particulate matter ($PM_{2.5}$) and its chemical components: coherence and public health implications. *CritRev Toxicol.*, 44(4), 299-347.

[10] Lu, F., Xu, D., Cheng, Y., Dong, S., Guo, C., Jiang, X. & Zheng, X. (2015). Systematic review and meta-analysis of the adverse health effects of ambient PM2.5 and PM10 pollution in the Chinese population. *Environ Res.*, 136, 196-204.

[11] Weinmayr, G., Romeo, E., De Sario, M., Weiland, S. K. & Forastiere, F. (2009). Short-term effects of PM10 and NO2 on respiratory health among children with asthma or asthma-like symptoms: a systematic review and meta-analysis. *Environmental health perspectives*, 118(4), 449-457.

[12] Tai, A. P., Mickley, L. J., Jacob, D. J., Leibensperger, E. M., Zhang, L., Fisher, J. A. & Pye, H. O. T. (2012). Meteorological modes of variability for fine particulate matter (PM 2.5) air quality in the

United States: implications for PM2.5 sensitivity to climate change. *Atmospheric Chemistry and Physics*, 12(6), 3131-3145.

[13] Baklanov, A., Molina, L. T. & Gauss, M. (2016). Megacities, air quality and climate. *Atmospheric Environment*, 126, 235-249.

[14] Seinfeld, J. H. & Pandis, S. N. (2016). *Atmospheric chemistry and physics: from air pollution to climate change*. John Wiley & Sons.

[15] Pöschl, U. (2005). Atmospheric aerosols: composition, transformation, climate and health effects. *Angewandte Chemie International Edition*, 44(46), 7520-7540.

[16] Kam, W., Liacos, J. W., Schauer, J. J., Delfino, R. J. & Sioutas, C. (2012). Size-segregated composition of particulate matter (PM) in major roadways and surface streets. *Atmospheric environment*, 55, 90-97.

[17] Baró Porras, F., Chaparro, L., Gómez Baggethun, E., Langemeyer, J., Nowak, D. J. & Terradas, J. (2014). Contribution of ecosystem services to air quality and climate change mitigation policies. *Ambio*, 43(4), 466-479.

[18] European Environment Agency (2016). *Air quality in Europe — 2016 report*. EEA Report No 28/2016. https://www.eea.europa.eu/.../ air-quality- in-europe-2016/download.

[19] Mateos, A. C., Amarillo, A. C., Carreras, H. A. & Gonzalez, C. M. (2018). Land use and air quality in urban environments: Human health risk assessment due to inhalation of airborne particles. *Environmental research*, 161, pp. 370-380.

[20] Baesens, B. (2014). *Analytics in a Big Data World: The Essential Guide to Data Science and its Applications*, 1st Edn. Wiley Publishing.

[21] Dhar, V. (2013). Data science and prediction. *Commun. ACM.*, 56, 64–73. doi: 10.1145/2500499.

[22] Provost, F. & Fawcett, T. (2013). Data science and its relationship to big data and data-driven decision making. *Big Data*., 1, 1. doi: 10. 1089/big.2013.1508.

[23] Greene, C. S., Jie Tan, J., Ung, M., Moore, J. H. & Cheng, C. (2014). Big data bioinformatics. *J. Cell. Physiol.*, 229, 1896–1900. doi: 10. 1002/jcp.2 4662.

[24] Blair, G. S., Henrys, P., Leeson, A., Watkins, J., Eastoe, E., Jarvis, S. & Young, P. J. (2019). Data Science of the Natural Environment: A Research Roadmap. *Front. Environ. Sci.*, 7, 121. doi: 10.3389/fenvs. 2019.00121.

[25] Philip Chen, C. L. & Zhang, C. Y. (2014). Data-intensive applications, challenges, techniques and technologies: a survey on big data. *Inf. Sci.*, 275, 314–347. doi: 10.1016/j.ins.2014.01.015.

[26] Lijteroff, R., Cortinez, V. & Raba, J. (1999). Urban development and air quality in San Luis City, Argentina. *Environmental monitoring and assessment*, 57(2), 169-182.

[27] Mazzeo, N. A., Venegas, L. E. & Choren, H. (2005). Analysis of NO, NO2, O3 and NOx concentrations measured at a green area of Buenos Aires City during winter time. *Atmospheric Environment*, 39(17), 3055-3068.

[28] Colman Lerner, J. E., Kohajda, T., Aguilar, M. E., Massolo, L. A., Sánchez, A. A., Porta, P., Opitz, G., Wichmann, O. Herbarth. & Mueller, A. (2014). "Improvement of health risk factors after reduction of VOC concentrations in industrial and urban areas." *Environmental Science and Pollution Research International*, 21, 9676-9688.

[29] Olcese, L. E. & Toselli, B. M. (1997). Effects of meteorology and land use on ambient measurements of primary pollutants in Córdoba city, Argentina. *Meteorology and Atmospheric Physics*, 62(3), 241-248.

[30] Tombesi, N., Pozo, K. & Harner, T. (2014). Persistent organic pollutants (POPs) in the atmosphere of agricultural and urban areas in the province of Buenos Aires in Argentina using PUF disk passive air samplers. *Atmospheric Pollution Research*, 5(2), 170-178.

[31] Puliafito, E., Gantuz, M. & Puliafito, J. L. (2010). Characterizing mobile emissions by on-board measurements. *International Journal for Applied Environmental Studies*, 5(2), 297-316.

[32] Wichmann, F. A., Busi, L. E., Cianni, N. F., Massolo, L., Müller, A., Porta, A. & Sly, P. D. (2009). "Increased asthma and respiratory symptoms in children exposed to petrochemical pollution." *Journal of Allergy and Clinical Immunology*, 12 (3), 632-638.

[33] Massolo L., Rehwagen M., Porta A., Herbarth O., Ronco A. & Müller, A. (2010). "Indoor-outdoor distribution and risk assessment of volatile organic compounds in the atmosphere of industrial and urban areas". *Environmental Toxicology*, 25(4), 339-49.

[34] Colman Lerner, J. E., Morales, A., Aguilar, M., Giuliani, D., Ditondo, J., Dodero, V. I., Massolo, L., Sánchez, E. Y., Matamoros, N. & Porta, A. (2014). "The effect of Air Pollution on Children's Health: a Comparative Study between La Plata and Bahía Blanca, Buenos Aires Province, Argentina." In: *WIT Transactions on Ecology and the Environment*, Vol. 181, Environmental Impact II, C.A. Brebbia, G. Passerini, Eds. 659-670. Print ISBN: 978-1- 84564-762- 9. 768 pp.

[35] Sosa, B. S., Porta, A., Lerner, J. E. C., Noriega, R. B. & Massolo, L. (2017). Human health risk due to variations in PM 10-PM 2.5 and associated PAHs levels. *Atmospheric Environment*, 160, 27-35.

[36] Achad, M., López, M. L., Ceppi, S., Palancar, G. G., Tirao, G. & Toselli, B. M. (2014). Assessment of fine and sub-micrometer aerosols at an urban environment of Argentina. *Atmospheric Environment*, 92, 522-532.

[37] López, M. L., Ceppi, S., Palancar, G. G., Olcese, L. E., Tirao, G. & Toselli, B. M. (2011). Elemental concentration and source identification of PM10 and PM2.5 by SR-XRF in Córdoba City, Argentina. *Atmospheric Environment*, 45(31), 5450-5457.

[38] Vasconcellos, P. C., Souza, D. Z., Ávila, S. G., Araújo, M. P., Naoto, E., Nascimento, K. H., Cavalcante, F. S., Dos Santos, M., Smichowski, P. & Behrentz, E. (2011). Comparative study of the atmospheric chemical composition of three South American cities. *Atmospheric Environment*, 45, 5770-5777.

[39] Puliafito, S. E., Allende, D., Pinto, S. & Castesana, P. (2015). High-resolution inventory of GHG emissions of the road transport sector in Argentina. *Atmospheric Environment*, 101, 303-311.

[40] D'Angiola, A., Dawidowski, L. E., Gómez, D. R. & Osses, M. (2010). On-road traffic emissions in a megacity. *Atmospheric Environment*, 44(4), 483-493.

[41] Allende, D. G., Puliafito, S. E., Fernández, R. P., Castro, F. H. & y Cremades, P. G. (2012). Herramientas para la evaluación integral de la calidad de aire en Argentina: modelos de escala urbana y regional. *Revista Proyecciones- UTN-FRBA*, 10, 31-44.

[42] Abril, G. A., Diez, S. C., Pignata, M. L. & Britch, J. (2016). Particulate matter concentrations originating from industrial and urban sources: Validation of atmospheric dispersion modeling results. *Atmospheric Pollution Research*, 7(1), 180-189.

[43] Venegas, L. E. & Mazzeo, N. A. (2006). Modelling of urban background pollution in Buenos Aires City (Argentina). *Environmental Modelling & Software*, 21(4), 577-586.

[44] García Ferreyra, M. F., Curci, G. & Lanfri, M. (2016). First Implementation of the WRF-CHIMERE-EDGAR Modeling System Over Argentina. *IEEE Journal of Selected Topics in Applied Earth Observations and Remote Sensing*, 9(12), 5304-5314.

[45] Carreras, H. A. & Pignata, M. L. (2002). Biomonitoring of heavy metals and air quality in Cordoba City, Argentina, using transplanted lichens. *Environmental Pollution*, 117(1), 77-87.

[46] Rodriguez, J. H., Weller, S. B., Wannaz, E. D., Klumpp, A. & Pignata, M. L. (2011). Air quality biomonitoring in agricultural areas nearby to urban and industrial emission sources in Córdoba province, Argentina, employing the bioindicator Tillandsia capillaris. *Ecological Indicators*, 11(6), 1673-1680.

[47] Estrabou, C., Filippini, E., Soria, J. P., Schelotto, G. & Rodriguez, J. M. (2011). Air quality monitoring system using lichens as bioindicators in Central Argentina. *Environmental monitoring and assessment*, 182(1), 375-383.

[48] Puliafito, S. E., Allende, D., Fernández, R., Castro, F. & Cremades, P. (2011). New approaches for urban and regional air pollution modelling and management. In *Advanced Air Pollution*. InTech.

[49] García Ferreyra, M. F. (2013). *Obtención de mapas de calidad de aire, a través de la implementación y primera aplicación del modelo de transporte químico CHIMERE sobre Argentina* [*Obtaining air quality maps, through the implementation and first application of the CHIMERE chemical transport model over Argentina*]. Tesis de Maestría en Aplicaciones Espaciales de Alerta y Respuesta Temprana a Emergencias. Facultad de Matemática, Astronomía y Física, Universidad Nacional de Córdoba.

[50] Licursi, M., Sierra, M. V. & Gómez, N. (2006). Diatom assemblages from a turbid coastal plain estuary: Río de la Plata (South America). *Journal of Marine Systems*, *62*(1-2), 35-45.

[51] Arkouli, M., Ulke, A. G., Endlicher, W., Baumbach, G., Schultz, E., Vogt, U., Müller, M., Dawidowski, L., Faggi, A., Wolf-Benning, U. & Scheffknecht, G. (2010). Distribution and temporal behavior of particulate matter over the urban area of Buenos Aires. *Atmospheric Pollution Research*, 1(1), 1-8.

[52] National Institute of Statistics and Census (INDEC). (2010). en línea. *Censo Nacional de Población, Hogares y Viviendas* [*National Population, Household and Housing Census*]. Disponible en: www.censo2010.indec.gov.ar.

[53] Masiol, M., Squizzato, S., Formenton, G., Harrison, R. M. & Agostinelli, C. (2017). Air quality across a European hotspot: Spatial gradients, seasonality, diurnal cycles and trends in the Veneto region, NE Italy. *The science of The Total Environment*, 576, 210-224.

[54] Sprent, P. & Smeeton, N. C. (2016). *Applied non-parametric statistical methods*. CRC Press.

[55] Ostro, B. & World Health Organization. (2004). *Outdoor air pollution: assessing the environmental burden of disease at national and local levels*.

[56] Carslaw, D. C. & Ropkins, K. (2012). "openair — An R package for air quality data analysis." *Environmental Modelling & Software*, 27–

28(0), pp. 52–61. ISSN 1364-8152, doi: 10.1016/j.envsoft.2011.09.008.

[57] *Air quality in Europe, report*. (2017). ISBN 978-92-9213-921-6, https://www.eea.europa.eu//publications/air-quality-in-europe-2017.

[58] *Defra's annual report and accounts 2017 to 2018*. https://www.gov.uk/government/publications/defras-annual-report-and-accounts-2017-to-2018.

[59] Li, F., Ginoux, P. & Ramaswamy, V. (2010). Transport of Patagonian dust to Antarctica. *Journal of Geophysical Research: Atmospheres*, 115(D18).

[60] Peel, M. C., Finlayson, B. L. & McMahon, T. A. (2007). Updated world map of the Köppen-Geiger climate classification. *Hydrology and earth system sciences discussions*, 4(2), 439-473.

[61] Venegas, L. E. & Mazzeo, N. A. (1999). Atmospheric stagnation, recirculation and ventilation potential of several sites in Argentina. *Atmospheric Research*, 52(1), 43-57.

[62] Ulke, A. G. & Mazzeo, N. A. (1998). Climatological aspects of the daytime mixing height in Buenos Aires city, Argentina. *Atmospheric Environment*, 32(9), 1615-1622.

[63] Gramsch, E., Cereceda-Balic, F., Oyola, P. & Von Baer, D. (2006). Examination of pollution trends in Santiago de Chile with cluster analysis of PM10 and ozone data. *Atmospheric Environment*, 40(28), 5464-5475.

[64] Ruiz, S., Fernández-Olmo, I. & Irabien, Á. (2014). Discussion on graphical methods to identify point sources from wind and particulate matter-bound metal data. *Urban Climate*, 10, 671-681.

[65] Henschel, S., Le Tertre, A., Atkinson, R. W., Querol, X., Pandolfi, M., Zeka, A. & Pascal, M. (2015). Trends of nitrogen oxides in ambient air in nine European cities between 1999 and 2010. *Atmospheric Environment*, 117, 234-241.

[66] Mavroidis, I. & Ilia, M. (2012). Trends of NOx, NO2 and O3 concentrations at three different types of air quality monitoring stations in Athens, Greece. *Atmospheric Environment*, *63*, 135-147.

[67] World Health Organization. (2006). *WHO Air quality guidelines for particulate matter, ozone, nitrogen dioxide and sulfur dioxide-Global update 2005-Summary of risk assessment, 2006*. Geneva: WHO.

[68] Resquin, M. D., Santágata, D., Gallardo, L., Gómez, D., Rössler, C. & Dawidowski, L. (2018). Local and remote black carbon sources in the Metropolitan Area of Buenos Aires. *Atmospheric Environment*, 182, 105-114.

[69] Mazzeo, N. A. & Venegas, L. E. (2010). An ambient air quality monitoring network for Buenos Aires city. *International Journal of Environment and Pollution*, 40, 184-194.

[70] World Health Organization. (1999), *Environmental Health Criteria*, No. 213: Carbon Monoxide (second edition), ISBN 92 4 157213 2 http://www.inchem.org/documents/ehc/ehc/ehc213.htm Checked July 2018.

[71] Zhou, S., Deng, Q. & Liu, W. (2012). Extreme air pollution events: Modeling and prediction. *Journal of Central South University*, 19, 1668−1672.

[72] Collini, E., Osores, M. S., Folch, A., Viramonte, J. G., Villarosa, G. & Salmuni, G. (2013). Volcanic ash forecast during the June 2011 Cordón Caulle eruption. *Natural hazards*, 66(2), 389-412.

[73] Ulke, A. G., Brizuela, M. M. T., Raga, G. B. & Baumgardner, D. (2016). Aerosol properties and meteorological conditions in the city of Buenos Aires, Argentina, during the resuspension of volcanic ash from the Puyehue-Cordón Caulle eruption. *Natural Hazards & Earth System Sciences*, 16(9).

[74] Reckziegel, F., Bustos, E., Mingari, L., Baez, W., Villarosa, G., Folch, A., Collini, E., Viramonte, J., Romero, J. & Osores, S. (2016). Forecasting volcanic ash dispersal and coeval resuspension during the April–May 2015 Calbuco eruption, *Journal of Volcanology and Geothermal Research*, doi: 10.1016/j.jvolgeores.2016.04.03.

[75] Ristori P., Otero, L., Pawelko, E. & Pallotta, J. (2010). Aerosol and atmospheric boundary layer temporal evolution in Buenos Aires, Argentina during May 12th, 2006. *Proceedings of the 25th International Laser Radar Conference*, 5–9 July 2010, St.-

Petersburg, Russia, Publishing House of IAO SB RAS, 2010. ISBN 978-5-94458-109-9. Vol. I, 684-686 (2010).

[76] Otero L. A., Ristori, P. R., Pawelko, E., Pallotta, J., D'Elia, R. & Quel, E. (2011). Biomass Burning Aerosol Detection over Buenos Aires City, August 2009. *Journal of Physics: Conference Series.*: IOP Publishing (doi:10.1088/1742-6596/274/1/012092), vol. 274 n°1, ISSN 1742-6596 [http://iopscience.iop.org/1742-6596/274/1/012092] .

[77] Otero L. A., Ristori, P., Pallotta, J. V., Pawelko, E. & Quel, E. (2012). The Puyehue-Cordón Caulle Volcanic Eruption, June 2011: Water Vapor, Atmospheric Boundary Layer and Aerosol Temporal Evolution in Buenos Aires, Argentina. *Revista Boliviana de Física*. [online]., 2012, vol. 21, no.21 [citado 20 Noviembre 2012], p.27-29. Disponible en la World Wide Web: <http://www.scielo.org.bo/scielo.php?script=sci_arttext&pid=S1562-38232012000100005&lng=es&nrm=iso>. ISSN 1562-3823.

[78] McConnell, J. R., Aristarain, A. J., Banta, J. R., Edwards, P. R. & Simões, J. C. (2007). 20th-Century doubling in dust archived in an Antarctic Peninsula ice core parallels climate change and desertification in South America. *Proceedings of the National Academy of Sciences*, 104(14), 5743-5748.

[79] Gassó, S., Stein, A., Marino, F., Castellano, E., Udisti, R. & Ceratto, J. (2010). *A combined observational and modeling approach to study modern dust transport from the Patagonia desert to East Antarctica*.

[80] Johnson, M. S., Meskhidze, N., Kiliyanpilakkil, V. P. & Gassó, S. (2011). Understanding the transport of Patagonian dust and its influence on marine biological activity in the South Atlantic Ocean. *Atmospheric Chemistry & Physics*, 11(6).

[81] Gassó, S. & Stein, A. F. (2007). Does dust from Patagonia reach the sub-Antarctic Atlantic Ocean? *Geophysical Research Letters*, 34(1).

About the Editor

Jorge Esteban Colman Lerner
Assistant Researcher of National Scientific and Technical Research Council of Argentina (CONICET, Argentina)
Associate Professor at the Arturo Jauretche National University (UNAJ, Buenos Aires, Argentina) and Teaching Assistant at the National University of La Plata (UNLP, Buenos Aires, Argentina)

Jorge Esteban Colman Lerner is a professor of Instrumental Analytical Techniques at the Arturo Jauretche National University (UNAJ, Buenos Aires, Argentina) and of Chemistry and Atmospheric Pollution at the National University of La Plata (UNLP, Buenos Aires, Argentina). He is also a postgraduate professor at the National Technological University (UTN Buenos Aires, Argentina) in the courses "Atmospheric Pollution", "Control of Gaseous Effluents" and "Tools for the Evaluation of Air Quality", and has a PhD (National University of La Plata in the area of Chemistry) in the determination of atmospheric pollutants and the application of technologies to mitigate their emission. He is an Associate Researcher of the National Council for Scientific and Technical Research (CONICET, Argentina), with extensive experience in analytical determinations in different matrices (air, soil, water and biological material) and synthesis, characterization and study of catalysts for the treatment of environmental pollutants in air such as volatile organic compounds (VOCs), semi-volatile organic compounds (SVOCs) and particulate matter (PM).

INDEX

B

C

D

E

F

G

H

I

J

L

M

N

Q

R

S

T

U

V

W

X

Y